THE
CONTEMPORARY
MEDITERRANEAN
WORLD

THE CONTEMPORARY MEDITERRANEAN WORLD

edited by

Carl F. Pinkele
Adamantia Pollis

PRAEGER

PRAEGER SPECIAL STUDIES • PRAEGER SCIENTIFIC

Library of Congress Cataloging in Publication Data
Main entry under title:

The Contemporary Mediterranean world.

  Based on a conference held at the Rockefeller
Conference Center in Bellagio, Italy, Aug., 1981.
  Includes index.
  1. Mediterranean Region--Politics and government--
1945-   --Addresses, essays, lectures. 2. Mediterranean Region--Foreign relations--1945-   --Addresses,
essays, lectures. 3. Mediterranean Region--Strategic
aspects--Addresses, essays, lectures. 4. World politics
--1975-1985--Addresses, essays, lectures. I. Pinkele,
Carl F. II. Pollis, Adamantia.
DE100.C67 1983   320.9182'2   82-16658
ISBN 0-03-060091-X

Published in 1983 by Praeger Publishers
CBS Educational and Professional Publishing
a Division of CBS Inc.
521 Fifth Avenue, New York, New York 10175 U.S.A.

3456789   052   987654321

Printed in the United States of America

CONTENTS

LIST OF TABLES

ix

LIST OF FIGURES

ACKNOWLEDGMENTS

This volume is to a large extent the product of a highly successful week of intense and productive discussions at the Rockefeller Foundation's Villa Serbelloni in Bellagio, Italy, during the week of August 24-28, 1981. The conference was directed by Carl F. Pinkele and, in part, was supported by Ohio Wesleyan University and the Rockefeller Foundation.

The editors wish to publicly thank everyone who participated in the Bellagio Conference, with special thanks to Professor Mumtaz Soysal of the University of Ankara, Turkey, and William Gaillard of the European Economic Community, who attended the conference but were unable to contribute to this volume. Our appreciation also to Barbara J. Pinkele and Bronislawa Lenczowski; to the New York and Villa Serbelloni staffs of the Rockefeller Foundation; to Betsy Brown at Praeger; to Janet King of the Ohio Wesleyan University Politics and Government Department, who corrected, adjusted, and typed through the planning and production of both efforts; and to Donald E. Sherblom, graduate student in the Department of Political Science, Graduate Faculty, New School for Social Research, for his great patience in putting the manuscript together.

INTRODUCTION
WHOSE AGENDA IS IT?
Carl F. Pinkele

Even a momentary glance at the contemporary Mediterranean region captures clearly the fact that a wide variety of policy making is simultaneously under way. [1] Well over twenty nations plus several regional and international agencies are actively involved in making public policies that pertain to contemporary Mediterranean political affairs. In addition to the more formal policy makers, numerous informal (nongovernmental) groups and movements are engaged in policy formulation and political activity, including many corporations and oil companies, the Palestine Liberation Organization (PLO), and the Muslim Brotherhood.

The general regionwide result of this quite sizable number of participants is a plethora of policy agendas arrayed across a broad spectrum of interests and perspectives. This introduction attempts to provide a lens setting through which the picture of the variety and tangle that characterize Mediterranean policy making can become more clear.

THE ANALYTICAL FRAMEWORK

The contemporary Mediterranean world is an extraordinarily interdependent environment. It is the case, as other chapters in this volume demonstrate, that few if any policy decisions (internal or external) are isolated or immune from structural patterns of the prevailing macropolitical economy. Kegley and McGowan have observed that "as nations' internal economic fates have become intermeshed, the dependence of each nation on others has risen. For many states, the advent of interdependence means that states cannot independently control their own fate." [2]

A fundamental distinction exists between policy interconnectedness and policy interdependence. Policy interconnectedness suggests a comparatively simple situation in which there are no structural costs, benefits, constraints, or opportunities; there is only interaction. Policy interdependence, by contrast, involves structural patterns of costs, benefits, constraints, and opportunities. Policy interdependence signifies reliance on others. In the case of policy interconnectedness one is observing and discussing the behavior of basically autonomous actors, while in the case of policy interdependence there is much less or no meaningfully autonomous policy making or political actors. Whether or not one is talking about policy at the level of micro or macro variables (people or nations, for example), interdependence involves restricted freedom of movement or choice among options, and, according to Keohane and Nye, "significant costs."[3]

A second aspect of policy interdependence requires that a distinction be made between symmetrical and asymmetrical interdependence. To find that a relationship between two policy makers representing two different nations is one of symmetrical interdependence is to find that the costs and benefits, particularly, and the policy-making options and restraints as well are roughly evenly distributed. Asymmetrical interdependence, on the other hand, occurs when costs and benefits, options, and restraints are not evenly distributed but are rather unevenly distributed. Thus, in the asymmetrical interdependent relationship one or more principals, parties, or actors receives considerably more benefit from it than does another (or others as the case may be).[4]

The matter of whether or not an interdependent relationship is a symmetrical one or an asymmetrical one is always a critically important judgment. It is an even more significant judgment when the question involves the determination of whether or not what one is observing is an instance or a pattern of symmetry or asymmetry. A structural pattern of one type of interdependence or the other suggests much about the future as well as the present policy-making potentials for a nation-state.

It is clear that the type of interdependence in a relationship is a matter largely of degree. Nation A, for example, may well enjoy more benefits than Nation B, who will suffer higher costs in an interdependent structure, and yet there will be some costs and benefits for each. It is the case, however, that as circumstances are altered both the quality and the quantity of costs will also change. An extremely enlightening Mediterranean example of just this sort of thing occurred when the Arab oil-producing nations flexed their united muscle toward the United States in 1973 as a result of American support of Israel in the Yom Kippur War. Irrespective of its both symbolic and real position of power in its Mediterranean relationships, the United States was indeed sensitive and vulnerable to the oil boycott policy decisions of the anti-Israel Arab states.[5]

The point to underscore is that in an interdependent relationship there is little autonomy for policy makers. And this is the case whether or not one finds symmetrical or asymmetrical interdependence. However, in the case of asymmetrical interdependence, because structural patterns result in a maldistribution of policy-making power, one actor (or set of actors) will have more policy-making latitude and leverage than will another. The degree of autonomy in

decision making is more often than not a reflection of one's structural position. The greater the degree of asymmetrical interdependence, the less power a nation and its formal decision makers have
over domestic and foreign-policy agendas.

Policy making begins with the creation of agendas, which are
themselves a form of more general policy making. Initially, agendas
reflect the fact that policy makers have translated their individual or
collective priorities into an action-directed plan in which, to para-
phrase Harold Lasswell, decisions are made concerning which items
will be dealt with, who will decide them, and when they will be decided upon.

Items which are viewed positively become the subject of policy
making in the sense that specific decisions are made, or plans are
prepared, to actively address the dimensions and specifics of the
matter at hand. They are also granted legitimacy in the sense that
by being held up as worthy of consideration and action, positively held
items receive a value-added aspect; they become, in a sense, valid
matters of concern, whereas issues and problems not taken into serious consideration are viewed, or at least broadcast, as being of
dubious merit. Items viewed negatively by policy makers are generally ignored, deferred, or held for recasting into some more acceptable mold. Just as items adopted for the policy makers' agenda attain a legitimacy, the opposite obtains oblivion for items not placed
on or appearing on that agenda. Items held to be not worthy of attention by the policy-agenda gatekeepers, for whatever reasons, are
either delegitimized or are not elevated to the status of being legitimate.

Combining an interdependence framework with a particular
focus upon policy-agenda building reduces the distinction between
politics and economics to relative insignificance while emphasizing
the importance of the fact that they are inextricably interwoven features of the contemporary world. This is not to say that in a specific
case, for example matters of naval strategy or restriction of nuclear
proliferation, more traditional political concerns do not occupy center stage, or, in the case of technical oil import-export matters,
that more traditional economic aspects do not dominate the immediate
scene. But in neither of these instances is it a matter purely of one
or the other (politics or economics). In far more instances than not,
there is no clearly identifiable line between what is of political and
what is of economic concern. Interdependency means complexity,
even ambiguity, because "economic dependence is positively related
to political compliance,"[6] while "military aid is the single most
important influence on the pattern of political compliance."[7]

When one links the matters of structural policy interdependency
and power, the question becomes, Whose agenda is more significant,

carries more weight, and exercises more leverage? In the instance of symmetrical interdependence, with power structurally distributed more or less evenly, policy agendas should reflect the fact that the interests of all concerned parties are represented and influence the shaping of policy outcomes. In the instance of asymmetrical interdependence, wherein power is structurally unevenly distributed, policy agendas will reflect the fact that the interests of one or more actors are more significant in terms of influencing outcomes than are those of the other actors in the relationship. Thus it is the case that policy agendas more or less accurately reflect power considerations, and power considerations in turn reflect the type and degree of policy interdependence.

One further ingredient needs to be added to round out this proposed analytical framework for comprehending the contemporary Mediterranean world. First, it is difficult to discuss matters of political economy without taking note of the important place that the pursuit of interests (either subjective-symbolic or objective-real) occupies in the behavior of political-economy actors. [8] Decision makers and would-be decision makers actively pursue the satisfaction of their perceived interests. This behavior occurs within structures that contour and prefigure both the mode in which interests are pursued and the likely success of the pursuit.

In an interdependent environment of either variety the interests of policy-making actors are structurally interlocked. Clashes, conflicts, agreements, and bargaining between and among interests are to a most significant degree in effect regulated by the character and extent of the circumstances of interdependence present. This is to say that the extent to which an actor's interests are satisfied or not satisfied, as well as shaped, will largely be a reflection of that actor's position in an interdependent structure. If it is the case that there is symmetrical interdependence, then one would expect to find that the actors in the relationship enjoy an evenness in the degree to which their own interests are reflected in the establishment of a common policy-making agenda. Furthermore, and in some ways perhaps more revealing, in a symmetrical environment each actor is able to formulate policy proposals (domestic and foreign) in a relatively autonomous manner. Quite the opposite from the situation resulting from a symmetrical environment occurs in the case of an asymmetrical one. In an asymmetrical situation, there is a significant degree of unevenness in the extent to which the respective actors influence policy agendas and policy outcomes. And in asymmetrical circumstances the interests of one actor or set of actors—the most powerful—penetrate the agenda-building framework of the weaker actors in both domestic and foreign policy-making arenas.

THE ASYMMETRICALLY INTERDEPENDENT
MEDITERRANEAN WORLD

The basic shape of the Mediterranean political economy is one of asymmetrical interdependence; although, as the chapters by Boxer and Nogee point out, there are areas in which the interdependence is more of a symmetrical nature. Interestingly enough, in the areas of pollution control and keeping the Mediterranean a nuclear free zone, the issues are so general and diffuse that interests are not galvanized toward clearly protectionist postures. But one can see glimpses in these areas also of how present and future power-interest considerations might well shift, and the pattern of asymmetrical structures of interdependency could penetrate here as well.

The Mediterranean policy-making world is a quite rigidly stratified one in which some nations—the United States, the Soviet Union, Western Europe, and Israel—have considerably more influence and autonomy in the shaping of events than do others.[9] The asymmetry found throughout most reaches of the policy area is a consequence of this systemic maldistribution of political, social, and economic costs and benefits.

Present Mediterranean policy makers in both domestic and foreign decision making find themselves either more or less constrained in their policy options depending upon the position they occupy in the interdependent structure.[10] No decision maker, whether a primary beneficiary of the asymmetrical process or not, has the luxury of acting autonomously. However, some national decision makers clearly have more leverage and flexibility in establishing policy agendas than do others. That is, some policy makers, the more powerful or favorably positioned, can and do exercise more autonomy—pursue their own interests more vigorously and with fewer structural inhibitions—than do their less favorably positioned counterparts. Less autonomous policy makers are in point of fact the victims of what is referred to as agenda penetration, wherein the agendas and hence the interests of the more powerful, penetrating actors weigh heavily in the decision-making calculus of the less powerful, agenda-penetrated actors. In a pattern of asymmetrical interdependence, as Nour and Pinkele argue was the case in the recent Egyptian situations, the external constraints placed upon the policy options for Nasser, Sadat, and Mubarak were such that when Egyptian actors constructed domestic and foreign policy agendas, the interests of non-Egyptian actors had a great and direct force in those activities.

Future Mediterranean decision makers will inherit the results of the contemporary play of generally uneven forces. Contradictions abound and the future looks grim.[11] The more powerful actors are in a very real sense bound to continue seeking advantages across a unified field that includes strategic, diplomatic, economic, and social factors. Challenges to the leverage of the more powerful will undoubtedly result in even more penetration, and there shall be challenges. [12] As the less powerful actors find themselves constrained in the setting of their own agendas, they will be not only frustrated but also unable to meet real problems with appropriate policies.

NOTES

1. As conceptualized in this volume, the Mediterranean world consists of those nations and peoples who physically are linked to the Mediterranean basin and those nations especially who are tied to that geographic area through a high level of interdependent policy interaction.

2. Charles W. Kegley, Jr. and Pat McGowan, The Political Economy of Foreign Policy Behavior (Beverly Hills: Sage, 1981), p. 12.

3. Robert O. Keohane and Joseph S. Nye, Power and Interdependence: World Politics in Transition (Boston: Little, Brown, 1977), p. 9.

4. Adrienne Armstrong, "The Political Consequences of Economic Dependence," Journal of Conflict Resolution 25, no. 3 (Sept. 1981): 401-28.

5. Power can be quite usefully defined by combining what Keohane and Nye suggest, on the one hand, and Armstrong on the other. According to Keohane and Nye, "power can be thought of as the ability of an actor to get others to do something they otherwise would not do . . . Power can also be conceived in terms of control over outcomes," The Political Economy, p. 11. Armstrong adds that "the idea of opportunity costs is central to the understanding of power," "The Political Consequences," p. 406. Political sensitivity refers to the matter of "degrees of responsiveness within a policy framework—how quickly do changes in one country bring costly changes in another and how great are the costly effects?" Political vulnerability "rests on the relative availability and costliness of the alternatives that various actors face," Keohane and Nye, The Political Economy, p. 12.

6. Armstrong, "The Political Consequences," p. 415.

7. Ibid., p. 416.

8. See the Smolansky, Smolansky, and Ginsburgs chapter in this volume for a discussion of the notion of interests.

9. Those chapters in this volume that make this point, as well as demonstrating how, especially for the major-power Mediterranean

actors, inherently non-Mediterranean concerns become a major part of Mediterranean-oriented and -exercised policy making, are Smolansky, Smolansky, and Ginsburgs; Kislov; Lenczowski; Dettke; Baade and Galloway; Schmidhauser and Berg; and Papademetriou. Warburg suggests that Egyptian foreign-policy makers have considerably more autonomy and freedom of movement than do Nour and Pinkele in their chapter "Camp David and After: Foreign Policy in an Interdependent Environment."

10. Those chapters making this point are Pollis, Woodward, Musto, and Nour and Pinkele. Jaber strikes a quite different posture by suggesting that the Arab Mediterranean, and Jordanian policy making specifically, is motivated largely by an internal elite's desires to modernize.

11. Russell Stone's chapter concerning Israeli attitudes toward the Palestinian issue and the Palestinian people is an important case in point. The present Israeli government, as well as past and probably future ones, been able to avoid making meaningful policies regarding the Palestinians because of what E. E. Schattschneider would term a displacement of conflict. The costs of not addressing themselves directly to the plight of the Palestinians seem to be catching up with Israeli policy makers, however.

12. See the chapter by Marilyn Waldman on Islamic resurgence.

PART I

INTERNATIONAL CONTEXT

1

A NATIONAL-INTEREST FRAMEWORK: THE UNITED STATES AND THE SOVIET UNION IN THE EASTERN MEDITERRANEAN

BETTY SMOLANSKY
OLES M. SMOLANSKY
GEORGE GINSBURGS

Any reasonable effort to analyze the present and predict the future foreign-policy actions of a nation must ultimately rest, either implicitly or explicitly, upon assumptions about the national interests on which policy decisions are based. Yet there is considerable reluctance among scholars to attempt explicit and systematic analyses of such interests. This hesitancy is readily understandable in light of the intellectual jungle that the very concept represents. However, the complexity of the issue does not justify its omission from overt consideration; otherwise, complex suppositions will be made implicitly, remaining hidden not only from the reader but, often, from the author as well.

Thus, in order to generate a discussion of Soviet and U.S. policy in the Eastern Mediterranean, a three-level framework for the analysis of their national interests will be presented, along with some basic postulates about the role such interests play in the determination and formulation of foreign policy. For the sake of clarity, these postulates will be incorporated as part of the commentary on the paradigm but will be emphasized by underlining.

The first level of the framework deals with the basic approach to the problem of interests. Though some scholars refuse to recognize the legitimacy of the distinction, others contend that national interests must be divided into two basic categories: objective and subjective (or perceived). While, admittedly, all interests are per-

3

ceived, this particular use of the term implies a distinction between separate groups of perceivers; <u>subjective</u> interests are defined as those based on the perceptions of the national leaders themselves, and <u>objective</u> interests refer to those posited by outside observers. The so-called subjectivist school would claim that there is no such thing as an objective interest, in other words, one which exists as an empirical entity apart from the perceptions of policy makers. The objectivists counter that such a position assumes a kind of omniscience among governmental leaders that simply does not exist. The first of the framework-attendant postulates on the role of interests in foreign-policy making allows the analyst essentially to avoid this controversy; it is simply that <u>exclusive recognition of objective interests would lead to ideal policy, while actual policies are based on perceived interests</u>. Thus, the possible existence of objective interests is granted but, for the purposes of comprehension and prediction, attention is focused upon perceived interests. Unfortunately for the scholar, the situation is further complicated by the fact that <u>perceptions of national interests must be inferred from some combination of the actions and utterances of policy makers</u> (assuming the relevant individuals can be identified). This second postulate points up an aspect of policy analysis that is made even more difficult by the fact that, while inferences are more easily drawn from verbal cues, verbal cues are a less reliable indicator than the typically more obscure meanings of actions resulting from policy decisions. Certainly, use of official government pronouncements alone, especially if taken at face value, tends to be misleading.

The second level of the framework contains three subcategories of interests: <u>intrinsic</u> (direct), <u>relational</u> (indirect), and <u>vested</u>. The first subcategory refers to those interests that grow out of some aspect of the nature of the geographic area under consideration, such as the availability therein of material resources needed by the state concerned. In contrast, those interests that originate not in the nature of the area itself but in its importance to other nations with which the government in question has a significant relationship (either amicable or antagonistic) are termed <u>relational</u>. Finally, <u>vested</u> interests are those that historically emerged as one of the first two types but have come to enjoy a life of their own, even though the original circumstances that generated them have changed or disappeared. Parenthetically, it is at this point that the paradigm lacks full consistency; that is to say, theoretically each of these three subcategories should be subsumable under each of the two more abstract classes of objective and perceived interests. However, logic suggests that an objective vested interest is an impossibility; indeed the very term is internally inconsistent. Thus, while all three subtypes are potential forms of perceived interests, objective interests must be either intrinsic or relational.

The final level of the framework concerns more substantive, concrete aspects of national interest. The subdivisions contained in it are therefore susceptible to disciplinary preferences; that is, while most analysts would probably agree with the inclusion of the suggested major categories, many might favor additions to them. Exercising the prerogatives of scholarly license, the authors have decided to add to the widely acceptable enumeration of military-strategic, economic, and political interests only the residual category of cultural-ideological interests. This is somewhat loosely defined as a desire to proselytize among the noncommitted on behalf of one's own central belief system. In Washington's case, this may involve a wish to promote "the American way of life," while for the Soviets it consists primarily of demonstrating the "superior virtues" of Marxism-Leninism.

The usefulness of the framework might be enhanced by further explication of some of the assumptions that underlie both its utilization and the limitations of its application to foreign-policy problems in a world dominated by superpower rivalry. The contest for influence between the Soviet Union and the United States is one of the dominant facts of the current international system. Thus, adversary assumptions continue to undergird most of the perceived interests of the superpowers. Even in those instances where the two decide on some forms of cooperation, as in the phase of "peaceful coexistence" (commonly referred to as détente), their collaboration all too often consists of the grudging, short-term type through which the parties seek the long-run enhancement of their respective positions vis-à-vis each other and the corresponding diminution of the other power's position.

As indicated earlier, if decision making in the foreign-policy sphere were perfectly rational, objective interests would supersede and sometimes negate perceived interests. Failing that perfection, but assuming a high level of realism about the world, among the subtypes of perceived interests, the intrinsic variety should enjoy priority over relational and both should outweigh the vested. However, the inclination of the human mind (even the "trained" variety) for emotionally generated rationalization more often than not so far outstrips its capacity for rational thinking that vested interests often come to assume priority over the other two categories. Among the vested interests most likely to enjoy an undue level of priority, the foremost is probably what could be termed regime survival. As Watergate demonstrated, the "engine Charley" mentality is all too common among politicians in democratic societies,[1] and the same is no doubt true of representatives of authoritarian governments as well.

While not as invariate as in the case of the three subcategories of intrinsic, relational, and vested interests, there appears to exist a hierarchy of importance among the four substantive types as well. Specifically, the military-strategic category should almost invariably take precedence over economic interests, and the latter over the political. It is at this point that actual policy decisions most typically violate the paradigm's rational approach. For while the transitory and ephemeral nature of most political successes would suggest that they are less worth pursuing than economic gains, ir- rationality in the decision-making process often accords them unwar- ranted importance. Moreover, military-strategic, economic, and political interests should be (and typically are) assigned greater sig- nificance than their cultural-ideological counterparts.

While it is hoped that the above categories of interest will stim- ulate further scholarly analyses, the authors recognize that the sys- tem's ultimate utility is limited by yet another factor, which must be attached to the paradigm as an additional postulate: specific policy decisions are typically based upon mixed, multiple motives and in- terests. Some of the latter are arational and are not even a conscious part of the decision-making process.

Finally, one more postulate that, too, illustrates the frame- work's limitations must be acknowledged for the sake of scholarly integrity. Specifically, while the framework is concerned with the interests that motivate policy, it ignores the other major factor that circumscribes the decision-making process, namely, the availability of alternative modes of policy implementation. Thus, even after agreement has been reached on the nature of the interests to be served, honest differences over the best available methods of policy implementation can (and usually do) remain. It is this fact, perhaps more than any other, that makes the understanding and prediction of specific actions by governments so difficult.

Keeping in mind the limitations imposed by these two postulates, we now turn to the framework's application to the foreign policies of the United States and the Soviet Union in the Eastern Mediterranean. Since our primary interest lies in the realm of present and future actual policy by the superpowers, the first two postulates suggest that the examination should focus upon the perceived interests that may be inferred from their recent and current behavior in the Medi- terranean.

The most striking fact that emerges from an attempted point- by-point application of the framework to Soviet and U.S. activity in the region is the existence of a lynchpin assumption held with appar- ently equal and unquestioning conviction by both superpowers. More precisely, both Moscow and Washington appear to accept as a tenet of faith the notion that the Mediterranean itself is an important key to the stability of the surrounding areas, particularly the Middle East. [2] The significance of this assumption emerges with ever-in- creasing impact in the course of the examination of the intrinsic in- terests that the superpowers apparently perceive themselves to have in the Mediterranean.

INTRINSIC INTERESTS

Interestingly, if solely intrinsic interests are analyzed and the assumption of the sea's importance to the superpowers' roles in the Middle East is accepted, the Kremlin appears to have a much stronger rationale for pursuing an activist policy in the Mediterranean than does the United States. Thus, if military-strategic considerations do in fact enjoy top priority in foreign-policy decision making, the sea's geographic proximity to the Soviet Union makes it appear to the Kremlin as an area that must be defended against potential strikes, which the United States could launch in case of a military showdown (general or local) between the superpowers. Conversely, its distance from the North American continent, along with the availability of intercontinental ballistic missiles armed with nuclear warheads, makes the Mediterranean a low-priority U.S. defense area. (Even if one accepts the proposition that "the best defense is a good offense," the Mediterranean, as a virtually enclosed sea, is no longer a particularly worthwhile forward strategic position.) In any event, if the Soviets believed that the U.S. military presence in the Mediterranean is, as advertised, a defensive one, then any attempt to counter it would not make sense as an intrinsic interest. Clearly, however, even though many Western planners have fallen victim to a good-guys-bad-guys mentality and thus fail to understand why anyone would doubt the purity of their motives, their counterparts in Moscow clearly do not share this vision of the world.

Moving from military-strategic to economic interests and from the Eastern Mediterranean to the adjacent Middle East, there can be no doubt that, in this category, the Western case is much stronger than that of the Soviet Union. In the particular case of the United States, there is no denying continuing dependence on Middle Eastern oil. In view of the post-1973 increases in the price of petroleum and the resulting accumulation of capital in the oil-producing nations, it is possible to speak also of a U.S. interest in broadening economic relations with them and in attracting their wealth for investment in the United States.

The Soviet Union, in contrast, has in the past derived some benefits from the importation of relatively limited quantities of Iraqi oil and Iranian natural gas. It has not, however, depended on them for its economic well-being and cannot, therefore, be said at this stage to possess an intrinsic economic interest in the area. Both superpowers, incidentally, benefit from stable and open access to transportation and communications facilities that the region provides, so that, in this respect, their interests tend to balance each other.

On the political level, in the case of the United States there appear to be no intrinsic interests that are separable from the economic

considerations referred to above. Thus, Washington is clearly interested in the political stability of the Middle East in that it creates
the necessary preconditions for maintaining and enhancing economic
ties that have become vital for the health of the U.S. economy. The
major problem in this connection, therefore, becomes that of the
best means to achieve this objective, especially in view of U.S. relational and vested interests in the area (see below). It might be
noted parenthetically in this connection that in the major regional military flare-ups of the past decade, such as the October 1973 war and
the Cyprus crisis, the U.S. Sixth Fleet proved impotent in maintaining regional stability.

As for the USSR, it could be argued that, as in many other parts
of the globe, Moscow is seeking U.S. recognition of its status as a
superpower with legitimate state (gosudarstvennye) interests in the
Middle East. It could also be said that economic interests and investments of various types in many other states of the area would dictate
the desirability of political stability along lines similar to those mentioned in the preceding paragraph in connection with the United States.
However, as long as the Sixth Fleet and Polaris threats remain in the
Mediterranean, the Soviets are likely to if not actually foment instances
of area strife, at least attempt to utilize them to undermine U.S. positions. It is probably not too far-fetched to argue that, in terms of
regional politics, this policy appears to the Kremlin leaders as the
most expeditious method of mitigating what is probably still seen by
some of them as a significant military threat emanating from the
Eastern Mediterranean.

Turning to the category of cultural-ideological interests, it
should be noted that the leaders of both the Soviet Union and the United
States, having been socialized in and conditioned by their respective
systems, probably feel a commitment, albeit an amorphous one, to
advancing their respective cultural-ideological interests. (Operatively,
however, such considerations become important only in ceteris paribus cases. Indeed, oral citations of such interests are most typically
useful in ex post facto efforts to justify decisions made for other reasons.) On balance, the Soviets probably assign a slightly greater overt
significance to such considerations because of the more elaborately
articulated nature of their ideology. [3]

RELATIONAL INTERESTS

This is neither the time nor the place for a review of the origins
of the cold war or for a detailed analysis of U.S. and Soviet foreign
policies of the 1940s and beyond. However, a brief account of some
of the events that took place in the Middle East and the Mediterranean

is necessary at this juncture to enable the reader to judge the merits of the case presented below.

The original decision to establish and maintain a strong U.S. military presence in the Mediterranean dates back to the early post-World War II period. It was made under the impact of what was perceived to be Stalin's determination to spread Soviet influence into Iran, Turkey, and Greece and of the growing weakness of Great Britain, long the dominant Western power in then Persia and the Eastern Mediterranean.

In view of the growing mutual distrust between Moscow and Washington after the death of Roosevelt, which resulted in part from Stalin's demands for control of the Turkish Straits area, Tripolitania, and the Dodecanese Islands, as well as from Soviet intransigence in Iran, it is difficult to disagree with the Truman administration's serious concern over and suspicion of Moscow's intentions in the Middle East and the Mediterranean. When to these problems was added the communist insurrection in Greece, the United States, convinced (erroneously as we know now) of Stalin's complicity in the affair, felt it had no choice but to help Athens put down the Marxist revolution. Among the actions taken was the proclamation of the Truman Doctrine and the subsequent deployment in the Mediterranean of the U.S. Sixth Fleet. These measures were reinforced by the extension of a U.S. naval-and-air-base network throughout the Mediterranean and the Middle East. Aware that he had overplayed his hand and preoccupied with other more pressing problems (consolidation of the empire, relations with the Chinese communists, domestic problems, and so forth), Stalin retreated into relative inaction in the Mediterranean world (interrupted briefly by active Soviet support for the establishment of Israel in 1947-48), and there the situation remained until Khrushchev's advent to power in 1955.

When reviewing the U.S. presence in the Middle East and the Mediterranean in the early post-Stalin period, the new Soviet leaders could not but be highly concerned by what had by then developed into a position of U.S. military-strategic dominance in the region close to the southern borders of the Soviet Union. The powerful Sixth Fleet, with support facilities in Spain, Italy, Greece, and Turkey, and with its two carrier task forces capable of delivering nuclear bombs to targets situated in the southern parts of the Soviet Union, was supplemented by impressive numbers of long-range bombers based in Morocco, Libya, Saudi Arabia, Turkey, and subsequently, Pakistan, which could launch nuclear strikes on most of Soviet Europe and central Asia.

There can be no doubt that what to Washington appeared as legitimate and justified containment of Moscow's expansion was viewed in the Kremlin as another manifestation of U.S. determination to encircle the Soviet Union. The threat was made even more formidable by the fact that throughout the 1950s and well into the 1960s the United States enjoyed an unquestionable superiority over the Soviet Union in terms of both nuclear weapons and their means of delivery.

In view of these considerations, Khrushchev's effort to dislodge the United States from its position of strength in this general area by all means available to the Soviet Union should have come as no surprise. On the political level, the chairman set out to undermine Western positions, manipulating to Moscow's advantage the Arab-Israeli and inter-Arab disputes. On the military-strategic plane, the determination to continue the development of a viable intercontinental ballistic missile (ICBM) force was accompanied by a far-reaching decision to reshape the Soviet navy into an instrument capable of neutralizing the U.S. navy in the Mediterranean, the South Norwegian Sea, and the Pacific (other major areas of U.S. strategic deployment). Because of the schism with Albania, resulting in the denial of the Valona submarine base to the Soviets in 1961, and even more important, because of the lead time involved in the construction of new types of naval vessels, the fruits of this decision did not become visible until 1964, when a permanent Soviet naval presence was finally established in the Mediterranean. In the meantime, Khrushchev's political and economic initiatives in the region not only reinforced Washington's suspicions of his ultimate intentions, both along the North Atlantic Treaty Organization's (NATO) southern flank and in the Middle East as a whole, but led to further strengthening of the U.S. position, as evidenced by the introduction into Italy and Turkey of Jupiter medium-range ballistic missiles and, in 1963, of the Polaris submarines into the Mediterranean. With the subsequent deployment of the Soviet naval squadron (eskadra), the issue was fully joined, and, in the ensuing years, the Soviet Union was hard at work improving its position in the Eastern Mediterranean. The high point was reached in the period between 1970 and 1972, when Soviet naval and air bases were, in fact, established in Egypt.

In the late 1950s and beyond, therefore, it mattered little as to who did what first and when; the superpowers were facing each other in the Middle East and the Eastern Mediterranean, determined to weaken each other's military capabilities and political positions. To sum up, although no intrinsic U.S. military-strategic interests were initially involved, perceived relational interests in this category developed in the course of the cold war. They entailed, in the context of U.S.-Soviet competition, the protection of NATO's southern flank and, in terms of Middle Eastern politics, an informal moral obligation to protect Israel against the possibility of annihilation at the hands of the Arab states, which, with the important exception of Egypt, have not yet reconciled themselves to its existence.

As in the case of U.S. intrinsic interests, however, the basic question became that of the best means for achieving these objectives. To begin with, it has been difficult for quite some time now to accept the gloomy prognostications concerning the Soviet military threat to the southern flank of NATO. For one thing, any encroachment on the southern flank would involve the Soviet Union in a confrontation with NATO as a whole, a course of action the Kremlin is not likely to take for at least one important reason: any overt moves against NATO would inevitably result in a head-on collision with the United States, a course of action that, in the final analysis, no Soviet leader of the post-1945 period has been willing to pursue too far. (The Cuban missile crisis is an exception that proved highly instructive.) Therefore, the argument that the U.S. military presence in the Mediterranean, especially in its current form, is essential to the security of NATO is very difficult to accept.

Turning to the Arab-Israeli sector, it is equally questionable, given Washington's clear-cut commitment to the preservation of Israel, whether a physical U.S. presence in the Eastern Mediterranean is required to ensure its continued existence. Based on previous experience, it is not too far-fetched to argue that, if kept sufficiently supplied with military hardware, Israel is perfectly capable of protecting its own national security. It is, of course, obvious that it could not do so if confronted directly by the Soviet Union, but, primarily for the same reasons that an anti-NATO move seems improbable, the likelihood of a major Soviet military action against Israel is hardly within the realm of the possible. Moreover, it might be useful to recall in this connection that the Kremlin has never challenged Israel's right to exist. On the contrary, this right has been explicitly recognized in all major Soviet statements on the Arab-Israeli conflict.

In terms of protecting other U.S. interests, it should be remembered that the Sixth Fleet and its support facilities in the Mediterranean had no impact on the Arab decision to attack Israel in October 1973 or to impose an oil embargo in its wake or on the Lebanese imbroglio. Furthermore, while the U.S. military presence paradoxically has served the neutralist Arabs by allowing them to play the superpowers off against each other, it is also regarded by many littoral states as a provocative manifestation of Western imperialism.

Turning to the Soviet Union, while a generally strong defensive position in the Middle East and the Eastern Mediterranean makes intrinsic sense owing to the region's geographic proximity, on the relational level, as argued, the chief Soviet interest remains the countering of the potential threat that U.S. naval and air power represents to the security of the Soviet Union. On the negative side, the Kremlin continues to favor the dislodging of the United States from its network of support facilities. In terms of positive action, the Soviet Union has established its own naval presence in the Mediterranean, temporarily bolstering its position by gaining access to naval facilities in Egypt and, more recently, possibly in Syria, although the acquisition of actual naval and air bases in Egypt was tolerated for only two years, 1970 through 1972; however, access to Egyptian naval facilities remained available until 1976.

In reference to economic interests, it is obvious that the main concern for both the Soviet Union and the United States is the enormous petroleum deposits of the Middle East. In the case of the United States, the problem is quite clear-cut: in addition to what has been said in the category of intrinsic interests, it is a fact that the United States' most important allies—Western Europe and Japan—cannot exist without access to Middle Eastern oil. For this reason, petroleum must keep flowing, and one may safely assume that Washington will not suffer any outside (namely, Soviet) interference with its extraction or transportation. Theoretically, the Soviet Union could attempt to exert its influence in this matter in one of three ways: it could take the oil fields by force; it could encourage the Arabs to adopt measures (such as embargoes, nationalization of Western oil companies, and so forth) detrimental to Western interests; or it could attempt to interfere with the shipment of petroleum on the high seas (particularly in the Persian Gulf and the Indian Ocean). Upon closer examination, all these possibilities must be rejected as highly unrealistic. As noted, seizure of the oil fields or of oil-carrying tankers on the high seas by military means would represent acts of aggression and would most likely meet an immediate U.S. response (at least the Soviets must assume that this probably would be the case). Soviet awareness of the West's vital dependence on Middle Eastern oil and thus of Washington's probable resolute response is sufficient to deter the Kremlin from any such precipitous action. As for the Arabs, they have proved to be singularly stubborn and independent in exercising their own judgment in the matter of petroleum extraction, processing, and sale. Thus, while the prospects of controlling the flow of Middle Eastern oil might indeed be alluring to some Kremlin leaders, Moscow's freedom of action is so severely circumscribed by the realities of the local, regional, and international situations as to render these possible aspirations rather meaningless.

In the political sphere, after a temporary lull (initiated at the Nixon-Brezhnev summit of 1972 and lasting till October 6, 1973) things were back to "normal." The superpowers, which in 1972 apparently agreed to limit their competition in the Middle East, were once again cast in the accustomed roles of limited adversaries by their respective (albeit, in the case of Egypt, estranged) clients. In October 1973, it was the Arabs who struck first, sensing correctly that, preoccupied with more pressing problems, Washington and Moscow were prepared to put the Middle East on the back burner. When jolted from their posture of benign neglect, both superpowers scrambled to rekindle the contest in an attempt to advance their own influence and prestige at the expense of the rival power. In so doing, both proceeded from the traditional assumption that political successes are worth achieving and disregarded the time-honored truth that, for the outside powers in the Middle East, political successes are usually short-lived and ultimately lead to failures. The most likely explanation for this unseemly return to the fray would appear to be that traditional thought habits are hard to break, particularly when they are reinforced by vested interests, which the superpowers have been busily accumulating in the decades past.

VESTED INTERESTS

On the military-strategic level, having previously established advanced positions of strength in and around the Mediterranean, the United States has found it impossible to abandon them unilaterally. Such a course of action became particularly unfeasible in the wake of the introduction into the Eastern Mediterranean of a Soviet naval squadron—unilateral withdrawal would be widely interpreted as a sign of weakness and indecision, or so, at any rate, have thought most of the responsible U.S. policy makers. Furthermore, Washington would throw away what might eventually develop into a useful bargaining chip in possible future mutual-force-reduction negotiations between the superpowers. Finally, there has been evident among members of the policy-making bureaucracy (both Democratic and Republican) an emotional commitment to maintain and enlarge the U.S. presence in the Mediterranean that makes sense only in terms of what might be described as career investment.

It would be rather surprising if the Soviets did not by now feel the same way about their eskadra and if the determination to maintain it in the Mediterranean were not being constantly reinforced by the substantial amounts of military aid delivered to a number of littoral states, most notably pre-1976 Egypt, Syria, Algeria, and more recently, Libya. To pull out unilaterally at this juncture would no doubt, as in the case of the United States, be construed as an admission that the extensive Soviet effort in the region has been a waste—a proposition no Soviet leader is likely to accept in public.

On the economic level, powerful interests in the United States (of which the oil companies are the most notable example) have typically reinforced the bureaucratic predilection for the policies designed to maintain a favorable status quo. Hence, the real, intrinsic interests of the United States have traditionally been presented as even more important than they actually were. This has been achieved by the subtle use of relational and vested-interest arguments.

Arguments presented above with regard to Soviet vested military-strategic interests apply to the economic category as well. More precisely, having invested heavily in terms of economic, technical, and financial aid, the Kremlin leaders cannot afford an actual or implied admission of failure. Yet it is imperative to remember that these investments were made in the expectation of reaping benefits (primarily military-strategic and political), many of which have not been forthcoming. The fact that the Soviets have not been alone in pursuing this kind of policy is hardly a consolation to them, nor is it a powerful enough inducement for unilateral disengagement.

Turning to the last subcategory, political interests, it should be noted that, in addition to Washington's moral commitment to Israel, one could speak also of a very substantial domestic vested interest shared by most responsible U.S. politicians of ensuring the survival of the Jewish state. To a somewhat lesser extent, Turkey, prerevolutionary Iran, and some conservative Arab states (above all Saudi Arabia, Kuwait, Oman, and Jordan) have enjoyed a similarly privileged status. (The money spent by some of them on an extensive and expensive public-relations effort in this country is, in part, responsible for this fact.)

Clearly, in the case of Israel, a full-fledged retreat from the U.S. policy of massive support would be pregnant with great risk on the domestic political scene for those who would choose such a course of action. On a purely pragmatic level, however, this kind of commitment reduces Washington's policy options in the Middle East and can run counter to solid intrinsic interests, as the 1973 oil boycott has demonstrated. All of this, incidentally, points not to the necessity of abandoning Israel but to the desirability of a negotiated peace settlement between Jerusalem and its Arab neighbors.

The Soviets, on the other hand, also have vested political interests that often run counter to their more rational, intrinsic concerns. Specifically, by consistently picturing Moscow as a foremost and steadfast champion of the Arab cause, the Kremlin leaders have painted themselves into a corner. For one thing, the Soviet Union cannot possibly back the Arabs all the way by fighting Arab wars because this would place them squarely on a collision course with the United States. This has resulted in reduced Soviet credibility not only in the Arab East but in the Third World as a whole. Moreover, having invested heavily and in many diverse ways in their Arab client states, the Soviets have not as a result acquired a degree of influence sufficient to enable Moscow to control their actions. Thus—and again the October 1973 war is a good illustration of this thesis—the Soviet

Union has, to a significant degree, found itself at the mercy of its clients. Sadat, for example, proceeded to improve dramatically relations with the United States while the Kremlin resentfully but helplessly watched from the sidelines. Finally, Iraq, Syria, and Libya have all pursued their respective interests without undue regard to Soviet sensibilities.

CONCLUSION

The authors are fully aware of the fact that an attempt to impose a rational analytical scheme on particular foreign-policy problems will be objected to by many people for various reasons. Even so, such efforts have much to recommend them—if nothing else, they might help all concerned to examine their basic assumptions, to identify rational priorities of national interest, and, hopefully, to make policy decisions accordingly.

In the particular case of the Eastern Mediterranean and the Middle East, for example, it appears in retrospect that U.S. efforts to establish a position of strength in the early stages of the cold war were counterproductive. They not only did not prevent the deployment of a Soviet naval squadron in the Mediterranean but made its appearance a matter of high priority for the Kremlin leadership as well. Moreover, the pressure exerted on the Arabs to adopt a strong pro-Western stand actually facilitated Moscow's entry into the Arab East and contributed to the erosion of Western positions in that part of the world. The intention of this exercise, however, is not to pass historical judgments or to assign blame but to prompt a review of the current validity of policies still pursued as a result of past decisions. More particularly, in this instance, the policy of blindly maintaining a strong U.S. presence in the Mediterranean appears misguided. An attempt was made above to demonstrate that the Sixth Fleet, its supporting facilities, and Polaris submarines stationed in the Mediterranean (the latter because of technological advances that enable the deployment of U.S. seaborne second-strike capability in the world's oceans) no longer serve a meaningful, positive purpose, except, in the case of the fleet, to neutralize the Soviet eskadra. For this reason, if both were to be removed as part of a negotiated mutual-force-reduction agreement, this state of affairs would benefit not only the superpowers but also most of the regional states as well.

This line of reasoning could be objected to on the grounds that the superpowers serve as a restraining element in local crises, and to some extent they do. However, as demonstrated by the Yom Kippur War and the Cyprus crisis, Washington and Moscow cannot prevent local conflicts from occurring. If willing to act in concert, they can in fact exert an enormous influence on the belligerents, but the important point in this connection is that their ability to do so is predicated not on a naval presence in the area but on their respective control over the supply of war material to those determined to fight it out. In short, a degree of aloofness is not only unlikely to harm the intrinsic interests of the superpowers but would, on the contrary, enhance them by helping to defuse a highly volatile and explosive situation.

NOTES

1. The term derives its name from the testimony of a cabinet officer designate in the first Eisenhower administration who, in his confirmation hearings, demonstrated the all too common inability of decision makers to distinguish between public and private interests. ("What's good for General Motors, is good for America," Charles E. Wilson, Chairman of the Board of General Motors, 1952.)

2. Such implicit belief in this type of questionable assumption is probably the kind of thing that leads some analysts to distinguish between objective and perceived interests.

3. While one could probably isolate separate cultural-ideological interests under each of the three headings of the second level of the paradigm (intrinsic, relational, and vested), their relative lack of importance in actual policy making suggests that their discussion in a paper of this limited scope be confined to these few general observations.

2

SOVIET PERCEPTIONS
OF THE
MEDITERRANEAN WORLD

ALEXANDER K. KISLOV

In the aggravated international situation of today, not only do intentions and goals of a state acquire great significance but also the
ways these intentions and goals are perceived by its main counterparts in the international scene. A misperception, let alone a pre-
meditated distortion, of these goals may today lead to further deterioration of the international situation or even cause crisis. Therefore, it seems useful to present a realistic picture of the Soviet view
of the Mediterranean world and of the Soviet Union's approach to the
region's problems. Such a picture might contribute to better international understanding and, thus, to a general improvement of the
international situation.

In its approach to Mediterranean problems, the Soviet Union
first of all proceeds from the fact that for millenia the Mediterranean
basin, which is the cradle of European and many other civilizations,
has been one of the few regions of the world constantly in the focus
of international politics. It remains so today, when this area amid
the continents of Europe, Asia, and Africa plays an important role
in world politics because of its geographic position, its military-
strategic and economic prominence, and the specific features of the
sociopolitical structures of the countries belonging to the region.

It is through the Mediterranean Sea, which washes the shores
of countries populated by about 300 million people, that pass the
shortest and most developed sea and air routes linking Europe with

Africa and a good part of Asia. Day after day about 2,500 large ships under the flags of all countries of the world have been ploughing its waters. Enormous reserves of oil and the main countries exporting this important source of energy lie on its shores. The military-strategic and economic prominence of the Mediterranean basin has constantly attracted the attention of pretenders to the role of ruler of the world as an exceptionally suitable and important springboard for possible widening of their influence in western Asia, Africa, southern Europe, and throughout the world. For a long time the region has been an object of sharp struggles between its peoples and different conquerors, as well as between old and new colonizers and imperialists fighting for redistribution of spheres of influence. In our time, the Mediterranean basin is regarded as one of the hottest flash points of our planet.

Moreover, among the nearly two dozen littoral states, there are developed capitalist, socialist, and developing countries that have rid themselves of colonial and semicolonial dependence relatively recently. Hence, it is only natural that the course of events in the region is a near-perfect reflection of all the local and global phenomena of the contemporary world (such as the juxtaposition of the two main social systems, interimperialist contradictions, acute internal political and social processes, local conflicts threatening to evolve into worldwide confrontations, and so forth). All these factors form the basis of international public opinion of and the considerable and fully legitimate interests of the Soviet Union in the Mediterranean situation.

The well-grounded concern of the Soviet Union for a positive solution to Mediterranean problems has its own specific reasons also. The Soviet Union being a Black Sea, and in this sense a Mediterranean, power, is concerned that the people of all the countries of the region should benefit from economic, scientific, technical, and cultural cooperation. This is particularly important for the Soviet Union, since a considerable part of Soviet territory is connected with the world's oceans through the Mediterranean Sea. It is also through this sea that lies the shortest year-round route connecting the western and eastern regions of the Soviet Union. Finally, the Mediterranean is a doorway to the southern border of the Soviet Union, a defensive line the importance of which has recently grown considerably in connection with the adoption by U.S. leadership of the so-called limited nuclear war doctrine. Another reason for the increased importance of the Mediterranean has been the continuing growth of U.S. and North Atlantic Treaty Organization (NATO) armaments and armed forces in the region and pressing demands (formulated in particular by Alexander Haig, formerly the NATO commander-in-chief in Europe and recently U.S. secretary of state, and by other other NATO leaders

including General Secretary J. Luns) to extend NATO's geographic
limits beyond the southern Mediterranean and even further to the
south and to the east. The Soviet Union has been doing and will continue to do everything possible in order to reverse this dangerous
tendency.

It is necessary to point out here that the Soviet Union does not
consider its geographic proximity to the Mediterranean Sea a fact allowing it to make any special claims to the region. The Soviet leadership does not think that it has any rights to the natural resources
of any world region or to the sea routes or other methods used to
transport those resources. Neither does it pretend to the role of
self-appointed guardian of the Mediterranean or any of its states,
nor does it seek advantages at the expense of others. As has been
stressed by the head of the Soviet state, L. I. Brezhnev, "We want
the Mediterranean Sea to become a sea of peace, good-neighborliness
and cooperation. We realize that it is far from easy to reach this
goal, since there exist too many knots of tension, too many contradictory interests of states."

Despite the dire intricacy of this problem, the urgent need for
turning the Mediterranean Sea into a zone of peace is augmented by
the fact that acute conflict and crisis situations, often beyond any
control and portending direct danger to peace in the region and the
world over, have existed here for many years. The Middle East
conflict, which has more than once brought the world to the brink of
global confrontation, is assuming evermore severe forms, as witnessed by the sharp aggravation of the situation in Lebanon and around
Libya, by the events in Egypt, and by U.S. response to these events.
The Iraq-Iran conflict persists, as does the conflict over the western
Sahara; the situation on the Horn of Africa has not been solved. Delay in the solution of the Cyprus problem also bears a threat to peace
in the region. Although from the point of view of geography all these
conflicts and conflict situations may be qualified as local ones, in
effect, their political, military, and economic significance and impact go far beyond the territorial borders of the immediately involved
states and, as such, have an impact on the interests of all the countries and peoples of the Mediterranean and of the world at large.

Regarding the Mediterranean as an area of considerable political, economic, and military-strategic problems, the Soviet Union
has repeatedly put forward important initiatives aimed at turning the
Mediterranean Sea and adjacent regions into a zone of peace and cooperation. Also directed at fulfilling this complicated task have been
the joint proposals by the Warsaw Treaty member states enunciated
at the meeting of the Political Consultative Committee of this organization in May 1980. Soviet peace initiatives have been further developed in the decisions of the Twenty-sixth Congress of the Com-

munist Party of the Soviet Union (CPSU) in February–March 1981 as well as in recent speeches of L. I. Brezhnev. The Soviet initiatives deal both with the settlement of specific conflict situations on a just and therefore permanent basis and with the amelioration of the situation in the Mediterranean region in general.

The gist of the Soviet initiatives is that they are aimed at removing military threat from the Mediterranean area by concluding wide-scale international agreements that take into consideration the interests of all the parties concerned. Agreements that ensure legitimate rights of the region's states and security of sea-lanes and other ways of communication linking the region with the rest of the world would open the door to creating in the Mediterranean basin a situation of stability and tranquility—an aim in which all humanity is interested. In particular, the Soviet Union put forward the idea of convening a new international conference to elaborate a comprehensive Middle Eastern settlement with the participation of all the interested parties, including naturally, the Palestine Liberation Organization (PLO). The Soviet Union has also more than once proposed to start serious and businesslike negotiations on extending to this region the confidence-building measures provided for in the Final Act of the Conference on Security and Cooperation in Europe signed in Helsinki on August 1, 1975, as well as on limiting and reducing military presence and military activities in those regions.

In support of this statement, please recall the quite concrete proposals in this sphere put forward recently by the Soviet Union. The Arab-Israeli conflict has for more than 30 years kept the Middle East and the Mediterranean region in tension and more than once brought the world to the brink of global confrontation. If one is to show a modicum of fairness, it must be admitted that this was not owing to any lack of sincere desire on the part of the Soviet Union to get an Arab-Israeli settlement. As Ambassador Charles Yost, former U.S. permanent delegate to the United Nations and representative at many negotiations dealing with various aspects of the Middle Eastern settlement said, "The Russians had for eight years negotiated really and in earnest with the United States to settle the Arab-Israeli conflict, and it was the USA, not they, that put an end to those talks." And it was not the Soviet Union but the United States that refused to implement the 1977 joint Soviet-U.S. statement on the Middle East aimed at a comprehensive settlement of the Arab-Israeli conflict and moved toward the separate Egyptian-Israeli agreement.

However, the historical experience of more than ten years testifies that the Middle Eastern knot, in which national and territorial problems have intertwined as well as problems of security and many others, can be untied only by a combined approach to them. Any attempt to pull only one thread out of this knot leads to its getting more entangled. That is why the Soviet Union has long been appealing for a comprehensive settlement in the Middle East. As L. I. Brezhnev stressed in his 1981 report to the Twenty-sixth Congress of the CPSU, "As for the substance of the matter, we are still convinced that if there is to be real peace in the Middle East, the Israeli occupation of all Arab territories captured in 1967 must be ended. The inalienable rights of the Arab people of Palestine must be secured up to and including the establishment of their own state. It is essential to ensure the security and sovereignty of all the states of the region including those of Israel. Those are the basic principles. As for the details, they could naturally be considered at the negotiations."

Reporting from the highest tribune, the head of the Soviet state appealed to set the wheels of a Middle Eastern settlement in motion. He further stated, "It is time to go back to an honest collective search for an all-embracing, just and realistic settlement. In the circumstances, this could be done, say, in the framework of a specially convened international conference. The Soviet Union is prepared to participate in such work in a constructive spirit and with goodwill. We are prepared to do so jointly with the other interested parties— the Arabs (naturally, including the Palestine Liberation Organization) and Israel. We are prepared for such a search jointly with the United States and I may remind you that we had some experience in this regard some years ago. We are prepared to cooperate with European countries and with all those who are showing a sincere striving to secure a just and durable peace in the Middle East. The UN, too, could evidently continue to play a useful role in all this."

This authoritative Soviet statement was made at a moment when the failure of all other efforts at Arab-Israeli settlement, even ignoring the basic principles of guaranteed rights of all the region's peoples, became absolutely clear. As such, this initiative was appreciated by many Arab countries as urgent and concurrent with the common Arab position on establishing a lasting and durable peace in the region. It is necessary to point this out to once again disprove the allegation often made in the past that Soviet proposals for a Middle Eastern settlement are not supported by the Arabs themselves, notably the Palestinians, who, it is claimed, are against a political settlement and demand all but the physical destruction of the state of Israel. In a special statement circulated by the Palestinian news agency WAFA, Yasser Arafat, chairman of the Executive Committee of the Palestine Liberation Organization, speaking on behalf of the PLO and the entire Palestine people, welcomed the Soviet proposals, saying that they had once more demonstrated most vividly the Soviet Union's sincere desire for a genuine settlement in the Middle East.

During recent months, the Soviet Union has put forward a series of broad initiatives aimed at an amelioration of the situation in the Persian Gulf region. Striving for creation of a calm and normal situation in the region, the Soviet Union has asked the United States and the other Western powers as well as China, Japan, and all the states that show an interest in this sphere to agree to the following mutual obligations:

Not to establish foreign military bases in the Persian Gulf region and on the adjoining islands, and not to deploy there nuclear weapons and other weapons of mass destruction;

Not to use and not to threaten the use of force against the Persian Gulf states, and not to interfere in their internal affairs;

To respect the status of nonalignment chosen by the Persian Gulf states, and to reject the attempts to draw these states into military blocs with the participation of nuclear powers;

To respect the sovereign right of these countries to their natural resources; and

Not to create obstacles and threats to the normal trade turnover and the use of sea communication lines linking the countries of the region with the entire world.

Naturally, the Soviet proposals have also provided that the countries of the region themselves would be equal participants in any possible agreements concluded on the basis of these obligations; all this, of course, would accommodate their vital interests.

In general, describing the recent Soviet initiatives in this sphere, the authors of a review prepared by the London-based International Institute for Strategic Studies and titled The Strategic Situation: 1980-81 have come to the conclusion that the Soviet Union repeatedly acknowledged the importance of the Persian Gulf area for the supply of energy resources to the West and that it does not want to put up obstacles for the West in this respect. Especially stressed is the Soviet proposal on holding an international conference that would guarantee normal trade exchanges and the use of sea lanes linking this region with the other countries of the world.

However, the very idea of the Soviet Union's participation in the system of measures aimed at guaranteeing stability and security in the Persian Gulf zone is not to Washington's liking (and Washington does not even try to conceal it). The Soviet Union, whose frontier lies just over 900 kilometers from the Persian Gulf, is being proclaimed "an external force" threatening the security of the Gulf, while the United States, more than 10,000 kilometers away from the Gulf, is unilaterally including this region in "the sphere of U.S. interests" and proclaiming itself a "guarantor" of the freedom of navigation in it. Such an approach can hardly contribute to real security in this region, even against the background of the calls to observe a

sort of code of behavior in relations with the newly free countries of
Asia, Africa, and Latin America.

As for such a code, the Soviet Union believes that the establishment of spheres of influence is unacceptable and contradictory to the
well-known principles of its policy. This country has always stood
for the strict and full observance of the principle of equality and
universally accepted standards of international law in relations among
all states. Today, in regard to the Persian Gulf countries and to all
newly free countries in Asia, Africa, and Latin America this stand
means

Recognition of the right of each people to solve its own internal
affairs by itself, without fear of outside renunciation or attempts to
establish any form of dominance or hegemony over nations or to include them into the sphere of influence of any power;

Strict respect for the territorial integrity of these countries
and the inviolability of their frontiers, with no outside support for
any separatist movement aimed at dismembering these countries;

Unconditional recognition of the right of each state to equitable
participation in international life and development of relations with
any countries of the world;

Complete and unconditional recognition of the sovereignty of
these states over their natural resources and of their full equality in
international economic relations, with support for their efforts to
eliminate the vestiges of colonialism and eradicate racism and apart-
heid, in keeping with relevant UN resolutions; and

Respect for the status of nonalignment and rejection of attempts
to draw these countries into military-political blocs.

As for the Mediterranean basin itself, the Soviet Union has repeatedly stated its point of view in a very clear and concrete form.
In the Soviet opinion, achievement of the following international agreements may contribute to changing the Mediterranean area from a region of military-political confrontations (and the littoral states know
many of them) into a zone of stable peace and cooperation:

Supply of the Mediterranean region with confidence-building
measures in the military sphere, which has already demonstrated its
effectiveness;

Coordination of reduction of the military forces in the region;

Removal of ships carrying nuclear weapons from the Mediterranean Sea;

Nondeployment of nuclear weapons in territories of the nonnuclear Mediterranean countries;

Assumption of obligations by nuclear powers to not use nuclear
weapons against any of the Mediterranean countries that do not permit

deployment of such weapons in their territories. Specifying this proposal, Comrade Brezhnev, in an answer to the question of Greek newspaper Ta nea in April 1981, stated,

> As the Soviet Union has declared more than once, it will
> never use nuclear weapons against those countries that
> refuse production and the acquisition of nuclear weapons
> and do not have them on their territory. This is a rather
> stable guarantee. But we want to go forward and at any
> time conclude a special agreement with any of the non-
> nuclear states, including Greece if it takes an obligation,
> in its turn, not to have nuclear weapons on its territory.

Putting forward this broad series of proposals, the Soviet Union does not claim any monopoly to peaceful initiatives. At the most responsible level it has repeatedly declared its readiness to discuss with all the states concerned any other initiatives and ideas leading to international détente, the widening and strengthening of international cooperation in very different spheres, and general amelioration of the world situation.

All this demonstrates that the basis of the Soviet approach to the Mediterranean area and adjacent lands is a desire not to turn the Mediterranean Sea into a sea of division, military danger, and confrontation, but to restore its historical role as the link between three continents, contributing to the economic, scientific, technical, trade, and cultural cooperation among the people inhabiting its shores and adjacent lands. The multiplicity and the combined nature of new Soviet initiatives demonstrate that they deal not only with ways of overcoming the recent conflicts but with arranging certain international preventative measures to guarantee the region against its use by some circles as a kind of detonator of international crisis.

While being very flexible in searching for solutions to the most urgent problems of the Mediterranean area, the Soviet Union at the same time strictly observes the agreed framework of the Final Act of the Conference on Security and Cooperation in Europe, including a special part on problems of security and cooperation in the Mediterranean area. In this part, the participants of the conference consider that the strengthening of security and the development of cooperation in Europe will stimulate positive processes in the Mediterranean area and express the striving to contribute a course of peace, security, and justice to the region in the interests of states, participants, and the Mediterranean countries not participating in the conference. Particularly, they state their desire to support and widen the contacts and dialogue initiated by the Conference on Security and Cooperation in Europe with nonparticipating Mediterranean states, including all

the Mediterranean states, in order to contribute to peace, arms reduction, security, and relaxation of tension and to widen the sphere
of cooperation—the tasks in which all of us have a common interest—
as well as to determine subsequent common tasks.

In the case of their realization, the Soviet initiatives concerning
the Mediterranean area will not only create favorable conditions for
strengthening mutual confidence but also facilitate development of
mutually beneficial economic, scientific, technical, and other kinds
of cooperation among all the Mediterranean countries, as provided
for by the final act signed in Helsinki. It should be noted that when
the Soviet Union puts forward its peaceful initiatives, fully recognizing
the interests of the Mediterranean peoples and countries, it strives to
incorporate into its proposals, to the maximum extent possible, the
ideas expressed by the broadest and most representative forums of
the region's peoples. This intention has been testified to in documents
of such mass forums as the International Conference against Imperialist Bases for Security and Cooperation in the Mediterranean, convened
in Malta in March 1980, and the International Conference for Liquidation of Foreign Military Bases in the Mediterranean Sea, which worked
on the island of Crete in June 1980.

Speaking of the Soviet initiatives for cardinal amelioration of
the situation in the Mediterranean area, it would be worthwhile to
dwell upon another important point. It is well known that the international situation in general as well as the course of events in any of
the world's regions are significantly dependent on Soviet and U.S.
approaches to the problems and on the state of relations between them.
The acuteness of international problems demanding solutions dictates
the need for an active Soviet-U.S. dialogue at all levels. Judging by
some statements, this point of view is not contradicted by U.S.
President Reagan, who has spoken about the importance of constructive cooperation between the United States and the Soviet Union in
order to maintain peace. As for the Soviet Union, its position was
formulated in a very precise manner. L. I. Brezhnev at the Twenty-
sixth Congress of the CPSU stressed, "We are prepared to have this
dialogue. Experience shows that the crucial factor here is meetings
at the summit level." Calling upon the U.S. government and the
governments of other NATO countries to think calmly and objectively
about the whole complex of international problems, including Mediterranean problems, in order to find solutions acceptable to all countries (and first of all to the countries of this region), the Soviet Union
lays down only one preliminary condition—there must be no preliminary conditions at all. Regrettably, Washington's deeds have so far
failed to testify to any real intent to find a speedy solution to the acute
problems facing the countries of the Mediterranean basin. Everyone
is supposed to know from early childhood that flammable materials

must be kept away from fire; all the more so if one deals with gunpowder. But in Washington and some other Western capitals, policy makers apparently think otherwise. On the fabricated pretext of defense of its vital interests against the "Soviet threat" to the oil resources in the Middle East, extremely distorting the Soviet peace initiatives, the United States is continuing to build up its armed forces and weapons in the Indian Ocean basin and to widen the network of its military bases, having concentrated in the Mediterranean Sea and around it an unprecedented naval and air armada. Such actions are totally unjustified because it is ridiculous to think that one can defend oil interests by turning oil fields into ammunition dumps.

This fact is quite clear to those in Washington who are persistently searching for pretexts to build up U.S. military power in the region. Thus, references are made to the situation in Afghanistan and around it. They persistently state, for example, that the limited Soviet military contingent temporarily stationed in Afghanistan at the request of the Afghan government appears to be a threat to the oil field in the Persian Gulf region and that the U.S. and some other Western armed forces deployed in the western part of the Indian Ocean are a sort of counterbalance to the Soviet contingent. However, it is well known that not only the very decision on possible deployment of U.S. armed forces in the western part of the Indian Ocean was already fixed in President's Directive 18 of August 1977 but that practical deployment began long before the events in Afghanistan. Long before those events, Washington unilaterally broke off the rather successful Soviet-U.S. talks on the limitation of military activities in the Indian Ocean region. It is significant that in spite of the special part of the Common Soviet-American Communiqué (signed during the meeting between the head of the Soviet state L. I. Brezhnev and U.S. President Jimmy Carter in Vienna in June 1979), which expresses the consent of the parties to the meeting of their proper representatives, without delay, to discuss renewal of the talks on the limitation of military activities in the region of the Indian Ocean, these talks (not because of the Soviet Union) have not been renewed. The United States has not given a proper response to the repeated Soviet proposal to limit military activities in the Mediterranean Sea itself.

As for the situation in Afghanistan and around it, the principle for the Soviet position on this problem is well known and clear. It proceeds from the fact that rejection of a people's right to revolution often means the rejection of their right to exist as a sovereign state. For example, the United States was born as a state as a result of revolution, receiving, as it is known, military assistance from the French army. It is an axiom that any revolutionary action clashes with counterrevolutionary reaction; the experience of Afghanistan has proved it again. A majority of the population, which was oppressed and had no rights, was vitally interested in the victorious completion of the April 1978 revolution. But a small minority—feudals, some tribal leaders, extreme nationalists, monarchists, and some Muslim clergymen—hated the revolution and launched a fierce struggle against it, managing to involve in the struggle temporarily a certain number of ignorant peasants, traders, and nomads. Nevertheless, we can undoubtedly state that this mixed counterrevolutionary conglomerate could not have fought long against a revolution expressing the interests of the overwhelming majority of Afghan people without wide support and interference from outside.

From the first months of the revolution, a real, undeclared war was launched against Afghanistan. It should be remembered that for about three years arms have been flowing into Afghanistan in large quantities, mainly through Pakistani territory. The fact of those arms supplies was directly confirmed by late Egyptian President Sadat shortly before his assassination. Thousands of mercenaries and bandits, equipped and trained by foreign military instructors, and other foreign agents were sent to Afghanistan. U.S. Defense Secretary Weinberger spoke about the usefulness of U.S. arms deliveries to the Afghan counterrevolutionaries, and his statement was supported by President Reagan. Thus, the U.S. president, for the first time, officially confirmed this dangerous course.

When in late 1979 the Afghan revolution became seriously endangered, threatening the loss of Afghanistan's independence, the Soviet Union in response to the repeated appeals of three successive Afghan governments, made a decision to send a Soviet military contingent to Afghanistan. Naturally, it was not an easy decision for Soviet leaders to make. First, it was necessary to support the Afghan government to repel the aggression from abroad. Second, there appeared to be a possibility of turning the friendly country, which has more than 2,000 kilometers of border with the Soviet Union, into a hostile strategic springboard for the oppression of progressive movements in Asia. Finally, we should not forget the human aspect of the problem. A Soviet refusal to respond to the Afghan government's appeal for assistance could have facilitated the forces of reaction in that country in doing what they did in Chile, where people's freedom was drowned in blood. All the facts indicated that the extent of bloodshed in Afghanistan might even have surpassed that in Chile.

Making this decision, the Soviet leadership acted strictly in accordance with the standards and principles of international law. The Soviet Union has no other goals but to help a friendly neighboring country repel and stop aggression from abroad and to prevent creation of a new point of tension on the southern USSR border. L. I. Brezhnev stressed, "We say to those who even from the very heart are appealing to stop Soviet military assistance to Afghanistan: firstly it is necessary to be worried about how to remove the reasons which made this assistance essential. A political solution can be achieved, and the essence of it is the guaranteed ending of the military invasion of counterrevolutionary forces into Afghanistan from the territories of the neighbouring states."

A well-grounded and concrete program opening the doors to this settlement was put forward by the Afghan government in May 1980. Permeated by a desire to change the course of events around Afghanistan and turn it toward the route of political settlement, that initiative, refined and supplemented in August 1981, with due regard to the opinions of a number of other countries, proceeds from an objective analysis of the real situation. It provides for holding bilateral Afghan-Pakistani and Afghan-Iranian talks in order to work out agreement on the wide range of questions of their relations on the basis of goodneighborliness, noninterference in each other's internal affairs, and mutual beneficial cooperation. Considering bilateral talks with Iran and Pakistan more fruitful and preferable, the government of Afghanistan at the same time expressed its readiness, given a willingness on the part of Iran and Pakistan, to hold trilateral talks. It does not object to the UN secretary-general participating in such talks. The Afghan government believes that the terms of the settlement, strictly providing for rejection of any interference in its internal affairs, should include reliable guarantees by some states, acceptable to Afghanistan and its neighbors, including the Soviet Union and the United States. Putting an end to and guaranteeing nonrenewal of military interference in the internal affairs of Afghanistan would help settle some of the problems dealing with a procedure and a timetable for the withdrawal of the Soviet military contingent. At the same time, as it was stressed in a Tass special statement, it must be definitely clear that as long as military interference and acts of aggression against the people and government of Afghanistan continue, the Soviet Union will keep supporting this friendly neighboring country on such a scale as will be necessary.

As L. I. Brezhnev stated in his report to the Twenty-sixth Congress of the CPSU, "The Soviet Union is prepared to negotiate the Persian Gulf as an independent problem. It is also prepared, of course, to participate in a separate settlement of the situation around Afghanistan. But also we do not object to the questions connected with Afghanistan being discussed together with the questions of Persian Gulf security. Naturally, this applies solely to the international aspects of the Afghan problem, and not to internal Afghan affairs. Afghanistan's sovereignty, like its nonaligned status, must be fully protected."

All these proposals have so far met with no response. However, if we approach them in an unprejudiced manner, they show the real way to political settlement in this region of our planet. As the Paris-

based magazine Africa-Asia reported, Indian Prime Minister Indira
Gandhi, receiving the foreign minister of one of the Asian countries,
stated, "We can settle the Afghan problem in several weeks. It is
Washington, which does not even want to examine the very possibility
of such settlement on the basis we consider constructive and durable,
that remains the main obstacle. We are confident that the Pakistanis
and Iranians are ready to examine the recent proposals of the Afghan
government. But the U.S. Secretary of State A. Haig responded with
a categorical 'No!'." According to I. Gandhi, the magazine went
on, "Washington believes that the so-called Afghan problem must
remain unsettled in order to use it in a cold war against the Soviet
Union."

One can say that the U.S. policy toward the situation in Afghanistan and around it is a special case and is not linked with the Mediterranean problems. Naturally, Afghanistan does not belong to the Mediterranean countries, but this question is raised because Washington
tries to justify many of its actions in the region by references exactly
to Afghan events. It also connects directly the concentration of its
armed forces in the region of the Eastern Mediterranean and especially
in the Persian Gulf with the deployment of the Soviet military contingent in Afghanistan.

Washington acts the same way in the Mediterranean basin itself.
There are many examples of this that none of those who are able to
see the facts and objectively analyze present realities can fail to note.
One of these is U.S. reaction to the latest Soviet proposals, discussed
above. The United States refuses even to discuss them or to put forward any counterproposals to serve the same goal, that of turning the
Mediterranean area into a zone of stable peace and friendly cooperation.

And what about Washington's efforts to pull Spain into the military
bloc of NATO? What about its policy of deployment in the Mediterranean countries, particularly Italy, of U.S. medium-range nuclear
missiles? One need only to glance at the map to understand that the
choice of the Sicilian town of Comiso as a base of U.S. nuclear missiles creates a serious threat to the countries of Africa and the Middle
East. There is another threat to the Mediterranean countries created
by the repeated maneuvers of Washington aimed at consolidating U.S.
military presence and proliferating U.S. military naval and air bases
in Spain and Greece, as well as by augmenting the U.S. Sixth Fleet,
constantly deployed in the Mediterranean Sea, with NATO naval units.
It is expected that besides Italian, Greek, and Turkish ships this
naval force will include a number of combat ships from the United
Kingdom, the Federal Republic of Germany (FRG), Holland, and
probably some other NATO countries. The recent agreement concluded by U.S. President Reagan and Israeli Prime Minister Begin,

which provides for so-called "strategic cooperation" between the two countries and gives to the Pentagon the possibility of using the territory and armed forces of Israel for its own purposes, can also be considered evidence of the further escalation of U.S. preparations for direct military involvement in this region.

Witness also the approach of the United States and the West as a whole to the problem of Cyprus. The U.S.-led NATO countries have ignored proposals supported both by the government and by the people of Cyprus for convening a representative international conference under UN auspices to discuss international aspects of the Cyprus problem with due regard to the interests and rights of both the Greek and the Turkish communities in Cyprus and on the basis of full respect for the independence, sovereignty, and territorial integrity of the Republic of Cyprus. Instead, NATO has been trying to resolve that complex issue, potentially very dangerous for peace in the Mediterranean, in a "family circle," with maximum gain for itself.

Certainly, it is quite clear that after the loss of the Iranian outpost by the United States and in connection with the development of a more aggressive strategy toward Middle Eastern countries, the importance of Cyprus as a possible springboard, from the point of view of the Pentagon, has increased. This may explain the reports appearing in Western media about deployment of foreign troops under the cover of NATO and the stationing of a military base in the northern part of the island in addition to those formally British but actually U.S. bases that have existed there for a long time. Naturally, both the old and the new bases will be used by the United States as an important stronghold for U.S. rapid-deployment forces.

All these facts can be understood and explained, but they cannot be justified. The key to the amelioration of the Mediterranean situation can only be found by common efforts of Mediterranean and other countries aimed at the liquidation of knots of contradictions, not by further intensification of military preparations in this region of the world.

The list of actions taken by the United States and other NATO countries that are far from being helpful to the amelioration of this situation could be continued. The common denominator is that Washington considers the Mediterranean area not as a link between the countries situated in the basin but as a southern flank of NATO; as the deployment area for the nuclear-equipped U.S. Sixth Fleet; and as the region controlling approaches to the Middle East, North Africa, and a broad area of the Indian Ocean, including the Persian Gulf. Dwelling upon the Soviet perceptions of the Mediterranean problem, it must be pointed out once more that, from the Soviet point of view, these problems cannot be settled militarily. Rejecting the imperialist policy of diktat and blackmail, the Soviet Union expresses its firm confidence that the most complicated international problems can and must be solved by way of negotiations. We can find reasonable solutions only if we act together in joint efforts. There is no other way. Only political means can produce a settlement of such dangerous problems as the Arab-Israeli and Iraq-Iran conflicts, the Cyprus problem, the situations in western Sahara and the Horn of Africa, and so forth. And one strict condition is to not permit anyone to aggravate states toward each other in order to plot and to benefit from their contradictions.

Categorically rejecting the "Soviet military threat" concept that is being persistently spread in the West, the highest organ of the Soviet state, the Supreme Soviet of the Union of the Soviet Socialist Republics, in its appeal to the parliaments and peoples of the world solemnly declared, "The Soviet Union is not threatening anyone, it does not seek confrontation with any state in the West or in the East. The Soviet Union has not been and is not striving for military superiority. It has not and will not initiate new spirals in the arms race. There is no type of weapon it would not agree to limit or to ban on a mutual basis by agreement with other states. The safeguarding of peace has been and remains the supreme aim of the foreign policy of the Soviet Union." These words fully apply to the Soviet approach to Mediterranean problems.

3

NEW DIMENSIONS OF
BIG-POWER RIVALRY
IN THE MIDDLE EAST

GEORGE LENCZOWSKI

As an oil-bearing, strategic, and conflict-ridden region, the Middle East continues to claim high priority in the policies of major powers, especially the United States and the Soviet Union. Today, at the beginning of a new administration in Washington, it seems appropriate to ask whether such changes as have occurred in recent years in the area have affected the fundamentals of big-power relationships or have signified mere moves on the chessboard of the basic pattern of rivalry for power, influence, economic benefits, and strategic advantage.

Perhaps it would be useful to define "recent years." The suggestion here is to focus on the entire decade 1970-80 as one replete with major developments in international politics of the region. The following ten events may be identified as relevant to this study:

British withdrawal from the Persian Gulf, 1971
Expulsion of Soviet advisers from Egypt, 1972
The Arab-Israeli war of 1973 and the oil embargo
The Cyprus crisis, 1974
Civil war in Lebanon and Syrian intervention, 1974-75
Sadat's peace initiative and the Camp David peacemaking, 1977-78
Communist takeover and Soviet invasion of Afghanistan, 1978-79
Revolution in Iran and the hostage crisis, 1979-80
The Iraq-Iran War, 1980-81
Upsurge of Islamic militancy from 1978 onward

One may inquire: Are these events symptoms, causes, or effects of any changes in big-power rivalry? Can they be invariably traced to that rivalry or may some of them be considered as phenomena independent of it? Has this rivalry undergone significant qualitative or quantitative changes?

Specifically, it may be useful to ask four major questions: (1) Has the position or policy of the Soviet Union toward the Middle East appreciably changed during the period under review? (2) Has the position or policy of the United States been subjected to significant reformulations and changes in implementation? (3) Has there been a major change or realignment in the position and power of regional countries? and (4) Has there occurred a significant change in the role of leading industrial nonregional powers in the Middle East?

THE SOVIET UNION

The world communist revolution and global extension of Soviet power and influence have never been officially repudiated as ultimate goals of Soviet policy, notwithstanding the advent of détente in the 1960s and 1970s. The Helsinki agreement of 1975 did not seem to bring about a real change in the global Soviet policy. Numerous statements regarding coexistence made by Khrushchev and his successors were always made with explicit or implicit reservations that (1) the ideological struggle with the West must continue, and (2) the Soviets consider aid to "liberation movements" an integral part of their policy. The Helsinki agreement could be summed up under two major headings: first, Western recognition of the Soviet sphere of influence in Europe, and second, the Soviet pledge to honor certain essential (from the Western point of view) human rights and freedoms. While the first part of the bargain—reaffirmation of Soviet dominance in Eastern Europe—was honored, it required no action but merely a hands-off posture on the part of the West. The second part, pertaining to human rights, remained essentially unfulfilled. As a ranking Soviet diplomat aptly described it, "It is silly for Washington to expect that by signing the Helsinki agreement the Soviet Union had pledged to dismantle its political system."[1] Subsequent international conferences in Belgrade in 1980 and Madrid in 1981 bore eloquent testimony on this latter point.

Furthermore, in 1968 Chairman Brezhnev pronounced his doctrine legitimizing Soviet military intervention in socialist countries to justify the suppression of Dubcek's "socialism with a human face" in Czechoslovakia. Again, there is no evidence that this basic tenet of Soviet policy has ever been repudiated.

Although both the Helsinki agreement and the Brezhnev Doctrine have focused on Europe, awareness of their existence and consequences is relevant to the study of the Middle East itself, because both documents have global implications.

The persistent Soviet objective in the Middle East has been to extend Soviet influence (or domination) to the area and, ipso facto, to remove Western presence or influence. The Soviets have used a variety of methods to accomplish this objective. Where possible and profitable they used military force to establish Marxist-oriented or puppet regimes at the expense of the sovereign states in the area, as was the case in Iranian Azerbaijan and Kurdistan in 1945-46 (both following an earlier pattern in Gilan in 1920-21). They formulated territorial demands toward Turkey in 1945 (Kars, Ardahan, and the Turkish Straits). These overt acts of expansionism, characteristic of the Stalin era, were under Khrushchev replaced by more flexible methods of courting the nonaligned states, identifying the Soviet Union with the national aspirations of postcolonial peoples and forming closer ties with such revolutionary regimes as had come into power in Egypt, Syria, Iraq, Algeria, Yemen, and South Yemen.

Soviet relationships with these "radical" governments were based on broad political support for their struggle against remnants of Western imperialism and their resistance to Israel. Aid and trade, substantial arms supplies, and the ready availability of Soviet advisers, civil and military, constituted the principal tools of this policy. Communist propaganda and organizational activity in those countries were generally adjusted to the needs of Soviet diplomacy. In Iraq, the Communist party made a bid for power in 1959, partly in Baghdad, by exploiting dictator Kassem's struggle with the Nasserites, but primarily in Mosul and Kurkirk. Later, Iraqi Communists tried to turn the Kurdish nationalist rebellion to their advantage. By contrast, in Egypt the local Communist party found the competitive force of Arab socialism too strong to contend with and, after some brushes with Nasser (in the course of which Khrushchev could not restrain his irritation), it chose self-dissolution while opting for individual penetration of the Egyptian political apparatus as a more desirable alternative.

Soviet diplomacy was seconded by the growth of Soviet military and naval presence. Soviet instructors—and sometimes pilots on active duty—found themselves stationed in Egypt, Syria, and Iraq; and the Soviet navy made its substantial appearance (45-50 ships on the average) in the Mediterranean as well through regular visits to the Persian Gulf.

Broadly, by the early 1970s, in terms of paramount influence, the Soviet Union replaced Western European powers in those key Arab countries in which the West had previously held a dominant position: this was true of the former British and French mandates of Iraq and Syria, respectively; of formerly French Algeria; the ex-British Crown Colony of Aden; and ex-Italian Libya.

In thus penetrating the Middle East, the Soviet Union benefited from three converging factors: nationalist struggle against colonialism and the general antiimperialist ethos this struggle had engendered, the Arab-Israeli conflict, and the often failing socioeconomic performance of local political systems.

The decade of 1970-80 marked a period of impressive Soviet advances into the area. A number of countries established such close relations with the Soviet Union as to qualify for client status. In a number of cases, this special relationship was formalized by means of a treaty. One may classify as client states Iraq (treaty of 1972), Syria (treaty of 1980), Somalia (treaty of 1974), Algeria, and Libya. And, following the Communist revolution and conquest of Afghanistan, three countries acquired the status of virtual satellites: South Yemen, Ethiopia, and Afghanistan itself. In each of these cases a treaty of alliance reaffirmed their close ties and virtual dependence on Moscow. The difference between the client and the satellite states focuses on Soviet ability to manipulate rather decisively the internal political process in these countries. While collaborating with the Soviet Union, clients maintain their sovereign prerogative to manage their domestic affairs by their own means; by contrast, the satellites fall into a state of pronounced dependence on Soviet decision making. Furthermore, the satellites adhere to Marxism-Leninism as their guiding ideology, while the clients can be broadly identified with one or another version of Arab socialism.

In addition to these political successes, the Soviet Union gained access to some eight military bases located in the territories of Iraq, South Yemen, Somalia, and Ethiopia. Similarly, the Soviet navy acquired anchorage facilities off the coasts of Algeria, Libya, and Crete. The Strait of Bab el Mandeb found itself within the close radius of Soviet ships operating from the naval base in Aden. Through the conquest of Afghanistan, the Soviet air force shortened its distance from the Strait of Hormuz by some 800 miles. In recent years, Soviet naval traffic through the Bosphorus and the Dardanelles averaged two to three passages a day.

To keep the Soviet position in proper perspective, it must be acknowledged that Moscow's advances were not a story of undiluted success. During the period under review, the Soviets experienced certain reverses and defeats, some perhaps due to their excessive self-confidence, some to the aggressive methods employed to promote their interests, and some to the reckless behavior of local communist parties.

Thus, in 1971 an abortive communist coup in Khartoum considerably reduced growing Soviet influence in Sudan. Serious miscalculation of Sadat's nationalist temper led to the expulsion of Soviet military advisers from Egypt in 1972. The Aden-supported guerrilla

warfare waged in Dhofar against the Sultanate of Oman suffered a
setback (perhaps only temporary) at the hands of the combined forces
of Sultan Qabus and the shah of Iran in 1975. Similarly, disenchanted
with the Soviet support of Marxist Ethiopia, the radical government
of Somalia abruptly severed its client relationship with Moscow in
1977 and, in due course, made the former Soviet base in Berbera
available to the United States. Furthermore, Soviet relations with its
client Iraq began to cool off appreciably after the revolution in neighboring Iran in 1979. This process was punctuated by the repeated
death sentences meted out to the communist leadership in Iraq and by
Soviet neutrality and virtual stoppage of arms supplies to Baghdad in
the wake of the Iraq-Iran War. Last but not least, the two conferences
of Islamic states in Islamabad in 1980 and Mecca in 1981 issued strong
condemnations of Soviet invasion of Afghanistan. Iraq, in fact, was
one of the promoters of the initial resolution of 1980. Only Syria
(because of its dangerous confrontation with Israel), Marxist South
Yemen, and Libya were critical of or reluctant to espouse this anti-
Soviet stance.

In comparing the decade of the 1970s with the preceding period
that followed World War II, is it possible to register some radical
departures from the established Soviet doctrinal positions and actual
behavior? Doctrinally, the 1970s did not seem to introduce major innovations. Since the abandonment by Khrushchev—for reasons of
grand strategy—of the Stalin-Zhdanov rigid view of the bipolar world
in favor of a more flexible approach to the Third World in 1955, theoretical foundations of Soviet policy appear to have remained fairly
constant, with only certain tactical modifications. The general corpus
of Soviet theory has become enriched by the addition of the earlier-
noted Brezhnev Doctrine in 1968.

As for actual Soviet behavior toward the countries of the Middle East, it is also difficult to discern any fundamental change.
There is a constant striving to achieve the same basic goals of their
foreign policy: expansion of Soviet influence by a variety of methods
and reduction of Western presence. The methods used vary according
to the circumstances, oscillating between sheer use of force and
milder diplomatic approaches. Theory—in this case the echoes of
the Brezhnev Doctrine—occasionally dictates a heavy-handed action,
as demonstrated in Afghanistan, and may prove counterproductive.
A single conquest of an independent Islamic country has the potential
of undoing many gains achieved in the Third World. Moreover, even
though Middle Easterners have never been noted for great solicitude
toward the plight of Eastern Europe since World War II, a mass
workers' movement against the communist system such as has
shaken Poland in 1980-81 cannot pass unnoticed in this age of mass
communications. If we add to it the Islamic political resurgence be-

ginning with the later 1970s, the Soviet or local communist task of persuading the elites and masses of the Middle East of the superiority of the communist system becomes more difficult.

On the other hand, it should be remembered that such successes as the Soviets have scored in the Middle East over the years could not primarily be ascribed to the impact of communism as an ideology. These successes, on the contrary, could be traced to the positive response of local leaderships—in Nasser's Egypt, in Syria, in Iraq, and elsewhere—to Soviet offers of arms, aid, and diplomatic support in their struggle against the real or imaginary Western imperialism or the assertive behavior of Israel. On the threshold of the 1980s, with memories of direct Western colonialism gradually fading away, the Arab-Israeli conflict has remained the single most important stimulant of actual or potential pro-Soviet attitudes in the Arab world.

Thus it is difficult to speak of a new dimension of Soviet policies in the qualitative sense. Although the Soviet Union has acquired new strategic strongholds in South Yemen and on the Horn of Africa, its conquest of Afghanistan is a costly venture not devoid of harmful consequences. Ultimately, it might prove to be an advantage if the Soviets succeed in "Finlandizing" Pakistan or securing control of the Strait of Hormuz.

As for Iran, the causes of its revolution were complex, partly internal and partly external, primarily U.S. misjudgments, and possibly indirect Soviet assistance to Khomeini. Regardless of the degree of presumed initial Soviet involvement, there is no doubt that Khomeini's revolution has opened new and enticing vistas to the Soviet leadership. In spite of its being a manifestation of Islamic resurgence, Khomeini's movement has not only brought discredit to Islam—at least in Iran—but also has caused such a disarray in the country as to seriously enhance the Tudeh party's and other Marxist groups' chances to seize power. Should that occur and should a nationalist reaction against such hypothetical communist rule take place, the Soviets might face a dilemma similar to that faced in relation to Afghanistan: to choose between accepting the defeat of their ideological allies or saving them from destruction by lending them massive military support.

Finally, with reference to Soviet setbacks in the Arab world, the defeats in Egypt, Sudan, and Somalia were painful. But is it possible to affirm with certainty that they were irreversible? With special reference to Egypt, in many ways a key country in the Arab world, its present estrangement from the Soviet Union is a function of its complex relationship with Israel and the United States. This relationship is based to a large extent on the expectation that both parts of the Camp David Agreement—those pertaining to Egypt and those pertaining to Palestine—would be implemented. Should this expectation be frustrated, Egypt's policy might undergo substantial or even dramatic changes. "The enemy of my enemy is my friend" is an adage too old and too often translated into practice to be safely ignored.

To conclude, while Soviet military threat to the area—especially after the takeover of Afghanistan—must figure as a possibility under certain circumstances, in any U.S. contingency planning it is the politically motivated receptivity of local regimes to Soviet proffers of support that constitutes the most important vehicle for Soviet advances in the area.

THE UNITED STATES

The aims of U.S. policy in the Middle East since World War II have often been identified as (1) access to the area's oil, (2) maintenance of peace and stability, and (3) resistance to Soviet expansion. These points are not mentioned here necessarily in their order of importance; they are rather ranged to correspond to the analysis that follows.

With reference to the first objective, access to oil, the nature of U.S. interest in it underwent a change in the 1970s. Initially, that is, in the period following World War II, interest in Arab and Iranian oil was based more on the natural tendency of U.S. free enterprise to secure new sources of supply and, through a worldwide network of transportation and refining and marketing facilities, ensure commercial advantage and profits. In those early days, the United States held primacy in production and exports on a world scale, and the hypothetical cutoff of Middle Eastern supplies did not carry with it the strategic and economic threat that it later acquired. In fact, when such a cutoff temporarily materialized during the Suez crisis of 1956, the U.S. capacity to produce and export was so substantial as to alleviate, without excessive effort, Europe's shortages.

Beginning in approximately 1970, the United States became an oil-deficit country. Its previous production peak of 11.5 million barrels per day (mbd) experienced a gradual shrinkage, to bottom out at about 8.5 mbd in the mid- and late 1970s, and this despite the Alaskan discoveries during that decade. In the meantime, until 1978-79 consumption of oil was steadily growing in both the United States and the major industrial democracies. Although the North Sea discoveries had tended to decrease Europe's relative dependence on Middle Eastern oil, the total Western European production, which reached 2.7 mbd by 1980-81, was not decisive enough to change the fact that some 60 percent to 65 percent of Europe's consumption had to rely on Middle Eastern sources of supply.

The war-generated oil embargo of 1973 and the subsequent quadrupling of oil prices by the Organization of Petroleum Exporting Countries (OPEC)—two distinct though related phenomena—brought forth, with greater urgency than ever before, the realization that the West's oil supplies from the Middle East are subject to two parallel dangers: a politically motivated denial of access and an economically motivated prohibitive price level. At the same time, concern began to be voiced in Washington regarding Soviet capacities and requirements in this sector. Whereas in the 1960s Sino-Soviet production (mostly Soviet) stood at the level of 8.5 mbd while U.S. production hovered around 11.5 mbd, in the 1970s not only did the roles become reversed, but by 1980-81 Sino-Soviet production climbed to 14.3 mbd. This substantial augmentation in output obviously denoted a higher demand in the Sino-Soviet bloc, including Eastern European satellites whose growing industrialization required increasing supplies of energy. Repeated studies issued on this subject by the Central Intelligence Agency (CIA) have been far from conclusive: from initial forecasts of a 3- to 4-mbd Soviet oil deficit expected in the mid-1980s, the agency lowered its predictions of the supply-demand relationship in the Soviet sphere to the point of almost repudiating its earlier estimates. Linked with this was the problem of whether the United States should issue permits for export of drilling equipment to the Soviet Union: the question was whether this course would lessen a possible Soviet interest in securing access to Iranian and Arab oilfields and yet, by increasing Soviet output capacity, strengthen their military potential.

The Arab oil embargo of 1973-74 gave rise to a spate of speculations—some in the form of articles by academic and defense analysts—about a possible seizure of the Persian Gulf oil fields by the U.S. armed forces. Inasmuch as some authors were known for their pronounced sympathies for Israel, some confusion was bound to ensue as to whether such proposals reflected U.S. or Israeli interests and objectives in the region.

Regardless of the precise answers to the problems of Western and Soviet access to the area's oil and its exact price level, it would be legitimate to say that, during the decade of the 1970s, Middle Eastern oil acquired a much more emphatic strategic connotation than it had had in the preceding period. Here, it would be justified to suggest that the change was of a qualitative nature.

The maintenance of peace and stability in the Middle East figures as the next U.S. objective. This was certainly an early aim, already expressed by President Truman and generally upheld by the successive presidents. To pursue this goal, Ralph Bunche acted as mediator in the first Arab-Israeli war in 1948; President Eisenhower and John Foster Dulles exerted efforts to put an end to the Suez crisis in 1956; U.S. troops landed in war-torn Lebanon in 1958; Kennedy's special envoy Ellsworth Bunker went on a peace mission to Yemen in the early 1960s; President Johnson conferred with Premier Kosygin at Glassboro in 1967; Henry Kissinger conducted his shuttle diplomacy in 1973-74; and President Carter launched the Camp David peace-making process in 1978.

While all these actions appear as a persistent U.S. striving to save the area from war and disruption, a question may be posed as to whether, in the 1970s, the original aim of peace and stability was not overshadowed by the ever-stronger U.S. commitment to Israel.

The U.S. government's attitude to Israel has from its very inception been characterized by friendliness and readiness to help. But during the first Arab-Israeli war in 1948 this did not include the supply of arms. In fact, at that time Israel had to fend for itself in this respect, either by buying arms from the Soviet bloc sources or by using clandestine methods to secure them from the West. Similarly, the United States adopted a reserved—or, more properly, negative—attitude toward Israel's territorial expansionism—regardless of its motivations and justifications—in the 1950s and 1960s. Israel's evacuation of the Sinai Peninsula and the Gaza Strip following the Suez War of 1956 was due first and foremost to the pressure of the Eisenhower administration, expressed in the quiet but decisive cutoff of further aid.

The Johnson and, at least nominally, the Nixon administrations were also on record opposing territorial aggrandizement. According to Johnson, "Boundaries cannot and should not reflect the weight of conquest."[2] In a similar vein, Nixon's Secretary of State William Rogers declared, "We do not support expansionism. We believe troops must be withdrawn as the [UN 242] resolution provides."[3] Both statements were in harmony with what Nixon's special envoy William Scranton had called on December 13, 1968, a need for "evenhanded policy."[4]

Although the change in U.S. policy in this respect seemed gradual, it was nevertheless so substantial as to warrant the question whether it did not constitute a de facto reversal of its previous stand. Beginning with the early 1970s, especially corresponding to the Nixon administration, even the president preferred to shun references to an "evenhanded" approach. While the United States officially held to the validity of the United Nations (UN) Security Council Resolution 242 of 1967 (which among other important points called for Israeli withdrawal from occupied territories), in practice the massive transfers of sophisticated weapons to Israel rendered the implementation of this resolution ever more unrealistic. And in contrast to the avoidance of military intervention during the wars of 1948, 1956, and 1967, the United States not only engaged in massive airlift to the hard-pressed Israelis in 1973 but also called for a worldwide alert of U.S. forces in preparation for a possible confrontation with the Soviet Union, itself a massive supplier of arms to Egypt both before and during the October hostilities.

Subsequent U.S. peace efforts leading to the Sinai I and Sinai II agreements in 1974 and 1975 could not erase the fact of a strong commitment bordering on military intervention in favor of Israel. It should be acknowledged, for the sake of fairness, that this commitment was not without reservations and that the path toward an ever-closer U.S.-Israeli relationship was not devoid of obstacles. During the Ford presidency, there was a brief period of U.S.-Israeli tension, expressed by means of a thorough reassessment of U.S. Middle East policy. The Camp David diplomacy, true enough, was an attempt to promote peace acceptable to both Israel and the Arabs. But the omission of important issues and the ambivalence of certain parts of the agreement in reality signified either the lack of will or the lack of power of the U.S. president to ensure that a long-range solid, comprehensive settlement be achieved.

In the meantime, Israel's policy seems to have firmed up in terms of insistence on the retention of the West Bank, the Gaza Strip, and the major portion of the Golan Heights and on the expansion of Israeli settlements in occupied territories. Furthermore, Israel appears to have developed a new doctrine in its strategy, that of enlarging de facto its security perimeter to include southern and even central and northern Lebanon, as well as Iraq, at least insofar as the latter's nuclear capacity is concerned. The U.S. response to all these moves and initiatives has been timid and hesitant and, if compelled by the circumstances (especially the mood in the United Nations) to join with other powers in censuring Israel, no more than lukewarm.

It should be pointed out that this discussion is not intended to enter into the merits of the Israeli policies. Israel's leaders, broadly supported by the electorate, have consistently advanced arguments focusing on security and survival to justify their military and territorial policies. They may or may not be correct in pursuance of such policies. The purpose of this analysis is to register the fact that U.S. commitment to Israel, even if the latter's policy is to achieve a "Greater Israel," seems in the 1970s, to have definitely overshadowed the initial priority given to peace and stability in the area. Or, perhaps this thought should be reformulated to say that, at least to important U.S. decision makers, U.S. interests in the Middle East have become increasingly identified with Israel's territorial acquisitions and virtual military hegemony in the region.

The third U.S. objective in the area—containment of Soviet advances—should, in terms of priority, be listed as the first, because if the Middle East as a whole were to acquire a satellite status, that is, become a part of a larger Soviet empire, all the preceding discussions regarding oil and Arab-Israeli relations would become largely academic. If we list the Soviet problem after our analysis of the two other problems, it is to stress that effective U.S. policy toward the Soviet Union in the Middle East is logically linked with the policies regarding oil and Israel.

The U.S. policy toward the Soviet Union could, since 1947, be described as a policy of containment. As a bipartisan policy, it focused first on the Northern Tier. The Truman Doctrine of 1947, inclusion of Turkey and Greece in the North Atlantic Treaty Organization (NATO) in 1951, the Baghdad Pact of 1955, and the bilateral security agreements with Turkey, Iran, and Pakistan in 1959 all served this basic and consistently pursued policy of containment. As a result, a seemingly solid regional security system was erected. Unfortunately from the U.S. point of view, this system suffered an almost complete collapse in the 1970s, mostly, but not entirely, during the Carter administration. The Cyprus dispute (in its two installments of the mid-1960s and mid-1970s) had the nefarious effect of distracting Turkey from the prime issue of its external security—that of Soviet proximity and latent hostility—and producing a disturbing estrangement from the United States, whose support and guarantees were so eagerly sought in the 1940s and 1950s. Even though Turkey (in contrast to Greece) chose to remain both diplomatically and militarily within NATO, the United States could no longer take Turkish cooperation for granted. Thus, while in 1958 during the Lebanese civil war, the U.S. air force could fly over and refuel in Turkey with Ankara's full concurrence, in 1973 during the Arab-Israeli war, such privileges were denied to Washington. Similarly, in 1979 the Turks made the use of their bases for U-2 flights over Soviet territory (designed to verify Soviet compliance with Strategic Arms Limitations Talks [SALT] agreements) conditional on Soviet agreement, thus in reality denying the United States their facilities. And, of course, one should not forget the temporary suspension of certain U.S. bases in Turkey in retaliation to the U.S. arms embargo in the wake of the Turkish invasion of Cyprus.

Simultaneously with this weakening of U.S.-Turkish ties (only recently improved after the military coup in Turkey in September 1980), Pakistan gradually withdrew from active participation in the Central Euorpean Nations Treaty Organization (CENTO). The causes of this process were complex, but, as in the case of Turkey, an issue not central to the CENTO alliance, the tense state of Pakistani-Indian relations, acted as a catalyst of this estrangement. It required a revolution in Afghanistan and its subsequent conquest by the Soviets in 1979 to produce a reexamination of Pakistan's policy toward the

United States. By 1980-81, Pakistan was sufficiently alarmed by Soviet advances to look again for U.S. support. But even this new turn in Pakistan's policy was not devoid of serious conditions and reservations.

It was, however, the 1979 revolution in Iran that truly destroyed the CENTO-based regional security system. Not only has Iran collapsed as an effective military power and as a rationally organized state, but it has also adopted a posture of ideologically motivated hostility toward the United States and the West in general. As a regional force, the shah's army could, before 1978, be counted upon to provide a stabilizing influence, as witnessed by its intervention in Oman. By 1980-81, with its generals murdered, its officer corps demoralized, and its ranks decimated, it could barely engage in a holding action against the invading forces of a much smaller Iraq. In short, from a center of regional strength, Iran transformed itself into an area of weakness, almost a military vacuum, and from an ally and to some extent a U.S. surrogate in the Gulf area, Iran turned into a prime agent of revolutionary hostility toward the United States and the regional status quo.

It was not surprising that, even despite its frequent reverses of policy and vacillations in the Middle East, the Carter administration felt compelled to clarify its stand toward the Persian Gulf in the wake of the Iranian and Afghan developments. The Carter Doctrine of January 23, 1980, used words strikingly similar to those of Lord Lansdowne in 1903, when it pronounced U.S. determination to defend the Persian Gulf region against any external assault by any means necessary, including the use of armed forces. The doctrine was, of course, a declaration of intent. To carry it out in practice would require much new military planning, an agreement on basic strategies, accessibility to regional bases and facilities, provision of manpower and equipment, and appropriate budgetary allocations.

The Carter Doctrine contained elements of continuity and innovation: continuity in that it was pronounced in the spirit of containment that guided the Truman and Eisenhower doctrines and other bilateral and multilateral arrangements; and innovation in that it focused on the Gulf, a subregion that had not been specifically mentioned in previous major policy statements.

The Truman Doctrine had as its immediate aim the salvation of Greece and Turkey from immediate Soviet threat. The Eisenhower Doctrine, while broadly addressing the entire Middle East, was particularly geared to the defense of the Arab world in view of the spectacular Soviet successes in identifying with Arab nationalism in the wake of the Suez War. By focusing on the Gulf, the Carter Doctrine was singling out a special oil-rich subregion where the Northern Tier concerns overlapped those of adjacent Arab states. As such, it con-

veyed a recognition that, in view of the increased dependence of the
West on the Gulf's oil, this particular subregion warranted a commitment of U.S. military strength to safeguard it from aggression or
subversion. And it also signalled serious concern that, in view of
the floundering of the old barrier provided by an allied and strong
Iran and neutral but independent Afghanistan, the Soviet menace to
the Gulf has become more immediate. This change of emphasis thus
reveals a new dimension of U.S. policy—a dimension expressed from
a shift from a broadly based Northern Tier security system (no more
tenable) to a more narrowly centered interest.

But this shift of emphasis does not and should not mean that the
defense of the Gulf can be treated in isolation from other concerns.
To ensure realistic implementation of the Carter Doctrine, U.S.
forces will have to rely to a greater or lesser extent on accessibility
to the region they are expected to protect and countries of the Gulf
region will have to cooperate with, or at least not oppose, the United
States in carrying out the task of defense.

Here is where the linkage between U.S. policies in other subregions of the Middle East and those in the Gulf becomes evident regardless of whether Washington would or would not want to separate
the issues. From Washington's point of view, for example, it may
appear desirable and convenient to separate the bombing of Iraqi nuclear plants by Israel from the question of defense of the Gulf. By
the same token, it probably would be more convenient to diplomatically
localize Israel's attacks on populated centers of Lebanon as something distinct from the Gulf problems. But despite this actual or potential preference for treating such issues separately, the linkage
does exist as a fact of life. Cooperation with Saudi Arabia, to be effective, must be based on reciprocity. And it is almost certain that,
while basically desiring U.S. support for its security, the Saudi government will be most reluctant to cooperate if it perceives U.S. policy as one of aiding—through its sophisticated arms—the enemy of
the Arabs in its military actions far beyond its borders.

Similarly, since the Khomeini revolution, opportunities had
arisen to normalize the long-strained relations between the United
States and the Bath regime of Iraq. In fact, one could notice several
direct or indirect signals from Baghdad that it was willing to reas-
sess in a positive sense its relationship with Washington. Again,
destruction of Iraq's nuclear facilities by weapons provided by the
United States, with only a reluctant censure (though a collective vote
in the UN Security Council) of Israeli action, has put a new and possibly formidable spoke in the wheel of U.S.-Iraqi relations. While
Saudi Arabia and Iraq are singled out here as major countries of the
Gulf region whose posture on the defense of the Gulf may materially
influence the success of the Carter Doctrine, examples of this sort
could be multiplied.

To ensure effective containment of Soviet advances in the area, the United States cannot rely on military force alone. If its policy is perceived as one of excessive identification with Israel, even if the latter is embarked on a policy of expansion beyond its original borders, it is likely to materially aid Soviet designs by enhancing the radicalization of the local elites and masses alike and by encouraging the existing governments, moderate or radical, to seek closer ties with Moscow as a counterweight to a U.S.-Israeli de facto alliance.

Although this basic linkage has existed for a long time, recent physical approach of the Soviet Union to the Gulf, the collapse of Iran as an element of stability, and the escalation of violence between Israel and its neighbors in Lebanon and Iraq have made the linkage between the Gulf security and the peace and stability of other subregions in the area more evident and more compelling. It is this reality that provides a new dimension in the Soviet-U.S. rivalry in the Middle East.

THE REGIONAL POWERS

The decade of the 1970s has also witnessed substantial changes in the position and alignment of a number of regional powers. The collapse of Iran and Afghanistan as viable and reliable political entities has already been noted in this chapter. Of a different nature, but no less weighty in its possible consequences, is the isolation of Egypt, a country that suffered an abrupt demotion from leadership in the Arab world to an object of ostracism. That this is a dangerous position for Egypt to be in is recognized by friends and foes of Egypt alike. From the U.S. point of view, it may be frightening to contemplate the possible alternatives to the late Anwar Sadat's regime in Egypt. It may be virtually taken for granted that such alternatives—either the Nasser- or Qaddafi-type radicalism or Islamic fundamentalism—would bring about a drastic reduction in U.S. influence not only in Egypt itself but in the region as a whole.

Not to be neglected are also the positions of Jordan and Syria. In the case of Jordan, there is demonstrable reluctance of King Hussein's regime to cooperate with Washington in its Camp David diplomacy. Even though the king's moves toward some dialogue with Moscow may appear more as symbolic gestures to convey his unhappiness with Washington's policy than as a real intention of establishing a firm Soviet connection, the very fact of such gestures taking place injects a new complication into the U.S.-Jordanian relations.

As for Syria, the Hafez Assad regime is currently exposed to severe domestic pressures aggravated by mounting external complications in its relations with Israel and Lebanon. Initially a moderate successor to its preceding left-wing regime, the Assad government has been pushed by circumstances into a more militant stance, partly by the Lebanese civil war and partly by the Camp David Agreement. At the root of both can be found the unresolved Palestinian problem.

Similarly, an evolution could be noticed in the attitudes of Libya. Initially staunchly anti-communist, Colonel Qaddafi subjected his country to a process of radicalization while effecting a rapprochement with Moscow. The result has been seen in massive transfers of Soviet arms to Libya and in a simultaneous policy of military and revolutionary expansion, be it in the form of advances in Chad, threats to Sudan and Egypt, or support for revolutionary movements on a global scale.

All these changes in the posture or policies of local powers have usually had an adverse effect on the position and influence of the United States. The area of diplomatic access for Washington has been substantially reduced. The United States has today no formal diplomatic relations with South Yemen, Iran, and Libya; it has a reduced diplomatic representation in Baghdad (under the flag of the Belgian Embassy) and Afghanistan; and because of the war or near-war conditions in Iraq, Syria, and Lebanon, the presence of U.S. interests, cultural or commercial, has shrunk to a bare minimum. Furthermore, such U.S. oil firms as continue their business in Libya do so at their own risk, without diplomatic protection. With two exceptions— Egypt and Israel—U.S. tourism in the area as a whole has also decreased considerably.

WESTERN EUROPE AND JAPAN

With regard to the position of leading industrial democracies, it was possible to record six major steps of Western European eclipse in the Middle East and North Africa: (1) retreat of the United Kingdom and France from their dominant positions in their erstwhile mandates of Iraq, Syria, Lebanon, Jordan, and Palestine; (2) full emancipation of Egypt following the Suez War of 1956; (3) expulsion of Italy from Libya; (4) the end of French rule in Algeria; (5) nationalization of oil and the ouster of British influence in Iran; and (6) British withdrawal from the Persian Gulf in 1971. While all these steps were indicative of a broader process of retreat from empire, the Suez crisis of 1956 injected into this process a special aggravation by closely identifying Franco-British policies with those of Israel. The late 1950s constituted probably the lowest point in the fortunes of the United Kingdom and France in the area.

From the beginning of the 1970s, however, both Western powers began gradually regaining their influence in the region. This evolu-

tion was expressed by increased arms deliveries to Middle Eastern
—mostly Arab—buyers, by expanding aid-and-trade policies, by increased oil purchases effected through bilateral intergovernmental
agreements, and by gradual disassociation from Israel. In the case
of France, which in the 1950s was the chief supplier of weapons to
Israel, a marked change occurred following De Gaulle's advent to
power. His policy, characterized by a pro-Arab tilt, has been pursued by his successors at least until the end of Giscard d'Estaing's
term. It is problematical whether, especially in view of the Iraqi
nuclear incident, it will undergo an appreciable change under Mitter-
and's new socialist dispensation.

Although Germany never exercised political dominion in any
part of the Middle East, its defeat during World War II temporarily
produced an eclipse of German cultural and commercial presence in
the area. The decision of the Bonn government to pay substantial
reparations to Israel as an expiation of Hitler's persecution of European Jewry further contributed to the shrinkage of German influence
in the Arab world. Gradually, however, in the 1960s and 1970s, West
Germany (and to some extent East Germany) began regaining its cultural and commercial position in the area. As a major importer of
Middle Eastern oil, Bonn had a tangible interest in the security and
stability of the region, not unlike the United Kingdom and France.

Thus Western European industrial democracies—and to a considerable extent Japan—developed a serious stake in the area that
led them to conduct two simultaneous policies: to forge closer economic links, especially with Arab countries and Iran under the shah,
with the purpose of ensuring safe and stable exchange of oil for their
goods, services, and technology; and to practice an evenhanded treatment of the Arab-Israeli conflict, expressed on the one hand in the
recognition of the right of Israel to secure existence and on the other
in the view that the Palestinian quest for a homeland or self-determination be fulfilled.

With this in view, Western European democracies issued at
one time or another statements of a collective nature. Moreover, at
the time of the massive U.S. airlift to Israel during the October war
of 1973, they generally refused to the United States the right to over-
fly their territories.

This Western European (and partly Japanese) attitude brought a
new dimension to the mutual relations of big powers in the area. In
1950 the United States, France, and the United Kingdom acted in concert by trying to regulate the arms flow to the Middle East. In 1956,
paradoxically, France and the United Kingdom acted in collusion with
Israel against U.S. support of Egypt. But in the 1970s, not only did
the roles become reversed, with the United States becoming more and
more alienated from the Arab world, but Washington went as far as

to adopt a reluctant attitude toward any ideas of concerted Western action vis-à-vis the Arab-Israeli conflict and the Palestinian question. This less-than-perfect harmony could also be noticed in the European response to U. S. boycott initiatives toward Afghanistan and Iran in the wake of the Soviet invasion and the hostage crisis respectively. Furthermore, Western Europe was effectively kept out of the Camp David peacemaking process by the United States.

CONCLUSION

In the foregoing lines an attempt has been made to demonstrate that, except for certain adaptations stemming from greater tactical flexibility, no major changes could be noticed in the Soviet policies toward the Middle East. These continued to employ a wide range of methods, from propaganda and diplomacy through aid and trade to an outright use of military force, to attain the basic objective of replacing Western, especially U.S., presence in the region. Persistent continuity rather than startling innovations has been the dominant feature of Soviet policy.

By contrast, no similar observation could be made regarding the United States. In terms of objectives—containment of the Soviet Union, access to oil, and promotion of peace and stability in the area—only the first two may be said to have remained consistent, with oil actually gaining in importance, owing to increased dependence on imports in the United States. Furthermore, in terms of implementation, containment of Soviet advances became seriously eroded through the collapse of the Northern Tier security system as exemplified by the Iranian revolution and the invasion of Afghanistan.

As for the third objective, promotion of peace and stability, it appears to have become overshadowed by a growing commitment to Israel. This commitment in the 1970s reached the point not only of ensuring Israel's right to exist but even of tolerating, at least implicitly, those military and demographic actions of Israel that were being undertaken in the name of Israel's security and survival. The effect of such U.S. policies on the region, especially its Arab part, were bound to have a destabilizing effect, both in terms of the domestic fortunes of various local governments and in terms of their attitudes toward the United States.

Furthermore, this U.S. posture contrasted with the policies of Western European nations, which individually and collectively, were seeking a solution based on a balanced approach to both protagonists in the Arab-Israeli conflict.

The U.S. dilemma in acting in isolation from other powers was further dramatized when, as a result of the Iranian and Afghan revolutions and Soviet expansion, the security of the Persian Gulf became a prime concern of U.S. diplomatic and military policy. The countries of the area insisted on a linkage of broader U.S. regional policy —primarily focused on Arab-Israeli relations—with that pertaining to the Persian Gulf. At the threshold of the 1980s, the United States was still groping for a satisfactory formula in the face of some unresolved inner conflicts: how to reconcile its commitment to Israel with its broader interests in the Arab and Islamic world, how to evade or ignore the linkage between the two while reluctantly acknowledging that the linkage existed, how to ensure cooperation with its Western allies while opposing their collective initiatives in the Middle East, and finally, how to contain Soviet advances in an area the major parts of which had become estranged from the United States.

NOTES

1. Statement made at a colloquium at the University of California at Berkeley by the Minister-Counselor of the Soviet Embassy in Washington, Spring 1980.

2. Ralph Magnus, ed., Documents on the Middle East (Washington, D.C.: American Enterprise Institute, 1969), p. 210.

3. New York Times, December 11, 1969.

4. Magnus, Documents, p. 223.

4

SECURITY IN THE CONTEXT
OF AN ENLARGED
EUROPEAN COMMUNITY

DIETER DETTKE

No doubt, Europe's security problems will grow in the near future as a consequence of growing instabilities in the Third World and a state of superpower relations that can only be described as dangerous. Therefore, as was stated in a recent International Institute for Strategic Studies (IISS) publication, "The 1980s will be marked by more, not less, conflict outside the defined regions of East-West Alliances."[1] Under these circumstances, caused by "internal strains within developing countries themselves and the concomitant decline of the structure of international order,"[2] it is difficult to see how the enlargement of the European Community (EC) can help to solve Europe's security problems. Quite the contrary, it is necessary to realize that enlargement increases rather than decreases our existing security problems. Amazingly, however, the security problems of enlargement are hardly discussed anywhere, including within the EC. For those who know a little about the thorny problem of defense and military problems in the framework of European integration, this is no surprise. There is no way to win laurels by taking up this subject. On the other hand, it is safe to predict that a group of nations like those in the European Community, with the economic and political weight that they can bring to bear, cannot permanently avoid including security issues in the decision-making process and ultimately developing some kind of security policy of their own, whether they like it or not.

SECURITY PROBLEMS OF THE ENLARGEMENT

The internal structure of the EC and also the way in which it looks at the outside world will necessarily change quite fundamentally when Greece, Spain, and Portugal are full members. Enlargement toward southern Europe is a fundamental process, and although it is difficult to determine what the final outcome will be, a number of internal and external consequences are clearly visible.

Integration of southern European agricultural states into a community of predominantly industrial middle and northern European nations is bound to cause not only institutional headaches but problems for European integration as such. The social, economic, and political structures of the new countries are closer to those in Third World countries; Pierre Hassner even talked about a "Third-Worldization of Europe" in connection with enlargement.[3] That may be an overstatement of the case, but it points to the important fact that the internal balance of the enlarged EC will be a new one, probably less stable than the existing balance. There will be increased heterogeneity and more disparity between societies, while differences of culture, religion, and tradition will add to the disparity and thus create more difficulties for a decision-making process that is already slow and delicate, not to speak of new financial troubles that will result as a consequence of the present system for handling agricultural products. The European agricultural market by nature has a tendency to overproduction, and that tendency will be reinforced. In addition, the ten EC nations already have a strained labor market, with possibly almost ten million people unemployed in the near future. With the inclusion of Spain and Portugal into the EC, this problem will intensify. In sum, enlargement for the European Community is an additional internal burden.

With the entry of Greece, Spain, and Portugal, the balance of political parties and forces will change. Greece, Spain, and Portugal have more or less conservative political parties in power, but they all have very strong socialist parties in opposition and with good prospects for coming to power. Finally, there are strong communist parties in each of the three new member countries.[4]

Even if the communist parties of the three new countries do not agree in their strategies, the fact remains that the communist movement in the European Community, whether orthodox or Eurocommunist, probably will grow stronger. It is less certain whether or not this automatically will mean more political influence. At least for the moment, the most striking phenomenon within the communist parties of the European Community is the appearance of two different groupings: the orthodox and Moscow-oriented French, Portuguese, and external Greek communist parties and the reformist or Eurocommunist Italian, Spanish, and internal Greek communist parties. Their divergences can hardly be exaggerated. The most interesting question today, of course, is what the long-term orientation of the French Communist party will be now that they share power together with the French Socialist party. No doubt, power sharing by the French Communist party is a delicate decision for both Socialists and Communists. One can only speculate as to who will profit from that move and whether the political concessions that the French Communists had to make in order to share power is a definite step toward a more reformist attitude and strategy.

The European Community, through enlargement, is on course toward inheriting an enormous potential for conflict. The EC is already directly involved in the conflict over Cyprus and the disputes between Greece and Turkey. In addition to that, the EC cannot avoid sharing responsibility for Turkey, a country of great strategic importance for the West. However, economic and domestic developments in that country, especially after the military coup of last October, prevent the EC from offering what ultimately might be the last chance to stabilize relations with the West, namely a closer relationship with the EC. Under normal circumstances, something less than membership but certainly more than association with the EC would be the ideal solution. Such a step is practically out of reach today, and as a consequence Turkey might well drift away from the West and invite foreign intervention. But enlargement does not only mean greater potential for conflict and instability in southern Europe and the Mediterranean. It also means greater involvement in adjacent regions that directly affect European security, namely the Middle East and the Persian Gulf. Greece and Spain in particular always had, and will continue to have, a close relationship to Arabic countries. Traditional relations to the Arabic world and new economic necessities come together here, and this will add momentum to efforts of the EC to seek a greater role in a region that is extremely explosive.

As a result of the process of enlargement, the Atlantic orientation of the European Community will decline. It is true that Greece and Portugal are members of the North Atlantic Treaty Organization (NATO); Greece just recently even returned into the military wing of NATO after half a decade of aloofness, and Spain might join NATO. But at the same time, one should not overlook that the security needs and concerns of the three new members are related to Mediterranean problems and the Third World. Nonalignment plays an important role in the foreign-policy perspectives of all three countries. If Spain joins NATO, for example, and it is doubtful whether under the present circumstances such a step is feasible domestically, it would probably seek a special status within the alliance (NATO). The status

of France or Denmark and Norway could be the model. Neutralism
has a long tradition in Spain, and seeking full membership in the alliance could split the country. Finally, in connection with the possible
membership of Spain in NATO, there might be repercussions in Eastern Europe, and, in this case, Yugoslavia might be a new strategic
target of the Soviet Union.

MILITARY CAPABILITIES OF THE
ENLARGED EUROPEAN COMMUNITY

The combined potential of the enlarged European Community,
compared with the superpowers, is very impressive even in military
terms. With a population of roughly 300 million (10 percent of world
population), a gross national product (GNP) of over 2,500 billion (in
U.S. dollars), almost 40 percent of world trade, and well over 2.5
million armed forces, the enlarged European Community matches the
two superpowers in almost every respect except nuclear weapons and
defense expenditures. The GNP of the Soviet Union, for example, is
probably not more than one-fifth of that of the EC. It is a legitimate
question why a group of nations with such a potential, enormous social and economic progress, and a high standard of living is militarily so dependent on the United States and vulnerable to the Soviet
Union.

The best way to answer this question is to look at the reasons
for the failure of so many attempts to create an independent European
federation including defense and foreign policy, the ultimate goal of
many Europeans ever since the process of European integration began. Such a militarily potent and independent Europe, the European
Defense Community, like the Loch Ness monster, comes up again
and again only to disappear with almost the same steadfastness.
There are obvious structural reasons that prevent Europe from merg-
ing into a political union with a military role, and enlargement in that
respect is certainly no panacea. First, there is one basic objective
strategic truth that Europe has to face, and that is, as Ian Smart
wrote some time ago:

> Given the relative size and characteristics of Soviet,
> British and French (nuclear) forces, there is no possi-
> bility that Britain or France will be able to launch a
> "disarming" first strike against the Soviet Union; what-
> ever damage they might do to Soviet missile silos or
> bomber bases, the Soviet Union would always have ample
> weapons left to inflict crippling destructions upon West-
> ern Europe. The only threat posed by British and

> French forces is thus a threat of retaliation, which may
> be presented in one of two forms: as a threat to retaliate
> for any Warsaw Pact military action which is held to
> threaten vital British or French interests, or as a threat
> to retaliate only for a strategic nuclear attack by the
> Soviet Union. <u>The former threat is one which entails
> suicide, the latter a blow from the grave.</u>[5]

Under such strategic conditions, Europe seems unlikely to be able to
develop a credible deterrent of its own. Second, and perhaps even
more fundamental, the European Community is today by character a
profoundly civilian entity with a high standard of living, the status of
an important trading partner, relatively stable democratic structures,
and enormous social progress. It is difficult to see how a civilian
community such as this can make sustainable efforts to project power
to the outside world. Part of Europe's attractiveness as a partner in
other regions is in fact its lack of power projection. Third, the degree of integration is still not sufficient for a political and military
union. There seems to be a solid public majority for European integration in the member states of the European Community, but it is a
permissive consensus, to say the least. In fact one should rather
acknowledge that fragmentation prevails. A European authority le-
gitimized by majorities in every country is almost unthinkable if you
look at the summit meetings and the fragile compromises they produce. It is possible that the process of integration could break down
if there were a serious attempt to build a defense community, partly
because the nuclear question is so controversial and divisive. The
smaller and nonnuclear countries of the European Community could
well live with U.S. dominance. It is doubtful whether they would be
willing to live with French and United Kingdom (UK) dominance that
would hardly fulfill their security needs, especially since there is no
credible conventional deterrent as an alternative. Fourth, Europe's
military, political, and economic vulnerabilities obviously contribute
to discouraging any attempt at military integration. As a result,
there is a strong need to influence events diplomatically, but without
any military effort. This is visible in the European Political Co-operation (EPC), but from there it is a long way to a defense community.

Political cooperation within the European Community has worked
quite successfully since the early 1970s and has produced common
foreign-policy positions in a number of areas, some even involving
security matters. For instance, on CSCE the European Community
from the beginning played a very active role and continues to do so.
For example, the European Political Co-operation led to a coordinated
position on the Conference on Disarmament in Europe. Other areas
of effective European political consultation have included the Middle

East, the dialogue with Arab states, developments in Portugal and Spain, a common attitude toward South Africa, trade with Eastern Europe (an agreement between the European Community and COMECON remains open), and energy. Also, the European Community is trying to speak with one voice in international organizations such as the United Nations. In spite of these successful efforts, it seems highly likely that the recent debate on a European Defense Community, as with its predecessors, will come to nothing. Those spokespeople and advocates in France who have tried to develop a concept of a greater European role within the alliance are attacked by Gaullists who denounce it as pro-American and as an Atlantic European union. The Guallists demand the dissolution of the North Atlantic Treaty as a condition for a true European Defense Community. If that is the ultimate objective, then such a defense community can certainly never get off the ground. For Germany, a greater role for the Eurogroup could be a possible approach to strengthen the European element within NATO, perhaps ending in greater autonomy in defense matters on this side of the Atlantic. The Eurogroup has been successful on a number of projects to improve capabilities. Of even greater significance has been the creation of an Independent European Program Group (IEPG) in armaments cooperation. Its success has not only been in armaments cooperation but in the inclusion of France in that cooperative effort. But again, it is a long road from armaments cooperation to a European Defense Community. The IEPG clearly reaches beyond the European Community, and it is difficult to see how a region stretching from Norway to Turkey can be militarily integrated.

Similarly, the Western European Union (WEU) could hardly be turned into a defense community. Among other problems, it lost any meaningful political role when the United Kingdom joined the European Community. The European Community itself, and even more so when it will be enlarged to 12 members by the inclusion of Spain and Portugal, is in many respects, impressive in its economic, political, and combined military weight. But it appears stronger from the outside than it really is given its present internal structure. As a focus for European defense, the Community is unable to solve the nuclear dilemma that Europe faces. Not only would this question be explosive internally but it would face obstacles from the outside, not the least from the Soviet Union. And quite frankly, part of the problem would certainly be the German role in such a community, or more precisely, the burden of history, clearly a vulnerability in this case. At least for the time being, the road toward a European military union and, therefore, any option outside the alliance seems to be closed. As a result, there will continue to be a striking discrepancy between what might be called European worldwide interests, objectives, and aspirations and the means at Europe's disposal to realize its objectives.

THE SECURITY POLICY OF THE
ENLARGED EUROPEAN COMMUNITY

There is widespread agreement today that as the 1980s begin
we live in a new strategic environment. The most important new element seems to be the possibility of confrontations arising in the Third
World and overlapping with the East-West relationship and/or with
economic security, or more accurately, with economic vulnerabilities,
partly as a result of economic systems that created dependence on oil
and raw materials. This new type of overlapping conflict can be seen
in the Gulf region, in the Middle East, in Africa, in Asia, and even
in Central and Latin America.

Can and should Europe, on the basis of an enlarged community,
play a greater role to cope with this new situation? Part of the difficulty of the transatlantic dialogue has to do with an uneasiness about
the consequences of greater European responsibilities. When these
were only aspirations of professional Europeans and mere pronouncements rather than imminent reality, there was no great problem.
However, increased U.S. interest and even pressure, particularly
from Afghanistan, for expanded European responsibility have prompted
controversy. Many people fear that Europe, having just rid itself of
its colonial past, might risk falling back into colonial attitudes again
as it assumes military tasks outside the NATO area.

As far as the situation in the Gulf region is concerned, it is
clear that Europe has a number of options. First, although NATO's
responsibilities cannot be expanded to cover that region, there is
scope for a flexible approach in which France and the United Kingdom,
as individual countries, could contribute toward securing Western interests, in addition to the United States, whose role there simply
cannot be replaced by the Europeans. Second, there must be a better
division of labor within the alliance, through an increased European
contribution to the central front, the North Atlantic, and through
planning for wartime host-nation support. Third, Europe can contribute, although not as a replacement for U.S. efforts, in seeking a
settlement in the Middle East. In fact, complete similarity in approaches among the allies might not even be an advantage in getting
the peace process there started anew. The most important new element in any Western (European or U.S.) Middle Eastern concept is
the role of the Palestine Liberation Organization (PLO). A process
of mutual recognition and acceptance of legitimacy is certainly necessary; otherwise the PLO and many Arab states will be pushed even
more into the arms of the Soviet Union. Fourth, there is a need for
greater European-Middle Eastern economic cooperation. A strictly
military approach does not work for the simple reason that military
means alone can hardly keep the oil fields working if a conflict breaks

out. The war between Iran and Iraq shows that in the event of a military conflict, the oil fields would be destroyed or at least paralyzed.

Related to these questions, however, is the larger concept of organizing better relations between the industrial countries of the West and the Third World. This is what appears to be perhaps the greatest weakness of the present U.S. administration. Particularly disturbing in this regard have been the dramatic cuts in U.S. foreign aid (generally including international organizations), the apparent unwillingness to accept the results of a decade's negotiations on the law of the sea, and the policies in El Salvador that can hardly be seen as strengthening the democratic forces existing on both sides of the civil war. On this latter point, efforts should be made to bring them together rather than exclusively stabilizing the junta. Last, the U.S. policies toward South Africa could have negative repercussions in the Third World. If the Soviets, on the other hand, have a chance to strengthen their influence in Africa, it will be because the West deprives black Africa of any hope of support in its efforts to change both the racist policies of the Republic of South Africa and the attempts to prevent a democratic solution for Namibia.

The West Europeans are trying to organize an improved dialogue with the Third World. A model institution for this dialogue is the Lomé Agreement, which gives more than fifty states of Africa, Asia, the Caribbean, and the Pacific better access to European markets. But there are more problems to be solved, including energy, monetary relations, and transfer of technologies, to name only a few. In any case, the West should avoid attitudes that give the impression of supporting outdated structures and relations.

East-West relations, from a European point of view, are still largely seen as a necessary long-term process of committing the Soviet Union to support stability, thereby encouraging it to remain a responsible and constructive world power. There is no doubt that détente has been successful in this respect. The process must shape a system of relations in Europe that could represent almost a kind of ersatz peace treaty if the developments in Poland can be controlled. This, by the way, is not only important for Europe. It is clearly a contribution to global peace and stability. Of course, the positive results of détente in Europe could be destroyed, for instance, through the Soviet intervention in Poland. The difference between the European reaction to such an intervention and the situation in Afghanistan is not only the result of geography, but by the even clearer commitment made by the Soviet Union in the Helsinki Final Act. The détente process never succeeded in creating similar global commitments by the Soviet Union, for instance in the Third World, where there is much greater instability. The real problem is not too much détente, but rather not enough détente commitments. This is true for economic cooperation as well. Economic cooperation has been, and still is, the primary Soviet motive for pursuing détente. To deny it this cooperation ultimately puts into jeopardy the whole détente process. In addition, those in the West who advocate the concept of economic deterrence should know that there must be cooperation before it can be used as a deterrent. The main problem, of course, is dependence. But here, the degree of cooperation so far is much too small to create a serious vulnerability for the West. There is at least a similar, if not even greater, risk to the East in this process. Détente adds to security, and thus if the West is not prepared to undertake negotiated settlements with the East, this would dangerously undermine the legitimacy of current defense policies. The theater nuclear force (TNF) discussion in the Netherlands and elsewhere shows that this effort for negotiations is in our self-interest if we want to avoid a situation where Western defense measures, even necessary ones, are confronted with increasing domestic opposition. In order to cope with this new strategic environment, Europe needs more flexibility within the alliance to maximize its influence in the world. As an imperial alliance, the chances of coping with the present world situation are not very bright. In a coordinated effort, with flexible methods, the West is certainly capable of meeting the challenges ahead.

NOTES

1. Christoph Bertram, Third World Conflict and International Security, Institute for Strategic Studies, Adelphi Papers no. 166 (London, 1981), pt. 1, p. 1.
2. Ibid.
3. Pierre Hassner, "Détente and the Policies of Instability in Southern Europe," in Beyond Nuclear Deterrence, ed. John Holst and Uwe Nerlich (New York, 1977), p. 46.
4. Since this writing the Panhellenic Socialist Movement (PASOK) won electoral victory in Greece, and with this development more salient questions about the future abound.
5. Ian Smart, Future Conditional: The Prospect for Anglo-French Cooperation, Institute for Strategic Studies, Adelphi Papers no. 78 (London, 1971), p. 3.

5

ENVIRONMENT AND REGIONAL IDENTITY IN THE MEDITERRANEAN

BARUCH BOXER

THE MEDITERRANEAN IN PERSPECTIVE

Environmental concern has been a significant feature of Mediterranean regional identity since the late 1960s. It has been expressed through subregional alliances on particular issues, adjustment to extraregional political and economic pressures, and the effective use of international institutions to further regional objectives. The Mediterranean situation is anomalous. Cultural diversity and economic disparity are characteristic of the region, yet initiatives for marine and coastal protection are persistent and adaptable. While these initiatives have led to only limited attainment of management goals, their unabated momentum sustains the legitimacy of environmental protection as a basis for regional dialogue. How has this come about? What makes it unusual?

The Mediterranean is not only of interest as an example of regional potential for international cooperation. Analysis of contributing factors may also help clarify the future place of environment as a focus for global negotiations. Barely a decade after the Stockholm

———————

Research for this chapter is based, in part, on earlier work under a Rockefeller Foundation Grant-in-Aid and Fellowship in Environmental Affairs. I gratefully acknowledge this support.

59

Conference on the Human Environment, environment and conservation issues have lost much of their distinctively apolitical appeal. Technical aspects of such issues as trade in toxic products are considered in intergovernmental councils, but efforts to gain government support for fundamental policy shifts (for example, the World Conservation Strategy) have foundered. [1]

One reason for this is that environmental issues are diffuse and constantly assume new dimensions. Scientific uncertainty and uncoordinated policy response in most countries confounds attempts to understand the physical and biological implications of man-induced environmental changes in relation to social and economic causes and effects.

Since the period of exuberant environmentalism in the early 1970s, overriding moral compulsions for wide-ranging solutions have been replaced by political and economic imperatives that more accurately reflect global and regional priorities. National and international bodies now at best pay lip service to the collective human responsibility for maintaining and preserving the ecological heritage.

Patterns of worldwide response to environmental problems during the 1970s reflect these trends. Two tendencies predominate: on the one hand, international legal mechanisms were sought for managing transfrontier pollution and pollution of commons (oceans, atmosphere, Antarctica), species and habitat loss, and such problems as stratospheric ozone depletion and tropical deforestation. Concurrently, environment and resource issues were addressed directly and indirectly in global debates on north-south trade, aid, fiscal and monetary policy, and technology transfer; on multinational corporations' role in pollution export to developing countries; and on the place and function of environmental assessment in bilateral and multilateral development assistance.

A dichotomy thus developed between issue-specific formulation of regulatory policy by governments and international agencies, and less clearly articulated, subsidiary treatment of environment and conservation questions as part of the global bargaining process on development, trade, energy, shipping, and national and international security. This effort continues in the context of law-of-the-sea deliberations, through negotiations sponsored by UN regional economic commissions and the European Community on transfrontier pollution and harmonization of national environmental standards, and in the activities of international secretariats responsible for marine mammal protection and trade in endangered species.

Most studies of the international policy dimensions of environmental issues reflect this dichotomy. Problem areas (soil degradation, ocean and coastal use, chemical hazards, and so forth) are commonly assessed either in terms of their national or international policy implications, or more abstractly, as several paths to problem resolution that may help clarify issues in international law. [2]

In contrast, there has been minimal consideration of the extent to which governments' international policy positions are representative of broad societal interests. Much is known comparatively about differences in the effectiveness of environmental regulations as a function of the role of interest groups (for example, shipping, processing, electric utility industries; public interest organizations) in standard setting and in challenging regulatory policies. We still know little, however, about the relationship between societies' perceptions of and response to problems and the choice of regulatory solutions. This, of course, is part of a larger issue: public understanding of the place of science in society. [3] But it also points out the need for closer scrutiny of historical and cultural factors. This chapter explores the phenomenon of Mediterranean environmentalism as it nurtures regional identity. Several analytical approaches will be developed that show the relationship between historical determinants and contemporary institutional response.

ENVIRONMENTAL CONCERN:
LEGACIES AND STRUCTURES

How can we begin to probe the meaning of "environmental concern" as it serves to strengthen Mediterranean regional identity? What, in fact, is the basis of environmental unity in the Mediterranean? To begin, intensified multilateral scientific and diplomatic activity in support of pollution control and environmental conservation in the 1970s can be seen as a manifestation of continuing social adjustment to the physical and economic opportunities and constraints of coast and sea. These initiatives focused regional attention on shared problems and provided institutional mechanisms for common effort toward specific assessment and management objectives. Through the Mediterranean Action Plan, a United Nations Environment Program venture, international legal efforts and scientific research mutually strengthen the region's commitment to environmental goals. [4]

But differential cultural response to the modification and abuse of sea and coast also define, in Mediterranean terms, what the extent of regional concern can be. One can leave it at that and assume with facile confidence that the multinational Mediterranean program proves the credibility of the Stockholm Conference formula for global solutions to problems through regional activities. Of far greater interest, however, is what we can learn from the Mediterranean case of how environmental problems come to have meaning for regionally grouped societies and states. Two dimensions of this process seem paramount: adjustment to the sea as a discrete physical and ecological entity, and the growth of urban and rural coastal societies specifically in response to pressures from contiguous hinterlands and extraregional influences. The characteristic interplay since the early 1970s of institutional and scientific initiatives that give form and substance to the Mediterranean regional effort reflects these trends, especially the importance of their historical antecedents.

First is the sea, which stands discretely as a semienclosed body of water extending roughly 2,500 miles from Gibraltar to the Levantine coast, with an average width of about 500 miles. Geographically and hydrologically, the Mediterranean is a complex system still poorly understood after more than a hundred years of modern scientific exploration. As a physical system, the Mediterranean delimits and energizes pollution and resource problems through the geophysical and climatological processes that govern currents, heat exchange, oxygenation, sedimentology, and tectonic change.

Until the completion of the Suez Canal in the 1860s, the sole outlet to the world ocean was the Strait of Gibraltar. For several millenia prior to this, physical isolation fostered contacts among coastal societies through trade, conquest, intermarriage, and warfare. This, in turn, engendered a distinctive Mediterranean cultural perspective that was expressed linguistically, ethnically, and through the development of literary and mythic traditions. Historical intermingling was also mirrored in architectural and building styles, cultural landscapes, and artisanal enterprises.

The sea itself, then, is one unifying element that has fostered dialogue among frequently contentious neighbors. The physical proximity of coastal societies to one another and the coastal orientation of maritime commerce ensured steady contact around the littoral. This gave rise to a Mediterranean cultural entity that could absorb outside influences without being sullied by them. Malta is living testimony to this.

The coast is the other common denominator of the Mediterranean environmental heritage. Coastal communities have, on the one hand, responded to the economic and military challenges of the sea. They also developed the urban configurations that set the coast apart from the hinterlands. There are similarities in the historical evolution of these coastal processes around the Mediterranean that underline the special character of the near seacoastal interaction. The sea always has been a waste sink and a source of food, as well as the avenue of commerce and warfare. There are fascinating parallels in the way Mediterranean coastal states today view the scale and severity of coastal and marine pollution. It is easier to obtain agreement on the regional as opposed to the subregional level. Also, governments more readily criticize ocean dumping than problems that stem from more intensive and competitive use of the near seas and the coastal land areas. There is a complex and kaleidoscopic dialectic here between regionalization/generalization of issues versus localization/specification. This is the tension of the seacoast interface, which simultaneously is a force for regional or subregional unity and conflict.

Although the formal actors in Mediterranean environmental affairs are ostensibly national governments, there is a more fundamental element that undergirds the regional matrix. This is the quality of Mediterranean social unity. J. Davis treats this issue quite effectively in his criticism of social anthropologists for their failure to recognize the distinctiveness of a Mediterranean social anthropology that is at the same time comparative and historical. His discussion is helpful in providing a social rationale for the coast-sea model of human adjustment that helps place current environmental problems in historical perspective.

He observes that "at a regional level . . . it makes sense to speak of general social characteristics . . . present as versions in each of the communities." This Mediterranean social character, he suggests, is found both in national societies and cultures that are conterminous with state boundaries, and through a Mediterranean social order that derives from individuals' commercial, familial, and conflictual relations. Thus, "the 'mediterranean' is not a 'society' or a generalization at a remote level of abstraction about the characteristics of its component societies. . . . The mediterranean must be understood as the result of the interaction [for various purposes] [emphasis added] of people from diverse societies: those institutions and processes which have been created to facilitate interaction; those relics which were created by it and now appear diffused in certain zones, if not through the area."[5]

As Braudel has pointed out, Mediterranean maritime communication traditionally was coast oriented, despite the sea's relatively narrow width.[6] This was a response to the presence of many subregional basins and seas, each with distinctive hydrographic regimes; the risks from unpredictable currents and storms implicit in trans-Mediterranean voyages; and the dissected physical geographic configuration of the northern coasts that fostered diverse but isolated economic life. Southward, the openness of the coast east of the Maghreb, with its exposure to the harsh desert climate, was also a deterrent to cross-sea traffic. Southern trade tended to focus on coastal ports that were widely separated. Trans-Mediterranean communication, whether north-south or east-west, has always been difficult. It still is. To travel today from Nice to Tripoli, for example, requires inconvenient connections via Rome or Malta. Similar difficulties confront the traveler from Cyprus to other Mediterranean or Near Eastern destinations.

What are the implications of these historically significant physical determinants for understanding contemporary Mediterranean environmental affinities? A basic point is that there has always been a contradiction between physical proximity and social distance in the Mediterranean. This contradiction has been evident throughout history and has been expressed at both regional and subregional scales. Physical factors deterred development of balanced trade among the north, the south, and the Levant that might have more successfully promoted regional autarky. Traditionally, localized specialized trade was conducive to political diversity, competition, and instability.

The areal compactness of the Mediterranean superficially implies possibilities for achieving social and economic cohesion through trade and communication. In fact, the interaction among peoples and places that has given identity to the Mediterranean is more a product of tension and conflict. Davis's sense of interaction as a key determinant of Mediterranean social unity takes on even greater credibility if we extend his argument beyond the anthropological realms of kinship, family, honor, and shame. Social unity was also nurtured in the ecological medium of the coastal sea, which through its biology, ecology, and hydrography, was the mechanism through which the threads of communication, conflict, economy, and society were interwoven.

The near seas, with their narrow continental shelves and low level of biological productivity, are areas where the impact of man on nature is especially intense, where economic risk is great, and where there is much pressure to diversify economic activities. For example, the Ionian Sea is now a focus of conflict among Tunisia, Libya, and Italy over fishing and offshore oil and gas drilling rights, and over territorial sea demarcation. Fishing disputes stemming from increased pollution in Sicilian waters force Italian trawlers into Tunisian- and Libyan-dominated areas. [7]

Intensity of coastal sea use, with the inevitability of competition in both the human and environmental realms, is due mainly to the narrowness of the coastal seas, their poor connections, and to their clear demarcation in most Mediterranean locations from the more distant open ocean. Here, geophysical processes have resulted in varied bottom topography, highly differential and contrasting depth gradients, and complex water column dynamics in both the western and eastern Mediterranean basins.

The coastal sea, then, has always been an important vehicle for definition and support of those "purposeful interactions" (to use Davis's term) that enrich and fill out the diverse facets of Mediterranean unity. But there is a landward dimension of this that is of equal importance. It was suggested earlier that the narrow coastal sea, physically and ecologically distinct from the deeper waters, is counterbalanced by similarly constricted lands bounded by steeply sloped mountains and inhospitable deserts. These strips and pockets of land are also impoverished by poor soils, inadequate and irregular water supplies, and limited opportunities for indigenous economic diversification.

Conflicts and tensions here in the interplay of man and nature parallel those of the coastal sea. The ecological consequences of these conflicts—erosion, surface runoff, destruction of the vegetative cover—persist. Now, however, the harmful effects of poor management are further aggravated by intensified large-scale commercial agriculture and industrial, urban, energy-related, transportation, and port development. This provides a coastal land-sea interface where historically evolved and more recent economic and social forces combine to create ever more serious pollution and conservation problems. The fact that response to these problems has assumed regional scale attests to the power of Mediterranean social unity that links the sea and coast.

Before discussing several aspects of contemporary environmental response that reflect historical precedents, it is necessary to look somewhat more closely at the coastal zone. What are the common features of ecological adaptation (in the sense of land use, economy, and settlement patterns) in the Mediterranean littoral that might support our social-unity hypothesis? Where do cities fit into this scheme?

Coasts and Cities

Mediterranean coastal societies have common features that reflect the unique quality of their social and economic adaptation to the physical world. Social unity results not only from interaction of people from diverse societies for certain purposes. Just as important is the unity that derives from interrelations of social organization and environmental use in the confines of the coastal zones. Specifically, two sets of variables operate: distinctive peasant, pastoral, and urban family structures and consistent and unchanging physical parameters. Other important features of the human ecology of the Mediterranean basin are the fact that characteristic social structures have a high degree of cross-cultural (and cross-boundary) durability, and that traditionally limited economic opportunities in the coastal strips have fostered a high level of adaptability and flexibility in response to changing political and economic circumstances. [8]

The facts of Mediterranean climate and physical geography are well known, but the depth and permanency of their impacts on the human geography of the coastal societies is not fully appreciated. Climate and geomorphology now obviously have less influence on economic choice than in the past. But then, as now, these physical constraints also generated shared perceptions among coastal peoples of the environmental costs and benefits of the use of land, water, forests, and soil.

Human use and modification of the coastal environment has always been governed by the spatial extent and temporal predictability of climatic cycles. Yet, contrasts exist between wet winters and dry summers, mountains and deserts, steep slopes and narrow coastal plains, densely settled urban nodes and sparsely populated rural areas, coastal agriculture and upland pastoralism. These contrasts in cultural ecology have always regulated and modulated the rhythms of Mediterranean life.

The shared environmental experience of culturally diverse but socially similar coastal groups also resulted in a mutually felt sense of marginality and distance (both self-perceived and externally imposed) with respect to shifting and uncertain external political and economic forces. In fact, Mediterranean peoples (and governments) seem to distance themselves psychologically from forces and events that impinge from the basin's periphery as well as from those that sweep through the center. This sense of independence and immunity has been clearly expressed in the context of the regional environmental protection effort. One has only to observe regional intergovernmental meetings to appreciate the significance of this. In the struggle for diplomatic advantage regarding location of coordinating units or program activity centers, allocation of research funds, or recognition of national scientific achievement, Malta and Lebanon speak with as much conviction and self-assurance as France or Spain.

The struggle of culturally diverse peasant and pastoral economies to maintain their viability under conditions of land deterioration and moisture loss, labor migration, and unstable markets for agricultural products has tended to foster common perspectives on economic matters. European Community dissension over agricultural pricing policies in relation to Greece's and Spain's entry is exemplary. More specifically in environmental terms, locally generated protests by farmers and fishermen have become commonplace in recent years. Protests frequently glorify the honesty and purity of the "simple" confrontation with nature while, at the same time, condemning the polluting and degrading effects of coastal development. Several case studies of the effects of technological modernization on Mediterranean fishing communities, for example, document the sense of loss, sorrow, and uncertainty of fishermen who, despite prospects of enhanced economic rewards, are forced to abandon artisanal practices. Similar feelings are voiced by Spaniards, Frenchmen, Italians, Tunisians, and Lebanese. [9]

In sum, such current resource and conservation concerns as soil erosion, wetland destruction, and plant and animal habitat loss can be seen as recent manifestations of a long-evolving pattern of past abuse and mismanagement. Coastal overdevelopment, for example, aggravates ecological imbalances from earlier eras while adding new stress from thermal pollution and industrial waste discharge. Thus, pollution and mechanical degradation of coasts and coastal sea bottoms and the aesthetic disfigurement accompanying recreational and tourist development is but the latest stage of a process that began with Greek, Roman, Arab, and Ottoman colonization. Most important, though, is the need to understand how, and in what characteristically Mediterranean contexts, these problems evoke contemporary response. A key element has been the role of the city in intensifying the problem and serving to generate and focus remedial action.

In their form and function, Mediterranean coastal towns symbolize the struggle for productive use of coast and sea that is at once a source of environmental unity and a threat to the ecological fabric of that unity. With the exception of a few larger cities such as Marseilles that are located on estuaries, most towns remain relatively small (in comparison with global urbanization trends); they tend also to be isolated from one another because of their insular locations, the indented and irregular continental margin to the north, and the narrow exposed coastal plain east of the Maghreb. They are also generally remote from centers of state power (with some exceptions, such as Algiers and Tripoli).

Counterbalancing this has been the historic economic interdependence of trade, shipping, agricultural processing and distribution, and maritime-oriented industry that has given coastal cities a shared sense of purpose and a common identity despite cultural differences. Several appeals for protection of the Mediterranean environment during the 1970s came from nongovernmental associations of cities. In fact, the first charter on the Mediterranean was issued in Beirut in 1973 by a conference sponsored by the United Towns Organization, representing 132 towns and cities in 16 Mediterranean coastal states. The charter appealed to urban coastal authorities to come up with an international pollution code for Mediterranean cities and urged governments to strengthen commitments to fight pollution. [10]

This historical legacy now takes on new dimensions in Mediterranean environmental affairs. Several elements remain crucial: competition among economic uses of the coast and near sea; pressure of population on ecologically fragile coasts and shallow coastal waters; and striving for political dominance among both coastal and insular centers, which reflects insecurity about strength and position relative to each other, to national polities, and to outside powers.

These factors have contributed in many ways to the growth of environmental awareness on the regional scale. Perhaps the most important parameter, however, is that expressed by the dichotomy between coast and interior (national) interests. The northern and southern Mediterranean coasts vary in the way these differences are revealed in attitudes toward environmental policy. In the north, there is a distinct dichotomy between the coast and national centers, such as Madrid, Paris, Rome, Belgrade, and Athens. In the south and the Levant, coastal communities' problems become national problems in regional discussions. It was hypothesized earlier that there is a contradiction between physical proximity and social distance in the Mediterranean, a tension that somehow links competing interests despite potential conflicts inherent in social, linguistic, and ethnic differences. This notion was developed initially to build a case for identifying an interconnected coast and sea as the key determinant of Mediterranean social, ergo environmental, unity. It is also meant to convey an image of horizontal linkage around the coast, where commercial competition and military and naval struggles for freedom of trade and communication served to unite societies at different times in shifting alliances for common purposes.

Another aspect of this, perhaps more relevant to understanding the dynamics of the contemporary scene, is the fact that urban areas along the coast often assume corresponding or complementary positions on environmental issues, sometimes even in transboundary situations. Very often, government representatives in regional intergovernmental meetings find themselves in the position of having to represent the interests of coastal communities and constituencies even though these interests conflict with national government policies. Especially on the northern coast, a vertical policy gap often prevails between coastal populations and national capitals that sometimes has the effect of strengthening regulatory policy ties among local authorities contrary to national objectives. Throughout the 1970s, there were numerous instances when Spanish and French; French, Monegasque, and Italian; Italian and Yugoslav; and Italian, Maltese, and Greek local or regional coastal authorities entered into formal or informal agreements on technical matters pertaining to harmonization of environmental standards or regulations. Areas of discussion have included waste-water discharge and treatment; enforcement aspects of territorial sea use; fishing rights; oil, gas, mineral, and material exploration and use; and oil spill pollution and cleanup. [11]

To the south, latitudinally expressed differences between coast and nation generally have not been as politically significant as in the north. Here, clear distinction between coastal and national policy objectives is evident, mainly in the case of Egypt, and to a limited extent in Morocco and Tunisia. The North African coast is interesting, however, in that intergovernmental antagonisms are most prominently focused on issue-specific demands or complaints. Here, antago-

nisms that result from competing uses of the coast and sea for tourism, energy-related development, shipping, and fishing in the several North African coastal states tend to arise from localized subregional conflicts. With weak links between coast and interior, intergovernmental confrontations thus focus on disputes arising in
coastal population centers. This is true as well in the case of Greek-
Turkish relations, although the disputes here tend to be associated
with law-of-the-sea-related territorial sea and economic zone jurisdictional questions that bear specifically on oil and gas drilling and
security zones in the Aegean Sea, rather than on pollution or resource
conservation.

In North Africa, urban coastal interests and national positions
in regional discussions often coincide. There is a dimension of the
hostility between Algeria and Morocco, Tunisia and Libya, and Libya
and Egypt that is hardly noticed outside of the Mediterranean. This
has to do with fishing rights and conservation measures, coastal
state enforcement rights to control vessel-source oil pollution in
coastal waters, coastal transport of pollutants, and technology sharing for waste-water treatment and recycling. On the North African
coast, the historical process of human adaptation to the coast-sea
interface that supported and gave meaning to the idea of Mediterranean unity now sustains a regional environmental dialogue. Governments find it to their advantage to support United Nations Environment Program (UNEP) efforts to share information and techniques for
purposes of environmental assessment, scientific research, and technical training.

A final word on contrasts between north and south in the way
urban sources of pollution and environmental degradation are seen
by coastal states of the region. The largest contributors of pollutants—Spain, Italy, and France—have always attempted to downgrade
the extent of their depredations. They have consistently pointed out
in regional meetings that, although their inputs to the Mediterranean
contain as large a proportion of industrial discharges as domestic
sewage, these effluents are more effectively treated prior to discharge
than the largely untreated domestic waste water entering the Mediterranean from urban centers on the south coast. This attempt by
the three most industrialized states to suggest that their more harmful inputs are tolerable because their treatment technology is superior
has united southern coastal states in response to a perceived patroniz-
ing stance taken by the three former colonial powers.

Tunisia has emerged since 1975 as a mediator between north
and south, representing the urban coastal interests of the so-called
developing Maghreb states and Libya in treaty-drafting discussions
on pollution sources. In this case, industrial versus developing state
dialogue on New International Economic Order issues takes on a

uniquely Mediterranean cast. The North African states take the position that their pollution is less harmful than that from the north because it is mainly nutrient-rich domestic sewage, in contrast to the more toxic industrial discharges from north coast industrialized states. They also insist that because pollution from the northern coast is more harmful, northern coastal states should bear the heaviest burden of cleanup and treatment costs.

International Organizations

The common perspectives of Mediterranean coastal states on the implications of environmental degradation and pollution for their marine- and tourist-oriented economies have created a highly favorable climate for the involvement of international organizations in Mediterranean environmental affairs. There are two main reasons for this. First, as we have seen, coast and sea have evolved symbiotically, and economic and political interests of Mediterranean societies have coincided at regional and subregional scales despite localized competition. This often led to a situation in which coastal communities developed a sense of independence from and contempt for political forces and trends over which they had little control. (This is strongly expressed today by universal resistance to U.S. or USSR participation in Mediterranean environmental activities.) Now, what we refer to as environmental identity, derives from the recognition by coastal societies that their constrained, coast-dependent economies require cooperation for conservation and management of environmental resources. This results in the distancing of local from national interests just described, a situation favorable for international organization entry. The authority of international organizations derives exclusively and directly from multinational governing bodies. In the Mediterranean, coastal and national interests compete. Therefore, international organizations can take advantage of this dichotomy to build regional institutions and programs.

Second, the geographical extent of the Mediterranean region includes Europe, Africa, and the Arab world. As a result, the region poses special problems for UN regional allocation formulas for development assistance, personnel policies, and international agency responsibilities. These formulas result from difficult political compromises. Mediterranean governments were able to use this unusually flexible situation to their advantage, particularly since the environmental protection focus was a new and unfamiliar area for international organizations. The breach was filled opportunely by the United Nations Environment Program, an agency created by the General Assembly in 1972 in response to the June 1972 Stockholm Conference.

Its mission was to stimulate and coordinate the environmental actions of UN bodies and to foster an awareness of the need for international environmental cooperation among developed and developing nations.

The Mediterranean in the early 1970s was perfectly situated for UNEP's mission. Regional marine pollution control was given high priority by the Stockholm Conference as a proper concern of international agencies. The UNEP developed a strategy (built on earlier Food and Agriculture Organization [FAO] initiatives) for treaty drafting, environmental assessment and research, financing and institution building, and education and training that emphasized service to governments. The responsibility for success or failure of the Mediterranean Action Plan was placed squarely in the hands of the coastal states. The UNEP, as the main coordinator of legal, planning, and scientific activities for governments and other sectoral UN agencies (such as the FAO, the World Health Organization [WHO], and the United Nations Educational, Scientific and Cultural Organization [UNESCO]), tried to assume a detached stance, but in fact the agency's global credibility still rests largely on the success of the Mediterranean effort.

The UNEP's coordinating function in the Mediterranean responded effectively to the historical tension between coast and interior and to the dialectic of tension and cooperation that is so characteristic of the Mediterranean coast-sea entity. It played on regional and subregional rivalries to develop an extensive network of research laboratories in 16 Mediterranean countries to implement a marine research and monitoring program. It also supported a regional oil combating center in Malta and has worked with the UN Development Program to encourage renewable energy use and aquaculture. An important indication of continued UNEP commitment to the Mediterranean regional effort was the establishment in 1982 in Athens of a permanent coordinating unit for the Mediterranean Action Plan. This unit is also strongly supported by the new socialist government, which is strongly committed to Mediterranean regional environmentalism in keeping with its support for domestic environmental improvement.

In closing, one can simply account for Mediterranean governments' continuing acknowledgment of the importance of regional environmental cooperation as a matter of economic necessity: fouled, polluted beaches threatening the viability of a tourist and second-home economy that brings 100 million people to the coast every summer. But this ignores the challenge of understanding what pollution and degradation have come to mean for the coastal societies in their assessment of their own position relative to the sea and coast. This has suggested some ways of thinking about the environmental basis of Mediterranean unity. These ideas need to be further tested in historical contexts as well as in terms of contemporary institutional change. We know enough, however, to recognize the continuing influence of a still-powerful heritage of coast and sea that challenges the acceptability of commonly held definitions of modernization, development, and security in relation to the Mediterranean.

NOTES

1. The World Conservation Strategy, published in March 1980, was prepared by the International Union for Conservation of Nature and Natural Resources in cooperation with the United Nations Environment Program and the World Wildlife Fund. Its aim is to "help advance the achievement of sustainable development through the conservation of living resources" by suggesting policy approaches to achieve "conservation efficiency" and integration of conservation and economic development.

2. An excellent analysis of the role of international organizations in dealing with various problem areas is David A. Kay and Harold K. Jacobson, eds., Environmental Protection: The International Dimension (Totowa, N.J.: Allanheld, Osmun, 1982). Jan Schneider, World Public Order of the Environment: Towards an Ecological Law and Organization (Toronto and Buffalo: University of Toronto Press, 1979), assesses relations between law and policy in international environmental affairs.

3. D. Allan Bromley, "The Other Frontiers of Science," Science 215 (February 26, 1982): 1043.

4. This is a multinational and multiagency program of scientific cooperation, treaty drafting, and education and training aimed at control of Mediterranean pollution. The program has been coordinated since the mid-1970s by the United Nations Environment Program. See Baruch Boxer, "Mediterranean Action Plan: An Interim Evaluation," Science 202 (November 10, 1978): 585-90, for an evaluation of scientific and policy aspects.

5. J. Davis, People of the Mediterranean: An Essay in Comparative Social Anthropology (London, Henley, and Boston: Routledge & Kegan Paul, 1977), p. 14.

6. Fernand Braudel, The Mediterranean and the Mediterranean World in the Age of Philip II, 2 vols. (New York: Harper Colophon, 1976), 1:chap. 2.

7. "Tunisia's Struggle to Develop Fishing," New York Times, January 1, 1982, p. 38.

8. Jeremy Boissevain, "Uniformity and Diversity in the Mediterranean: An Essay in Interpretation," in Kinship and Mediterranean Society, ed. J. G. Peristiany (Rome: American Universities Field Staff, 1976).

9. Oriol Pi-Sunyer, "Two Stages of Technological Change in a Catalan Fishing Community," and Paul D. Starr, "Lebanese Fishermen and the Dilemma of Modernization," both in Those Who Live from the Sea: A Study in Maritime Anthropology, ed. M. Estellie Smith, American Ethnological Society Monograph no. 62 (New York: West, 1977). See also Bernardo Cattarinussi, "A Sociological Study of an Italian Community of Fishermen," in Seafarer and Community, ed. Peter H. Fricke (London, 1973).

10. "Drive on to Save the Mediterranean," New York Times, June 9, 1973.

11. The wider political significance of these local agreements is discussed in Baruch Boxer, "Mediterranean Pollution: Problem and Response," Ocean Development and International Law: The Journal of Marine Affairs 10 (1982): 320-23.

6

NUCLEAR PEACE
IN THE
MEDITERRANEAN

JOSEPH L. NOGEE

Nuclear peace refers to the absence of a war involving nuclear weapons. Nuclear peace exists as long as the countries of the Mediterranean basin (and the rest of the world) are at peace or, if they are at war, as long as none of the belligerents use nuclear weapons. Should nuclear weapons be used against any member of the Mediterranean basin or should any of the 19 states in the region or the Mediterranean Sea be the location from which a nuclear attack is launched, then nuclear peace would cease to exist. [1] This chapter examines the factors working for and against nuclear peace in the Mediterranean. Admittedly, any analysis that treats the Mediterranean basin as an integrated or self-contained area imposes upon the region a unity that is to some extent arbitrary. The Mediterranean Sea is the hub of three continents and several cultures. Each of the states in the region is more politically involved with one or more states outside of the region than it is with many of the countries within the region. Thus, the maintenance of peace in the Mediterranean basin depends at least as much upon conditions outside of the region as it does upon intraregional relations.

The author thanks James E. Trinnaman for helpful ideas and materials, the Strategic Studies Institute of the Army War College where this was written, and the University of Houston MAGIC Center and Mary Kershner for the manuscript preparation.

TABLE 6.1

Comparative Analysis of Mediterranean Demographics

	Population, 1980 (in millions)	Armed Services, 1980 (in thousands)	GNP, 1979 (in billions of U.S. dollars)
Albania	2.770	41.00	1.10[a]
Algeria	19.500	101.00	32.00
Cyprus	0.835	9.00	—
Egypt	40.460	367.00	16.50[b]
France	54.000	494.73	566.00
Greece	9.530	181.50	32.50[c]
Israel	3.900	169.60	16.40[b]
Italy	57.100	366.00	317.00
Jordan	3.104	67.20	2.69
Lebanon	2.800	23.00	2.90[d]
Libya	2.933	53.00	19.00[b]
Malta	0.338	0.80	0.76[c]
Morocco	20.000	116.50	15.20
Portugal	9.900	59.54	21.80
Spain	37.720	342.00	165.00
Syria	8.800	247.50	9.20[b]
Tunisia	6.390	28.60	6.99
Turkey	45.500	567.00	45.30[c]
Yugoslavia	22.130	264.00	45.00[c]

[a]1974 figures.
[b]GDP (gross domestic product).
[c]1978 figures.
[d]1977 figures.

Source: International Institute for Strategic Studies, The Military Balance, 1980–1981 (London, 1980).

75

There are two major threats to nuclear peace in the Mediterranean: (1) a war between the North Atlantic Treaty Organization (NATO) and Warsaw Pact members and (2) nuclear proliferation among the states involved in the Arab-Israeli conflict followed by the outbreak of war. These by no means exhaust the possibilities for conflict in a region as volatile as the Mediterranean. Among the potential interstate wars one might mention are a Soviet invasion of Yugoslavia in support of competing factions in Yugoslavia; Greece versus Turkey over Cyprus or rights in the Aegean; Morocco versus Algeria fighting over the Sahara; Libyan aggression against Tunisia, Egypt, or some other African states; Syria versus Jordan or Iraq in competition for regional leadership; or a power struggle between Ethiopia and Egypt or the Sudan. None of these conflicts would likely lead to a nuclear war unless, in the case of Yugoslavia, the issue broadened to include NATO and the Warsaw Pact. Table 6.1 shows the populations and size of the armed forces of the Mediterranean countries.

NATO, THE WARSAW PACT, AND NUCLEAR WEAPONS

Under present conditions, the probability of a war in Europe is low. Neither East nor West wants it, and there are no outstanding crises likely to trigger an unintended conflict. However, should NATO and Warsaw Pact countries become involved in a war, for whatever reason, the chances are significant that nuclear weapons would be used and even higher that the countries along the Mediterranean would become involved. The Mediterranean is the south flank of NATO, not the anticipated locus of the outbreak of a NATO-Warsaw Pact conflict (assumed to be central Europe), but it is almost certain to be the scene of intense struggle for control of vital sealanes of communication. The strategic importance of the Mediterranean in U.S. and Soviet strategy has been summed up by Abraham S. Becker.

> In no other area beyond European territory proper has the strategic military balance between the United States and the USSR been a factor in international affairs as it has been in the Mediterranean. American military power has been deployed elsewhere around the rim of the Soviet Eurasian expanse, but its strategic significance, in the sense of capability of striking at targets in the USSR, has been small [Vietnam] or it has not been a major issue of regional politics, as has been the case with the ballistic missile submarines deployed in the Pacific and Indian Ocean. In contrast, the nuclear

power of the sixth Fleet has figured intermittently in
the SALT talks and has been one of the reasons Moscow
has sought air and naval base privileges among its Mediterranean clients. The Mediterranean is not only the
back door to Western Europe: as Moscow so frequently
reminds us, the Mediterranean leads to the USSR's
back door, the Black Sea.[2]

Nuclear weapons now exist in substantial numbers in Mediterranean countries as well as in the sea itself. They include tactical battle-
field weapons having a range under 160 kilometers (km), theater nuclear
forces (TNF) for use against military targets but at distances ranging
above 160 km (these are divided into long-range and short-range TNF
with 1,000 km as the dividing line), submarine-launched ballistic missiles (SLBMs), as well as warheads discharged from aircraft.

No Warsaw Pact member borders the Mediterranean Sea (Albania withdrew its participation in 1961 and formally left the Pact in
1968). Among the NATO members on the Mediterranean littoral,
France, Greece, Italy, and Turkey each maintains nuclear weapons
on its territory. Although France withdrew from military participation in NATO in 1966, it is generally assumed that in a NATO-Warsaw Pact war its military forces would join with the NATO powers. Greece in 1974 withdrew from the NATO military command
structure on the grounds that the Alliance had failed to stop the Turkish invasion of Cyprus. In November 1980 Greece rejoined NATO's
military command, but in December 1981 the socialist government
of Andreas Papandreou suspended his government's return to NATO
because of differences with Turkey over the extent of Greek military
authority in the Aegean Sea.[3] Difficulties between Turkey and the
United States developed in 1975, when Turkey partially suspended
military cooperation in response to the U.S. arms embargo of that
year. When the embargo was lifted in late 1978, Turkey resumed
military cooperation with the United States.[4] NATO's southern flank
is weakened because both Greece and Turkey consider each other a
greater threat than the Soviet Union.[5] Spain recently became the
sixteenth member of NATO. For several years, U.S. nuclear submarines armed with nuclear weapons were stationed at Spain's Rota
naval base under special agreement with the United States. In 1976
the United States agreed to denuclearize the U.S. presence in Spain
by 1979.[6]

Tactical nuclear weapons were introduced in Western Europe
when the NATO council in 1954 decided that they were necessary to
compensate for Soviet superiority in conventional weapons. Nuclear
warheads currently in Lance and Pershing missiles are subject to
"dual key" control, requiring the authorization of both the United

States and the host country before they can be used. By contrast, the long-range theater nuclear forces (LRTNF) scheduled for deployment in 1983 will, following the preference of the Europeans, be exclusively under U.S. command and control.[7]

A major decision to modernize NATO's nuclear arsenal was made in December 1979 because of the deployment of the Soviet SS-20 intermediate-range ballistic missile (IRBM), which has no counterpart in Western Europe. The SS-20, a mobile missile capable of being reloaded, can strike anywhere in Western Europe from Soviet territory.[8] To maintain what it considers a necessary balance, NATO plans to deploy 572 ground-launched missiles capable of reaching Soviet territory from European sites. These LRTNF will consist of 108 Pershing-2 ballistic missiles and 464 ground-launched cruise missiles, all with single nuclear warheads. The only portion of this force to be deployed in the Mediterranean is 28 launchers, with a total of 112 ground-launched cruise missiles scheduled for Italy. However, the possibility that the Netherlands and Belgium—which had tentatively agreed to the stationing of LRTNF on their territories —may renege on their commitment may lead to a change in the distribution of the missiles in other NATO countries. At this time the fate of NATO's TNF modernization remains uncertain. There is strong objection to the program among segments of European public opinion, and implementation of the plan will depend in part on the outcome of negotiations between the United States and the Soviet Union on strategic and theater nuclear arms limitation.

Nuclear weapons are also located within the depths of the Mediterranean Sea. Shortly after the Cuban missile crisis, the United States began deployment of the Polaris submarine-launched missile in the eastern Mediterranean. It is no coincidence that this deployment was followed in 1964 by the appearance of a sizable Soviet naval force. The Soviet naval force (the Fifth Eskadra) increased substantially following the Arab disaster in the June 1967 war with Israel. Undoubtedly the Soviet buildup was in part designed to restore some of its lost political influence in the Arab world. However, according to Oles Smolansky, "It is no exaggeration to state that eskadra's primary initial function was to try to neutralize the nuclear threat created by Polaris submarines and thereby to enforce the credibility of Moscow's claim to nuclear parity with the United States."[9] Strategic deterrence has always been a major function of U.S. naval power in the Mediterranean. This function is probably declining, not only because of the growth of Soviet naval power but also as a consequence of changes in military technology. Improved accuracy of intercontinental ballistic missiles (ICBMs) and the lengthening of SLBM range through the Trident submarine have diminished the necessity for stationing strategic nuclear weapons close to the land mass of the Soviet Union.[10]

Although the NATO-Warsaw Pact confrontation constitutes one of the two central threats to nuclear peace in the Mediterranean, there are powerful factors acting to constrain the superpowers. Perhaps most important is the mutual recognition that there are no gains for either side to be had in the Mediterranean worth the cost of a nuclear war. Competition for influence in the Mediterranean Third World continues, but it is clear that Moscow and Washington are determined to avoid a military confrontation if at all possible. For NATO the main arena is Europe, and for the Soviet Union it is the homeland and the territories adjacent to it; the rest is sideshow and both sides recognize that.

NUCLEAR PROLIFERATION IN THE
MIDDLE EAST

A second major threat to nuclear peace in the Mediterranean arises from the possibility of another Arab-Israeli war with one or more of the parties in possession of nuclear weapons. One cannot exclude the possibility that nuclear weapons might be used in an intra-Arab war or a war between an Arab and a non-Arab state not involving Israel, though such a nuclear war would be unlikely.

The first question relevant to the issue of nuclear proliferation in the Middle East is, Does Israel now have the bomb? The answer is probably yes, though Israel has never tested the weapon. Certainly Israel has the capability to produce the bomb. Israel operates a nuclear reactor at Dimona that has never been subjected to international —or indeed any foreign—inspection since it was started up in 1963. It is estimated that the Dimona reactor can produce up to 8 kilograms of plutonium a year. In a recent analysis, Henry Rowen and Richard Brody calculated that by the end of the 1970s the Dimona reactor could have produced enough plutonium for as many as 20 bombs. [11] In order to extract the plutonium from the spent nuclear fuel, Israel would need a reprocessing plant, which, according to reports of the International Atomic Energy Agency (IAEA), Israel has. [12] The Israeli government has never admitted to possession of the bomb, though authoritative Israeli sources have acknowledged that Israel could assemble a bomb in short order if it were necessary. Officially, Israel's position is that it will not be the first country to introduce nuclear weapons into the Middle East. This policy has at times been described as one of deliberate ambiguity or "deterrence through uncertainty." [13] A document released by the U.S. Central Intelligence Agency (CIA) a few years ago stated, " We believe that Israel has produced nuclear weapons." [14] But it should be noted that Israel has never exploded a bomb or nuclear device; there must be some uncertainty in Israeli (as well as others') minds regarding the reliability of the weapon. No strategy can be based upon an untested weapon.

Would Israel's security be enhanced by the open acquisition of nuclear weapons? The principal rationale for an Israeli bomb is that of deterrence: it would keep the nation from being overrun by its Arab enemies. It would also deter a future Arab nuclear power from using nuclear weapons against the Jewish state. Generally, those Israelis viewing the future pessimistically will see in atomic weapons a compensation for weakness. Fears that the Israeli economy may not indefinitely support large conventional forces will lead some to look to nuclear arms as a means of providing more "punch for the pound." The prospect of an erosion of support in the West (particularly in the United States), or the growth of Soviet power in the Mediterranean, or even further increases in Arab economic power would all induce Israelis toward the idea of an atomic weapon. At this point, no one can be certain how much dependence upon Arab oil in the future will put the industrial world at the mercy of Arab political demands.

The arguments against a strategy of nuclear deterrence may be even more compelling. First among them is the almost certainty that a known Israeli weapon would stimulate several Arab governments to try to acquire the bomb for reasons of prestige as well as military consideration. A known Israeli bomb would undermine—and possibly destroy completely—any chance of creating an effective nonproliferation regime in the Mediterranean or Middle East. The question then is, How much better off is Israel in a situation of nuclear standoff? Would not Israel itself be deterred from ever using the weapon? It is hard to imagine how Israel could benefit from any war in which nuclear weapons were used in the Middle East. A small country with "a population concentrated in a narrow area which could be laid waste with a few bursts," Israel is simply too vulnerable. [15] Israel, of course, may believe that it can possess the bomb while denying it to its adversaries. In 1965, Labor Minister Yigal Allon, in announcing Israel's intention not to be the first to introduce nuclear weapons to the region, noted significantly, "May I add that Israel will not permit any of its neighbors to start this destructive race." [16] The destruction of Iraq's Osirak reactor on June 7, 1981, is evidence both of Israel's capability and Israel's determination to carry out this policy Still, as will be discussed below, the prospects are infinitely greater of keeping the region nuclear-free with the cooperation of the states involved than if it is against their wishes. One final point: Israeli acquisition of the bomb would run strongly against the U.S. policy of nonproliferation. President Reagan, continuing the policy of his predecessor, has announced that the United States will "seek to prevent the spread of nuclear explosives to additional countries as a

TABLE 6.2

Nuclear Programs in the Middle East

Egypt

Research reactors

WWR-C-Cairo; 2 megawatts; 1961; 10 percent enriched uranium.

Power reactors

Two 600-megawatt (electric) reactors have been under negotiation for several years with Westinghouse (with a letter of intent signed for one); this deal has been delayed by lack of congressional approval. There have also been press reports that two 150-megawatt (electric) reactors will be provided by a German-Austrian consortium for installation in Suez. None of these plans seems likely to be implemented in the immediate future.

Iraq

Research reactors

IRT-2000; 2 megawatts (electric); in operation; 10 percent enriched uranium.

Osiris type (under construction by France; also Italian participation); 70 megawatts (electric); 93 percent enriched uranium.

Power reactors

600 megawatts (electric) (apparently under negotiation with France); pressurized water reactor; late 1980s?; 3 percent enriched uranium.

Israel

Research reactors

IRR-1; 5 megawatts (thermal); 1960; 90 percent enriched uranium.

IRR-2; 26 megawatts (thermal); 1964; natural uranium.

Power reactors

None on order. A 950-megawatt (electric) light water reactor has been under negotiation with various supplier states. If unable to import one under conditions they deem acceptable, officials assert Israel may build one of its own design.

Kuwait

Research reactors

None.

Power reactors

None. But interest has been expressed in obtaining four to six 600-megawatt (electric) dual-purpose units by the year 2000, starting in the late 1980s.

Libya

Research reactors

None.

Power reactors

440 megawatts (electric) (negotiated with the USSR) - dual purpose: power and desalinization.

600 megawatts (electric) (under discussion with France) - pressurized water reactor.

Syria

Research reactors

None.

Power reactors

None planned. Syria contemplates a feasibility study for a 600-megawatt (electric) nuclear power plant.

Source: Reprinted with permission from Nonproliferation and U.S. Foreign Policy, Joseph A. Yager, ed., pp. 208-10. Copyright © 1980 by The Brookings Institution, Washington, D.C.

fundamental national security and foreign policy objective . . . and also view any nuclear explosion by a non-nuclear-weapon state with grave concern."[17] Thus, open possession of the bomb would risk serious alienation between Israel and its main ally.

Nuclear proliferation among the Arab states is a definite possibility. Already in several Middle Eastern states peaceful nuclear programs have begun or are planned. Many will have a potential for nuclear weapons. There are two basic processes in the peaceful production of atomic energy that can be utilized for the acquisition of fissionable fuel for military purposes. One involves the enrichment of natural uranium (by diffusion, centrifuge, or laser techniques) to make it usable (fissionable) in a power reactor. The same plant that enriches uranium for power reactors can produce fuel for atomic bombs. Currently, there are no enrichment plants in the Arab world, though Israel is reported to be working on a laser isotope separation plant.[18] The other potentially dangerous operation is the reprocessing of spent nuclear fuel rods to extract the plutonium in them. Plutonium is a man-made element created as a by-product of uranium fission, which can itself be used as a fuel for nuclear reactors or the explosive material in nuclear weapons. Thus, every government that operates a nuclear reactor is producing the ingredients for atomic bombs. Though no Arab government yet possesses a plutonium reprocessing plant, there is nothing to stop them from importing the technology to build them.

Table 6.2 is a summary of existing or planned nuclear programs of several of the countries of the Middle East, compiled by Henry Rowen and Richard Brody.[19] Recently, the United States announced approval of the sale of two large atomic reactors to Egypt. As reported in the press, the reactors would have a generating capacity of about 2,000 megawatts.[20] Egypt signed a similar agreement in March 1981 with France. Rowen and Brody believe that if these Arab nuclear programs are implemented, the countries of the Middle East will move closer to the acquisition of nuclear weapons and that in the long run the asymmetry between Israel and the Arabs will not last. "It seems more and more likely," they conclude, "that by 1990 several states in the Middle East will have nuclear explosives or be in a position to acquire them quickly."[21]

The acquisition of nuclear weapons is a function of capability and determination. While the capabilities of a state can be ascertained with some precision, its intentions remain easily concealed. Libya, for example, is a question mark. Colonel Qaddafi apparently sought to purchase nuclear weapons from the Chinese when he sent his deputy, Major Abdul Jalloud, to Peking in 1969. Since then Libya has signed the Nonproliferation Treaty (NPT), but it is believed that Colonel Qaddafi still desires to acquire nuclear weapons.[22]

Another country believed to be a nuclear risk is Iraq. Before the June 7 Israeli raid, there were suspicions about Iraq's nuclear intention. [23] These suspicions were fueled by several events: Iraq's insistence upon highly enriched uranium for Osirak, when a denatured fuel unusuable in bombs was available; Iraq's purchase from Italy and Brazil of "hot cell" facilities, which are used to shelter technicians from radioactivity during the separation of plutonium from spent fuel; the unexplained purchase of 200 tons of uranium from Portugal and Niger above and beyond that needed to operate Osirak; and finally, the refusal to permit IAEA inspection of Osirak for a period following the outbreak of war with Iran. On the other hand, Iraq is a signatory to the NPT, its government has denied an intention to build the bomb, and Iraq did permit IAEA inspections in January 1982. Charles N. Van Doren, in a recent study, concluded, "My research has uncovered no evidence of actual Iraqi efforts to develop or manufacture a nuclear explosive device, as distinguished from its acquisition of materials and equipment that would in time have been capable of yielding enough special nuclear material to do so."[24]

Iraq is of particular concern to Israel because of its oil wealth, close ties to the Soviet Union, and especially the intense hostility expressed by the government of Iraq toward Israel. Furthermore, statements by Iraqi authorities have on occasion given the impression that eventually Iraq will acquire the bomb. What, for instance, can one make of this statement made by President Saddam Husayn during a cabinet meeting in Baghdad on June 23, 1981?

> Irrespective of Iraq's intentions and of its present or future capabilities, I believe that anyone or any state in the world which really wants peace and security and which really respects peoples and does not want them to be subjugated to foreign forces should help the Arabs in one way or another to acquire atomic bombs to confront the actual Israeli atomic bombs, not to champion the Arabs and not to fuel war, but to safeguard and achieve peace. Irrespective of the Arabs' intentions and capabilities and even if the Arabs do not want them and are unable to use them, I believe that any state in the world that is internationally and positively responsible to humanity and peace must tell the Arabs: Here, take these weapons in order to face the Zionist threat with atomic bombs and prevent the Zionist entity from using atomic bombs against the Arabs, thus saving the world from the dangers of using atomic bombs in wars. [25]

It will be several years before Iraq's nuclear program recovers from the June 7 Israeli blow, but no one doubts that the effort will be resumed. Newspaper accounts report that Saudi Arabia has offered to compensate Iraq for its loss.

Presently, the potential nuclear power of most concern to Israel (and others) is Pakistan. Though not in the region of concern to this study, Pakistan could well provide a nuclear weapon to one of Israel's hostile neighbors. In a letter to UN Secretary-General Waldheim on May 27, 1981, the Israeli Permanent Representative to the United Nations charged that "there is abundant evidence indicating that Pakistan aims at producing nuclear weapons."[26] U.S. officials also believe that Pakistan is covertly building an atomic weapon, and U.S. intelligence sources believe that it now has the capability. It is known that Pakistan's principal security concern is India, not Israel. There is no clear evidence that Pakistan intends to share its arsenal should it acquire one. Libya has approached Pakistan with the intent to purchase nuclear weapons, apparently with no success. Shai Feldman believes that Pakistan is unlikely to provide Libya with the bomb.[27]

The only other Arab country that could be capable of building a nuclear weapon in the 1980s is Egypt.[28] In the past Egypt has given evidence of a desire to be a nuclear power. However, after the Camp David accords Anwar Sadat associated Egypt closely with U.S. efforts to bring stability into the Middle East—including the goal of nonproliferation. Recently, in a reversal of Egypt's earlier position, Sadat agreed to ratify the NPT even without a corresponding action on the part of Israel. Undoubtedly, the desire to obtain nuclear assistance from the United States was a factor in Sadat's decision to ratify the NPT, though, even beyond that, Sadat and his successor Hosni Mubarak have stamped Egypt's foreign policy vis-à-vis the Arab-Israeli conflict with the mark of moderation. It is the conclusion of Rowen and Brody that neither Israel nor any of the Arab countries are likely to adopt overt nuclear weapons programs. "But in the long run many of these nations seem likely to at least drift closer to having nuclear weapons."[29]

Not all analyses of the problem of nuclear proliferation agree that the growth in the number of nuclear-weapons states is a bad thing. Some argue that nuclear proliferation will have a positive effect on Middle Eastern politics, that it will produce prudence among Middle Eastern elites as it has done with the superpower elites, and that it will add to the stability in interstate relations.[30] Others see merit in proliferation in compelling the parties to the Arab-Israeli conflict to come to terms with each other. Paul Jabbur, for example, sees the prospect of nuclear war as providing "the requisite stimulant for an honorable negotiated solution" of the Arab-Israeli conflict.[31]

A case can be made that nuclear deterrence has been a factor in the reluctance of the Soviet Union and the United States to go to war against each other in the post-World War II period. But it seems

very questionable to make the same assumptions about political relations in the Middle East that one makes about the superpowers. The
United States and the Soviet Union, unlike many of the Middle Eastern
governments, are stable regimes. Joseph Nye argues, for example,
that the stability produced by nuclear weapons in great-powers relations is not replicable in unstable areas. "The transferability of prudence assumes governments with stable command and control systems;
the absence of serious civil wars; the absence of strong destabilizing
motivations, such as irredentist passion; and discipline over the temptation for pre-emptive strikes during the early stages when new nuclear weapons capabilities are soft and vulnerable. Such assumptions
are unrealistic in many parts of the world."[32] Most analysts would
agree that they are unrealistic in the Middle East under existing conditions. Thus, one must conclude that nuclear proliferation does not
add to overall stability in the Middle East and certainly does not contribute to nuclear peace in the Mediterranean.

THE NONPROLIFERATION TREATY AND
NUCLEAR-FREE ZONES

Efforts to promote nuclear peace in the Mediterranean cannot
be separated from broader policies for nuclear peace globally. There
is a general consensus within the international community that nuclear
proliferation increases the possibility of a nuclear war and should
therefore be halted. But the consensus dissipates over the question
of how to keep non-nuclear-weapons states from acquiring the bomb.
For analytical purposes, the approaches to nonproliferation can be
classified into three different types: the legal, the technical, and the
political.[33] The legal approach refers to obtaining as wide an adherence to the NPT as possible, ideally encompassing every state on
earth. Non-nuclear-weapons states adhering to the NPT legally bind
themselves not to obtain nuclear weapons and to place all their nuclear
installations under international control (through the IAEA) to provide
assurances that peaceful nuclear facilities will not be used covertly
to manufacture the bomb.[34] It should be noted that IAEA controls
consist essentially of accounting and inspection techniques to detect
a diversion of fissionable fuel from peaceful to military uses. They
provide a warning of illegal activity. IAEA controls do not consist
of physical control over nuclear plants, nor do they provide any means
to stop whatever illegalities the inspectors may find. As one authority put it, "They are analogous more to burgler alarms than to locks
on a door."[35]

The technological approach focuses upon preventing non-nuclear-
weapons states from acquiring what are now called sensitive facili-

ties, that is, those that give direct access to fissionable fuel. There are today two types of facilities that are considered to be sensitive: plants that enrich natural uranium (U-238) to make it fissionable and reprocessing plants that extract plutonium (another fissionable fuel) from spent uranium rods.[36] Until recently, only the nuclear-weapons states possessed sensitive facilities. Non-nuclear-weapons states did not need uranium enrichment plants so long as the nuclear-supplier states provided enriched uranium for power-producing nuclear reactors. Outside of the nuclear-weapons states, there are only a few states that now have the technical capability to manufacture sensitive facilities and thus go nuclear. In the Mediterranean region these include Italy and Israel.

The political approach seeks to stop proliferation by removing the underlying causes that induce nations to acquire nuclear weapons. Essentially, that involves satisfying the security needs of potential proliferants. A few of NATO's non-nuclear-weapons states, notably West Germany and Italy, could and probably would go nuclear were it not for the nuclear umbrella provided by the United States. Indeed, it is believed that non-NATO countries such as Switzerland and Sweden are influenced in their determination to eschew nuclear weapons by being implicit beneficiaries of the Western alliance.[37] The sense of security, which is essential to overcome the pressures to go nuclear, is based upon broader considerations than just formal guarantees.[38] Alliances can play a role in nonproliferation only to the extent that they are credible. These three approaches are not at all exclusive. In fact, the policies pursued by the nuclear powers to dampen the spread of nuclear weapons have involved elements of all three approaches.

Adherence to the NPT is one of the most reliable assurances that a state does not intend to become a nuclear power. As indicated above, the IAEA safeguards that are a part of the NPT cannot guarantee that a signatory will not violate its commitment and attempt to divert fissionable fuel for military purposes. Also, with only 90 days notice, an NPT signatory can legally disavow its commitment not to go nuclear. Nevertheless, no country that has signed the treaty has yet violated it or denounced it. The fact is that every country acquiring nuclear weapons to date has done so by manufacturing them from unsafeguarded facilities; that is, those countries that are known to have a weapons capability (such as India) or believed to have it (such as Israel) have refused to sign the NPT. Every nuclear-weapons state has built the bomb clandestinely in facilities not subject to international control. Thus, the position a state takes toward the NPT is important in assessing the prospects for proliferation.

Table 6.3 summarizes the actions of each of the Mediterranean states with regard to three international treaties relevant to prolif-

TABLE 6.3

Nuclear–Treaty Involvement in the Mediterranean

	Limited Test Ban Treaty	Seabed Treaty	Nonproliferation Treaty
Albania	—	—	—
Algeria	Signed	—	—
Cyprus	Ratified	Ratified	Ratified
Egypt	Ratified	—	Ratified (February 26, 1981)
France	—	—	—
Greece	Ratified	Signed	Ratified
Israel	Ratified	—	—
Italy	Ratified	Ratified	Ratified
Jordan	Ratified	Ratified	Ratified
Lebanon	Ratified	Signed	Ratified
Libya	Ratified	—	Ratified
Malta	Acceeded	Ratified	Ratified
Morocco	Ratified	Ratified	Ratified
Portugal	Signed	Ratified	Ratified
Spain	Ratified	—	—
Syria	Ratified	—	Ratified
Tunisia	Ratified	Ratified	Ratified
Turkey	Ratified	Ratified	Ratified
Yugoslavia	Ratified	Ratified	Ratified

eration. The limited test ban treaty of 1963 prohibits nuclear weapons tests in the atmosphere, outer space, and underwater. This treaty is an important link in the nonproliferation effort, because a government must test a weapon before it can put it into production. The Seabed Treaty outlaws placing nuclear weapons on the seabed or ocean floor. While relatively minor as arms control agreements go, this is a step toward the gradual denuclearization of the globe, commencing (along with the Antarctic and Outer Space Treaties) with the less-inhabited regions of the world.

Only two countries—Albania and France—have totally avoided any legal commitments in the nonproliferation realm. In both instances, the action—or lack of action—is more symbolic than substantive. Albania is a small and poor country not known to have either nuclear capabilities or intentions. Its refusal to endorse these measures is probably more a function of its policy of isolationism than it is of any nuclear aspirations. [39] France already is a nuclear power and has agreed to act with regard to the NPT as though it were a signatory. Of the remaining states, Algeria, Egypt, Israel, Libya, Spain, and Syria have either not signed or failed to ratify one of the other two treaties. It is difficult to imagine that the nonsignatories to the Seabed Treaty have refused because of any intention to implant nuclear weapons off their coasts. Not even the superpowers or any of the nuclear-weapons states have indicated a desire to carry the arms race onto the ocean floor. Presumably some Arab opposition to signing the treaty derives from the failure of Israel to do so.

That brings us to Algeria, Israel, and Spain as the only nonsignatories to the NPT. Algeria is considered to be far removed as a proliferant, both from the perspectives of capability and of intention. Spain, on the other hand, could, if it so chose, build a bomb within a decade, but is unlikely to do so. Its distance from the Soviet Union offers a protective barrier not available to most of Europe. Now that it is a member of NATO, Spain falls under the U.S. nuclear umbrella and will be more subject than ever to U.S. antiproliferation policy. Furthermore, Spain's domestic politics do not include sizable elements pushing for the nuclear option. [40] Israel, as has been indicated above, already has the capability if not the bomb itself. The Israelis give as their reason for rejecting the NPT the fear that the protections against an Arab violation of the NPT are inadequate and the belief that without the deterrent of its nuclear capability Israel in the future would be subject to Arab blackmail. They place no assurance whatsoever in the guarantees provided by the superpowers through the United Nations Security Council. [41] In short, the NPT is a useful step in the direction of denuclearizing the Mediterranean, but without the concurrence of Israel the entire structure falls apart.

THE MEDITERRANEAN AS A NUCLEAR-FREE ZONE

An alternative strategy for denuclearizing the Mediterranean basin is the creation of a Nuclear-Weapon-Free Zone (NWFZ) in the Mediterranean region. The idea of NWFZs is older than the NPT, though with one exception it has not been notably successful. The first proposals for an NWFZ were made in the 1950s with reference to central Europe, the most important being the Rapacki Plan to prohibit nuclear weapons in the two German states, Poland, and Czechoslovakia. Subsequently, proposals were made for denuclearizing the Balkans, Latin America, Africa, southern Asia, the Pacific region, the Middle East, northern Europe, the Adriatic Sea, Scandanavia, the South Pacific, the Indian Ocean, and the Mediterranean basin. Some of these proposals have been supported by the United Nations General Assembly, including NWFZs for Latin America, Africa, the Middle East and southern Asia. At its Tenth Special Session, devoted entirely to disarmament, the General Assembly gave general endorsement to the idea of NWFZs, noting that in the formulation of specific plans "the characteristics of each region should be taken into account."[42]

So far the only NWFZ to be created is the Latin American Nuclear-Weapon-Free Zone established by the Treaty of Tlatelolco in 1968. Although NWFZs will inevitably differ from region to region, the Treaty of Tlatelolco can be taken as a model of what an NWFZ is likely to be. The signatories agree to prohibit the testing, use, manufacture, or production of nuclear weapons in their territories. Also prohibited is the receipt, storage, installation, or deployment of such weapons. In order to maintain verification of the agreement, the parties must put their nuclear facilities under IAEA safeguards and accept special inspections by a regional organization created for that purpose. In separate protocols, the nuclear-weapons states (which are not parties to the treaty) have agreed to respect the terms of the treaty and to apply them to any territories in Latin America under their jurisdiction. Another commitment by the nuclear powers, which is significant because it is not contained in the NPT, is the obligation not to use or threaten the use of nuclear weapons against the parties to the treaty.[43]

The Treaty of Tlatelolco stands out as an exception to the generally dismal record of zonal denuclearization. Today there are no negotiations going on anywhere by any regional group to advance the scope of NWFZs. There are several explanations for the Latin American success. Most important is the fact that Latin American denuclearization did not affect the existing military balance either between the superpowers or any major alliance system. The initiative came from within the region itself and was not directed against any particular country or security arrangement. Since no Latin American country possessed nuclear weapons, the complicated issue of reversing a nuclear buildup never had to be faced. Finally, the countries of the zone agreed to proceed with the effort without the full participation of all the states in the hope that ultimately the recalcitrants—now including only Cuba—could be won over. In other words, the success of the endeavor was not made contingent upon unanimity at the beginning.[44]

Efforts to denuclearize the Mediterranean have to date foundered on the problem of maintaining a strategic balance between East and West. The first conference on denuclearizing the Mediterranean basin convened in July 1964 at the initiative of an Algerian peace committee. It adopted the Algerian Appeal, which called for the evacuation of all atomic weapons from the Mediterranean countries "as well as bases from countries committed to NATO."[45] Several attempts were made by some nonaligned countries during the 1960s to revive the idea of turning the Mediterranean into a zone of peace, but none were taken seriously by the NATO governments because of what was perceived as the imbalanced character of the proposals.[46] At least as far back as the 1950s, the Soviet Union has advocated the removal of atomic weapons from the Mediterranean region.[47] Soviet efforts intensified during the early 1960s, when the United States began the deployment of Polaris nuclear missiles in submarines in Mediterranean waters and during the period when the United States was promoting creation of the NATO multilateral nuclear force (MLF). On May 20, 1963, the Soviet Union proposed "that the whole area of the Mediterranean Sea should be declared a zone free from nuclear missile weapons. It is prepared to assume an obligation not to deploy any nuclear weapons or their means of delivery in the waters of this area provided that similar obligations are assumed by the other Powers."[48] Moscow's proposal was rejected by the United States as "designed precisely and solely to change the existing military balance at the expense of the United States and its allies."[49]

Leonid Brezhnev on several occasions sought to link denuclearization of the Mediterranean with his general program for détente. In 1971, 1974, and 1976 he proposed measures to remove nuclear weapons from the Mediterranean, particularly naval forces with nuclear weapons. As recently as June 10, 1981, Brezhnev reiterated his desire for turning the Mediterranean into a "zone of stable peace and cooperation." He proposed reducing armed forces in the area, withdrawal from the Mediterranean of ships carrying nuclear weapons, and renunciation of the deployment of nuclear weapons by Mediterranean countries that lack them.[50] Moscow's proposals for the Mediterranean have always been very general, and there is some question how seriously they were meant to be taken. For example, some Soviet statements imply that they favor making illegal the transportation in ships of nuclear weapons anywhere in the Mediterranean Sea.[51] Yet, Soviet authorities insist that NWFZs cannot violate international law, which includes freedom of the seas. Speaking before the United Nations Conference of the Committee on Disarmament, a Soviet delegate stated,

> An important factor in concluding agreements on the establishment of nuclear-weapon-free zones is the problem of the boundaries of those zones. The boundaries of nuclear-weapon-free zones must be determined in accordance with generally recognized rules of international law, including the principle of freedom of navigation on the high seas and in straits used for international navigation. An agreement on a nuclear-weapon-free zone cannot extend the nuclear-weapon-free status of a zone to the territory of States not situated in the zone or to the high seas, since that would constitute a violation of generally recognized rule of international law, including the principle of freedom of navigation on the high seas.[52]

Interestingly, on this position the United States and the Soviet Union are in full agreement. Recent Soviet statements regarding an NWFZ for northern Europe reveal something of an inconsistency with their position toward a Mediterranean NWFZ. While endorsing the prohibition of nuclear weapons from the Scandinavian countries, Moscow is refusing to remove nuclear weapons from the Kola Peninsula adjacent to Scandinavia on the grounds that "the military potential on the Kola Peninsula is part of the global strategic balance between the United States and the U.S.S.R. and is not aimed at the Nordic countries."[53] Furthermore, Moscow has rejected the idea of including in the NWFZ the Baltic Sea, where the Soviets maintain a fleet of submarines carrying nuclear weapons. The Soviet position regarding northern Europe and the Baltic is the mirror image of the U.S. position in the Mediterranean. Whether the Western powers simply took the Soviet proposals for propaganda only or whether they were convinced that the military advantage of a denuclearized Mediterranean was too advantageous to the Soviet Union, the fact is that the Soviet proposals have failed to result in serious negotiations between the two sides.

The position of the Western powers has been that NWFZs in the Mediterranean (or any area) can be created only if (1) they do not upset the existing military balance, (2) they are initiated by the states in the region, (3) they include all militarily significant states in the region, and (4) there are effective controls to verify observance.[54]

These conditions will clearly be difficult to apply to the Mediterranean. It is extremely difficult to imagine any scheme for denuclearizing the Mediterranean that does not leave the Warsaw Pact better off than NATO. One is tempted to conclude that the idea is unworkable as long as the factors that produced the two pacts continue to exist.

From the perspective of the states involved in an NWFZ, the basic security interest in participating is to avoid being attacked or threatened by a nuclear-weapons state. For the NATO members in the Mediterranean there are only two types of guarantees that would be meaningful. One would be to expand the territorial base of the zone to include the Soviet Union. No one seriously believes that this is feasible. The other is for the Soviet Union (and the other nuclear-weapons states) to give a formal commitment not to use or threaten any member of the NWFZ with nuclear weapons (as is provided for in Protocol II of the Treaty of Tlatelolco). Moscow is prepared to give this commitment, but not without reservations. In debate in the Disarmament Committee of the United Nations, Soviet spokesmen have stated that

> the Soviet Union, for its part, will be prepared to as-
> sume the obligation to respect the status of nuclear-
> weapon-free zones provided they are genuinely free of
> nuclear weapons, and the other nuclear States will as-
> sume similar obligations. At the same time, the Soviet
> Union reserves the right to review its obligations con-
> cerning respect for the nuclear-weapon-free status of
> the zones if a State in respect of which it has assumed
> such obligations engaged in aggression or becomes a
> participant in aggression.[55] [Emphasis added]

During the course of the debate, the Swedish delegate raised the obvious point, "Who would make that assessment and what would be the value of a commitment not to use nuclear weapons, if it is withdrawn in case of war?"[56] Clearly France, Italy, Greece, and Turkey are going to demand greater assurance from Moscow before they are prepared to abandon NATO or NATO's nuclear defense.

There are two NWFZ proposals in regions tangential to the Mediterranean that may have better prospects than a denuclearization of the entire Mediterranean. They are an NWFZ for Africa and for the Middle East. Support for an NWFZ in Africa arose in 1960 following the first nuclear test explosion in the Sahara Desert by France. During the 1960s, the General Assembly, the Organization of African Unity (OAU), and the Conference of Nonaligned Countries all endorsed the principle of a denuclearized Africa. In December 1974, the General Assembly for the first time unanimously passed a declaration

denuclearizing Africa. Although the United Nations has frequently
reiterated its support for the idea, nothing concrete has been done by
the states of Africa to implement these resolutions. General Assembly resolutions are not legally binding. It will take a multilateral
convention to establish a legal NWFZ in Africa.

A major barrier to a continent-wide agreement is South Africa,
which is the only country in Africa that is close to having a nuclear
capability. South Africa operates several nuclear reactors, is one
of the world's largest uranium-producing countries, and, if its claims
are to be believed, has a secret process for enriching uranium.
Though South Africa has never claimed to have the bomb, high authorities do claim to have the capability of making a bomb. Also, South
Africa is not a party to the NPT.[57] Recent General Assembly resolutions have supplemented calls for denuclearizing Africa with condemna-
tion of South Africa's nuclear program. In many respects, the circumstances in Africa are better for a consensus on denuclearization
than anywhere outside of Latin America. It is clear, however, that
until South Africa's relationship with the rest of the continent is nor-
malized, the majority of the countries are going to have to take the
initiative without it. The OAU, for example, could take steps to organize an African conference to draft a treaty. Ratification could
even begin without South Africa by including in a treaty the equivalent
of Article 28 in the Treaty of Tlatelolco. Article 28 specifies that
the treaty does not enter into force until all the eligible countries in
the region have ratified the treaty. By that mechanism Latin America
moved forward without Cuba. Admittedly, Cuba is not a potential
nuclear power, while South Africa is. But the African states would
not be legally bound until South Africa was also.

The prospects for a Middle Eastern NWFZ cannot be rated high
at the present time. Here, as elsewhere, states are reluctant to
abandon a potential weapon in the face of perceived security vulnerabilities. In the mid-1970s, Iran and Egypt took the initiative in proposing before the United Nations that steps be taken to establish an
NWFZ in the Middle East. With the concurrence of most of the states
of the Middle East and all five nuclear-weapons states, the General
Assembly endorsed the idea in December 1974. Israel, arguing that
an NWFZ could be brought about only by means of direct consultations
among the states of the region, abstained. In recent years, Israel's
position has softened on this issue. At the Thirty-fifth Session of the
General Assembly, Israel submitted its own proposal for an NWFZ in
the Middle East. The Israeli draft called for a regional conference
of all states to negotiate a multilateral treaty establishing an NWFZ
in the Middle East.[58] Several Arab states rejected this initiative,
presumably because of the implications for the issue of Arab recognition of Israel. Indeed, it is possible that the Israeli initiative was

introduced as a ploy just to obtain that political advantage. In spite
of the rejection of its draft, Israel did join with the General Assembly
in a consensus decision on December 12, 1980, calling for creation
of an NWFZ in the Middle East. While the General Assembly resolution makes no provision for negotiations among the states of the region leading to a treaty, it does invite the Secretary-General to "ex-
plore the possibilities of making progress toward the establishment
of a nuclear-weapon-free zone in the region of the Middle East."[59]
The issue will again be before the General Assembly.

CONCLUSION

One can be reasonably sanguine that nuclear conflict in the
Mediterranean is not likely to occur in the near future. Whether or
not it ever happens will depend in large part on events that take place
outside of the region. Nuclear peace in the Mediterranean, in other
words, depends substantially on peace in Europe and the Middle East.
The efforts to avoid nuclear war by denuclearizing (either via the
NPT or an NWFZ) the Mediterranean are useful but peripheral. These
efforts are hampered by political antagonisms that must be resolved
before denuclearization can take place and which, if resolved, would
probably make denuclearization unnecessary. There is an element
of gamesmanship in the campaign for denuclearization. For some
it is a propaganda or manipulative ploy; but for others it reflects considered conviction. Those who are serious about denuclearization
must inevitably come to grips with the necessity of all the states in
the region not only to accept the existence of each other but to eschew
major aggression.

Denuclearization may not be able to save the Mediterranean,
but it cannot hurt. A useful first step might be the creation of a committee representing all 19 Mediterranean states to examine means
to promote denuclearization. Whether or not denuclearization is possible depends on how much unity there is among the peoples of the
region. Several years ago, Jesse Lewis, Jr., observed that "a Mediterranean consciousness is emerging in America."[60] The vital
question today is, Is there a consciousness emerging in the Mediterranean? Have the peoples of the Mediterranean developed a sufficient
commonality among themselves to produce a genuine community?
Few would agree that such a community exists now, but some believe
that it is emerging.

NOTES

1. For the purposes of this discussion the Mediterranean is defined as the 17 countries bordering the Mediterranean Sea plus Portugal and Jordan.

2. Abraham S. Becker, The United States and the Soviet Union in the Mediterranean (Santa Monica, Calif: Rand Corporation, 1977), p. 11.

3. New York Times, December 9, 1981, p. 1; see also U.S., Congress, Senate, Perspectives on NATO's Southern Flank: Senate Delegation Report to the Committee on Foreign Relations, 96th Cong., 2d sess., June 1980, pp. 39-41.

4. Senate, Perspectives on NATO's Southern Flank, p. 10.

5. Alvin Z. Rubinstein, "The Soviet Union and the Eastern Mediterranean: 1968-1978," Orbis 23, no. 2 (Summer 1979): 305.

6. Instituto Affair Internzaionali, The Mediterranean: Politics, Economics, Strategy, Papers, vol. 1., p. 40.

7. U.S., Congress, House, Subcommittee on Europe and the Middle East, The Modernization of Nato's Long-Range Theater Nuclear Forces: Report to the Committee on Foreign Affairs, 96th Cong., 2d sess., December 31, 1980, p. 24.

8. Soviet IRBMs are situated in the northeast Crimea and in the southern fringes of the Transcaucasus. See CSIA European Security Working Group, "Instability and Change on Nato's Southern Flank," International Security 3, no. 3 (Winter 1978-79): 151-76.

9. Oles Smolansky, The Soviet Union and the Arab East under Khrushchev (Lewisburg: Bucknell University, 1974), p. 301.

10. Becker, The United States and the Soviet Union, p. 13.

11. Henry S. Rowen and Richard Brody, "The Middle East," in Nonproliferation and U.S. Foreign Policy, ed. Joseph A. Yager (Washington: The Brookings Institution, 1980), p. 206.

12. It is reported to be a "pilot" reprocessing facility; ibid.

13. Faud Jabber, Israel and Nuclear Weapons: Present Option and Future Strategy (London: Chatto and Windus, 1971), pp. 124, 127.

14. Rowen and Brody, "The Middle East," p. 221.

15. Jabber, Israel and Nuclear Weapons, p. 129.

16. Ibid., p. 122.

17. New York Times, July 17, 1981, p. A4.

18. Rowen and Brody, "The Middle East," p. 209.

19. Ibid., pp. 206-8.

20. New York Times, June 30, 1981, p. 6.

21. Rowen and Brody, "The Middle East," pp. 177, 208; see also Shai Feldman, "A Nuclear Middle East," Survival 23 (May-June 1981): 107.

22. Rowen and Brody, "The Middle East," p. 206; Feldman, "A Nuclear Middle East," p. 110; and Charles N. Van Doren, "Iraq,

Israel, and the Middle East Proliferation Problem" (Report prepared for the Arms Control Association, Washington, D.C., June 25, 1980), p. 19, mimeographed.

23. See International Institute for Strategic Studies, Strategic Survey 1980-1981 (London: IISS, 1981), pp. 114-15. A thorough analysis of Iraq's nuclear capabilities is contained in Van Doren, "Iraq, Israel, and the Middle East."

24. Van Doren, "Iraq, Israel, and the Middle East," p. 2.

25. Foreign Broadcast Information Service 5, no. 121 (June 24, 1981): E3.

26. Van Doren, "Iraq, Israel, and the Middle East," p. 45.

27. Feldman, "A Nuclear Middle East," p. 110.

28. New York Times, July 14, 1981, p. A6.

29. Rowen and Brody, "The Middle East," p. 234.

30. See, for example, Feldman, "A Nuclear Middle East," admitting that the absence of a second-strike capability could lead to nuclear instability, while believing that "the principal source of stability in a nuclearized Middle East will be the enormous difficulties each nuclear state will face when attempting to destroy its rival's nuclear forces," pp. 112-13.

31. Quoted in Rowen and Brody, "The Middle East," p. 189.

32. Joseph S. Nye, "Sustaining Nonproliferation in the 1980's," Survival 23 (May-June 1981): 104.

33. Joseph L. Nogee, "Soviet Nuclear Proliferation Policy: Dilemmas and Contradictions," Orbis 24, no. 4 (Winter 1981): 755-59.

34. The text of this treaty can be found in U.S., Arms Control and Disarmament Agency, Arms Control and Disarmament Agreements, (Washington, D.C.: ACDA, 1980), pp. 90-94.

35. Rodney W. Jones. Nuclear Proliferation: Islam, the Bomb and South Asia, The Washington Papers, vol. 9. (Beverly Hills: Sage, 1981), p. 12.

36. It takes a much higher degree of enrichment for uranium to be weapons-grade than it does for use in reactors producing electricity. For a good summary of the technical aspects of nuclear energy and international controls see ibid., pp. 11-23.

37. Ibid., p. 19.

38. Ian Smart, "European Nuclear Options," in NATO, the Next Thirty Years: The Changing Political, Economic and Military Setting, ed. Kenneth A. Myers (Boulder, Colo.: Westview, 1980), p. 129.

39. An interesting popular description of Albania's foreign and domestic policies is Mehmet Biber, "Albania Stands Alone," National Geographic 158, no. 4 (October 1980): 530-37.

40. Smart, "European Nuclear Options," p. 124.

41. Jabber, Israel and Nuclear Weapons, pp. 126-27; Rowen and Brody, "The Middle East," p. 226.

42. "Annex", UN Monthly Chronicle 15, no. 7 (July 1978): 6.

43. For the text of the treaty see Arms Control Agency, Arms Control, pp. 63-76.

44. Argentina has signed but not ratified. The ratifications of Brazil and Chile do not go into force until all eligible countries also ratify the treaty.

45. Milorad Mijovic, "Denuclearization of the Mediterranean," Review of International Affairs 15, no. 344-5 (August 1964): 7.

46. See J. Djerdja, "Misled Opportunities," Review of International Affairs 19, no. 434 (May 1968); V. Vladisavljevic, "Mediterranean Confrontation," ibid., no. 449 (December 1968); and Radovan Vukadinovic, "Europe and the Mediterranean," ibid. 26, no. 639 (November 1976).

47. U.S., Department of State, Documents on Disarmament, 1945-1959, vol. 2 (Washington, D.C.: Government Printing Office, 1960), pp. 1434-36.

48. U.S., Arms Control and Disarmament Agency, Documents on Disarmament, 1963 (Washington, D.C.: Government Printing Office, 1964), p. 193.

49. Ibid., p. 243.

50. New York Times, June 10, 1981; see also John C. Campbell, "Communist Strategies in the Mediterranean," Problems of Communism, May-June 1979, p. 14.

51. A. Vanin describes as an important move aimed at reducing military tensions in the Mediterranean "the Soviet proposal [put forward in July 1974] to withdraw all Soviet and U.S. ships and submarines carrying nuclear weapons from the Mediterranean." International Affairs, February 1978, p. 116.

52. United Nations, Committee on Disarmament, Comprehensive Study of the Question of Nuclear-Weapons-Free Zones in all its Aspects: Special Report of the Conference of the Committee on Disarmament, 1976, p. 77.

53. New York Times, July 24, 1981, p. A3.

54. UN, Comprehensive Study, p. 20.

55. Ibid., p. 77.

56. Ibid., p. 75.

57. See William Epstein, "A Nuclear-Weapon-Free Zone in Africa," "Roderic Alley," and "Nuclear-Weapon-Free Zones: The South Pacific Proposal," Occasional Paper 14 (Muscatine, Iowa: Stanley Foundation, 1977), pp. 5-24.

58. New York Times, June 26, 1981, p. A26.

59. United Nations, General Assembly, Resolution (35/147), December 12, 1980.

60. Jesse W. Lewis, Jr., The Strategic Balance in the Mediterranean (Washington, D.C.: American Enterprise Institute for Public Policy Research, 1976), p. 117.

7

ISLAMIC
RESURGENCE
IN CONTEXT

MARILYN ROBINSON WALDMAN

We have all learned from everyday experience the vital link between the context in which we place a phenomenon and the meaning
we are able to attribute to it. If we want to demonstrate that the rose
is the most beautiful flower in the world, we do not display it in a
bouquet of roses; if we want to show that a particular rose is especially lovely, we do.

In trying to grasp the significance of the so-called resurgence
of Islam in the late 1970s and early 1980s, we have lacked not only
a complete account of the phenomenon, but more than that, the contexts that could make it fully meaningful. Although the resurgence
of Islam can, like any other phenomenon, be placed in a very complex conceptual matrix, two perspectives are essential: the role of
religion in the contemporary world as a whole and in historical Islamic traditions themselves.

By isolating and focusing on what is happening to religion in the
Islamic world alone, or in the Mediterranean part of it, we particular-
ize and defamiliarize what are in fact two quite general and familiar
twentieth-century developments: the crisis of meaning associated
with accelerated secularization and the dramatic shifts in traditional
premodern loyalty patterns occasioned by the rise of nationalism.

The crisis of meaning has derived from what the doyen of religious studies Mircea Eliade describes as the progressive desacral-
ization of a once thoroughly sacralized cosmos.[1] This process has

been accompanied and encouraged by an expansion of the role of science and scientific standards to a point at which they became the preeminent, if not exclusive, means of evaluating truth, a truth concomitantly devoid of any ultimate meaning. When in modern societies a strictly limited means of verification became the only one, questions that were formerly crucial came to be bracketed or relegated to the private sphere, their traditional answers condemned to the appearance of irrationality among all "thinking" people.

Simultaneously, the rise of the secular national state, the political form closely associated with modernization, caused religious allegiance, as a source of identity and focus of loyalty, to function in new and ever more problematic ways. In traditional empires, where citizenship was so much more vague and less demanding and exclusive, religious adherence had coexisted with and often been superseded by multiple focuses—kin, occupation, neighborhood, town, region, voluntary association. Or religious loyalty itself might be transnational, as in the Roman Church or Muslim Ummah (community of the faithful).

These observations are as apt for traditional Islamic states as for their contemporaries in Europe, the Mediterranean, India, or China, except that Islamic states tended to institutionalize and routinize intercommunal relations in unusual ways, to the extent that legitimate and imitable Islamic regimes in Spain (711-1492) and India (1000-1750) could remain predominantly (50 percent to 90 percent) non-Muslim throughout their history. The source of this seeming anomaly was the concept of <u>dhimmah</u>, a contract of mutual obligation presumed to exist between Muslim rulers and their non-Muslim subjects. The idea was introduced in the Koran itself and worked out as a central principle in the Sharia, the law of Islam as derived from canonical sources.[2] Furthermore, the general pre-modern absence of a level of allegiance comparable to that of nation, reinforced by Islamic multicommunalism, and the haste with which most Muslim peoples had to develop one, severely tested and altered the strength and significance of old loyalties.

In this process, one of the most tragic and destructive features of modern politics began to emerge—the fusion, or confusion, of religious with national identity. Eventually, this feature began to intertwine with an accentuated version of an always common tendency for religious loyalty to serve as an expression of social, economic, occupational, ethnic, and cultural concerns. This explosive combination leaps out at us from India and Pakistan to Lebanon and Israel to Northern Ireland. It, not some inherent Islamic principle, helps to explain the decline of the fortunes of long-lived non-Muslim communities in Morocco, Turkey, Egypt, Syria, Iraq, and Yemen ever since World War II, as well as the catastrophe befalling Bahais and to a lesser extent Jews in Khomeini's Iran.

In the Muslim world, as in other colonial areas, colonizers' manipulation of the educational system and the identities of newly trained indigenous elites helped set the stage. For when loyalty, expressed largely through the paying of taxes to a distant absolute monarch ruling an international multicommunal empire with uncertain borders, was replaced by loyalty to a hastily created abstract idea executed by not particularly legitimate leaders ruling over artificially determined fixed but problematic borders, religious loyalty acquired new significances of many sorts. Nationalism itself can be a constructive force; but it has also been destructive, especially when linked with religion, creating new conflicts, making latent conflicts manifest, or charging old conflicts with new symbolic vitality.[3]

In Muslim lands in particular, the old pluralistic legal pattern could not be maintained. Whereas in prenationalist days built-in inequalities had been accepted and, even when the Sharia was the law of the land, separate legal systems functioned for different religious communities, in the new secular states with their new European-inspired legal systems, all citizens had to be subject to the same law, a situation comfortable for neither minority nor majority. With ethnic and language identities more or less difficult to develop, to be, for example, Yemeni and Muslim could be equated (or even Arab and Muslim, even though many Arabs are Christian), whereas to be Turkish and Muslim could be almost completely dissociated. Where communal relations were too complex, as in Syria, it was difficult to use Islam as a national attribute, whereas Islamic identity was vital to the creation of Pakistan.

Before one can describe Islam's own peculiar history, it is necessary to remember that any ideological system, which religions are in part, functions as a source of public and/or private values by providing three social necessities: (1) a language, especially of symbols, for talking about and explaining the world, for conveying one's assumptions and values; (2) an identity, for placing oneself and one's group in the world; and (3) techniques or norms for managing the world and living in it. But it is also important to remember that no complex society is ever ideologically monolithic, and that any given ideological system has to compete with one or more others for what we might call the "mental space" of its audience. The degree to which any given system competes successfully for that space is a subtly complex comparative topic that needs much more scholarly attention and consideration. It depends not only on the relative strength or weakness of competitiors but also on the shifting saliency of the given system for different groups as their material and nonmaterial circumstances change, as well as on the talents of what Eisenstadt would call its entrepreneurs in using it to address those conditions.[4]

The recent resurgence of religion as a strong competitor is a global phenomenon that is appearing not only in the Islamic world but also in such diverse places as the United States, Latin America, Israel, Poland, and Ireland. In analyzing the increasing significance of religious value systems in the contemporary world, one must be aware that all high religious traditions carry within them certain dynamic tensions, for example, between this-worldliness and other-worldliness and between stability and spontaneity.[5] One of the things that distinguishes all world or high religions from primitive religion, which is largely this-worldly in its aims and methods, is a tension between this-worldliness and otherworldliness, a tendency to put this world to an otherworldly use and to use the other world to trans-value this one.

This tendency is particularly pronounced in all the historicistic, monotheistic religions and, among them, markedly so in Islam, which sees itself as a complete social experiment in which God's omnipotence requires that everything be pervaded by his will. The choice for Muslims has not historically been between religion and politics or church and state (there are no words for religion and politics in traditional Islamic languages and no church in traditional Islamic states) but between such pairs as din and dunya, which correspond roughly to living in this world with moral referents (both for spiritual and mundane purposes) and living in this world without moral referents. For Muslims, what we call politics, even social action in general, has frequently been an expression of faith.

With no church to define orthodoxy, the whole community became the church; its law defied separating out the secular and focused on orthopraxy rather than orthodoxy, calling the ruler to account along with his subjects. Thus the major "theological" splits in Islam have ensued from disagreements about who should be the leader. The very calendar of the Islamic peoples begins not with Muhammad's first revelation from God (610), but rather with his immigration to the place where he established his comprehensive model community on earth. The mosque, not the palace, was the monumental building that announced the power of the ruler; and his overthrow might well be plotted there, too. Clearly, then, the relationship between what we call religion and what we call politics is difficult to conceptualize in our terms when we are talking about Islamic societies.

In looking at the second tension, between stability and spontaneity, it is helpful to use Marshall Hodgson's notion of the "dialectic of a cultural tradition":

> In general, then, but especially in the high culture of
> preModern citied societies, which has been the primary
> milieu of Islam, we may describe the process of cul-
> tural tradition as a movement composed of three mo-
> ments: a creative action, group commitment thereto,

and cumulative interaction within the group. A tradition originates in a creative action, an occasion of in-
ventive or revelatory, even charismatic, encounter.
. . . The second moment of a cultural tradition is
group commitment arising out of the creative action:
the immediate public of the event is in some way institutionalized and perpetuated; that is, the creative action becomes a point of departure for a continuing body
of people who share a common awareness of its importance and must take it into account in whatever they do
next, whether in pursuance of its implications or in
rebellion against them. . . . This group commitment
retains its vitality through cumulative interaction among
those sharing the commitment; above all, through debate and dialogue, as people work out the implications
and potentialities latent in the creative event to which
they are bound. [6]

As this process has occurred in the monotheistic traditions,
the revealed scriptures, strictly canonized and time-limited as they
are, have exerted a profound stabilizing force, which finds strong
expression in the literalist schools that have sprung up in all of them,
in the quasi-canonical scripturally derived legal codes they have all
produced, and in the religious specialists they all utilize.

At the same time, spontaneity is persistently sought and found.
On the fringe and sometimes beyond the pale, spontaneity can be produced by an actual reopening of direct revelation, as by the Church
of the Latter-Day Saints, the Bahai faith, or certain types of extreme
Shiism: more commonly, revitalization issues from the immediate
confrontation with the Divine sought in mystical endeavors. Frequently, routinized or nonroutinized expectations for change provide
its vehicles, as in the messianic or chiliastic traditions also common to all three monotheisms.

That is to say, Islam, like its "siblings," developed its own
mechanisms for private and public internal rejuvenation. Like them,
it has strong and long-lasting reform traditions of its own, of which
current dialogues are but the latest stages. Such reform has taken
many shapes; but until modern times, most of them partook of a
concept of change common to many, if not most, traditional societies
—a past-oriented change in which what we would call innovation occurred, to be sure, but in the guise of re-novation. To say, as is
commonly done these days, that Islam has scorned innovation, is
misleading. The word often used for innovation in Arabic—bidah—
tended historically not to have the connotation of change per se but of
change that was socially disruptive and incapable of being digested.

In such a situation, change did, however, occur slowly; and the force of received or inherited patterns and ideals, never completely absent from any society, was very, very powerful. Change was tolerable to the extent that it made or kept the past present, or appeared convincingly to do so. Progress could be measured in terms of proximity to an already established norm, not in terms of departure from it. Flexibility, logically, came from the possibility of reinterpreting that past in light of contemporary circumstances, or perhaps redefining its sources.

One could look at the reform movement within the Roman Church that culminated in the sixteenth-century Protestant Reformation in just that way. As a reformer bent on re-novating, Luther paved the way for a radically different type of change by returning to the gospels themselves instead of relying on commentaries accumulated over the centuries. Such an apparently conservative return to the past can in fact radically improve one's ability to reinterpret it flexibly. Along the same lines, it is interesting to note the quite old form of Islamic rethinking that returns to scripture (the Koran) precisely so as not to be bound by the accumulated constraints of centuries.

In Islamdom, expectations for renovation were both institutionalized and not. One form of institutionalization was connected with the figure of the mujaddid (literally, re-newer), who was expected by some Muslims from very early Islamic times on to appear in one or more persons at or around the turn of every Islamic century, in order to reduce any angular deviations from the fixed point of Muhammad's community that would in time, if left unchecked, lead Muslims far wide of their base. Such a figure could be a warrior, a ruler, a thinker, or a judge, working alone or with a group. Al-Ghazali, the Sufi theologian of the sixth Islamic century, was viewed as such, but so were ibn-Tumart (sixth-century North African military leader), ibn-Taymiyyah (seventh-century legal reformer), Jibril ibn-Umar (thirteenth-century militant West African reformist), and more recently Jamal al-Din al-Afghani (fourteenth-century thinker, writer, and political agitator). It is probably not accidental that Khomeini's return to Iran coincided with the turn of the fifteenth century of Islam; and even if it was, the significance was not lost. Not surprisingly, mujaddid-related activity around the turn of the fourteenth Muslim century, when most Muslim lands were under European control, was much less evident than around the turn of the thirteenth and fifteenth, when they were independent.

Renovation has, however, also taken place outside the concept of mujaddid, in march warrior ghazi movements like the Ghaznavids (tenth century, of our era), Safavids (sixteenth century), or Ottomans (fourteenth century); or through messianic figures, like the nineteenth-century Sudanese Mahdi, who might or might not be associated with the mujaddid.

As we turn to a survey of the recent history of Islam's reform traditions, let us note that religious systems theoretically have had, at least in traditional societies, a built-in competitive edge in their claim to have the inherent and exclusive right to dominate an adherent's mental space. Nevertheless, Islam as an ideal system and a system of ideals has had to compete with other sources of values (religious and nonreligious) from its very inception, or incorporate them. If we stand at a great distance and survey the whole 1,400-year sweep of Islamic history, we see the saliency of Islam as a source of public values waxing and waning periodically. Periods of waxing would include

1. From the latter part of Muhammad's life (A.D. 624) through the second civil war (fitnah) and the death of Husayn (685);
2. From the beginning of active pious opposition to the Umayyad dynasty (730s) through the reign of the caliph al-Mutasim (830s);
3. From the beginning of the Shii century (945-1055) through the end of the Sunni revival under the Seljuqs (1055-95) or even the Muwahhidun (Almohades, early twelfth century);
4. From the establishment of the Ottoman state (early 1300s) through the stabilization of the Safavids of Iran (1550);
5. From the Wahhabi movement (1780s) through the West African jihads and the heyday of the late nineteenth-century reformers;
6. From the establishment of the state of Pakistan (1947) to the present, with intensified saliency during the last five or ten years.

Although it is beyond the scope of this chapter to analyze the configurations of circumstances that fostered premodern surges, one can list here the conditions that seem to be stimulating the current one:

1. An improving material situation in a few parts of the Muslim world, sparked by oil and accompanied by increased international influence and pride;
2. A deteriorating material situation elsewhere and a related disenchantment with other ideological systems;
3. The differential impact of modernization, alienating rulers from certain citizens through economic and educational change;
4. A widespread growth of anticolonial rhetoric throughout the developing world;
5. The relative peripherality (and thus unassailability) of Islam as a source of public policy while other ideologies were being tried and discredited.

The most recent period of saliency, together with its strong roots in the one before, can be seen as a continuation of an ongoing dialogue; but both must be analyzed also in the context of the dramatic alteration in the material circumstances of Muslim peoples that began to take place in earnest around A.D. 1750 and constituted the first widespread permanent reversals of Islamic expansiveness and vitality in Islam's history.

Buttons can provide us with a memorable metaphor for these reversals. When Europeans first began to make their presence felt in the Muslim world in the sixteenth century, they were often made fun of because their buttons popped off when they tried to assume the civilized cross-legged sitting posture of their hosts. Later, when Muslims began to adopt the frock coat and tight pants themselves, the laugh was on them, because they often did not button them right.

Saint-Exupéry captured the far-reaching cultural significance of this transformation of identity in his story about the discoverer of the Little Prince's birthplace.

> I have serious reason to believe that the planet from
> which the little prince came is the asteroid known as
> B-612. This asteroid has only once been seen through
> the telescope. That was by a Turkish astronomer, in
> 1909. On making his discovery, the astronomer had
> presented it to the International Astronomical Congress,
> in a great demonstration. But he was in Turkish cos-
> tume, and so nobody would believe what he said.
> Grownups are like that. . . . Fortunately, however,
> for the reputation of Asteroid B-612, a Turkish dictator
> made a law that his subjects, under pain of death,
> should change to European costume. So in 1920 the as-
> tronomer gave his demonstration all over again,
> dressed with impressive style and elegance. And this
> time everybody accepted his report. [7]

In the eighteenth century, however, this significance was not yet completely apparent, and Muslim reaction to the European challenge took three fairly positive and optimistic forms.

1. Restoration of existing political systems (as accumulated over time) and even exaggeration of their characteristics, as in the Laleh Devri (Tulip Period), 1703-30, in the Ottoman Empire;

2. A highly selective imitation of Western systems, with no certain awareness or acknowledgment of the power or superiority of their totalities, as in the creation of European-style army units to fight alongside traditional ones—it could even be said that a form of this style persists in Saudi Arabia's selective modernization;

3. A radical, in fact antitraditional, revival of an authorita-tive past, as in the Wahhabi movement or the numerous West Afri-

can jihads of the nineteenth century, a style that continues to be expressed by such odd bedfellows as the Muslim Brotherhood of Egypt and the Mujahidin-i Khalq of Iran (of whom more will be discussed later).

It is important to pause at this point for an exploration of what is meant by an "antitraditional revival of an authoritative past," a phrase that sounds like a contradiction in terms but is essential to any understanding of intra-Muslim discussion and conflict. The concept rests on a specific definition of tradition—as the accumulated cultural heritage of a people transmitted from one generation to the other and recognized as such. Use of this definition makes it readily apparent that tradition varies from group to group and from time to time.

Among Sunni Muslims, the vast majority, tradition could include the Koran (interpreted literally or figuratively), various commentaries thereon, and Hadith (reports about the exemplary speech and behavior, or Sunna, of Muhammad), and several classical legal schools, but also jurisprudential precedents and customs associated with being Muslim, like wearing the veil, celebrating Muhammad's birthday, or behaving in certain ways during visits to Muhammad's tomb (traditional behavior that the so-called traditionalist Wahhabis violently abhorred and condemned).

For Shii Muslims, Islamic tradition can include not only the Koran and numerous (other) commentaries thereon and (other) Hadith and (other) early legal schools, but also the sayings and judgments of the imams and their representatives as well as the status and powers traditionally assigned to religious specialists (ulama or mullahs) say in the Safavid or Qajar periods, as well as historically generated customs firmly associated with being (Shii) Muslim, like the wearing of the chador by some groups of Iranian women or the making of pilgrimages to the birth or burial places of the imams or the annual ritual acting-out (taziyyeh) of the passion of Husayn.

For either, the possibility of further internal variation also exists. And for both, the chance of a Sufi (mystical) dimension to tradition is not absent, as indicated by the growing popularity of the (Sunni) Mevlevi Order in Turkey (misnomered Whirling Dervishes in the West) or the persistence of the spiritual athleticism of the (Shii) Zurkhanehs in Iran, which have, among other things, trained Iran's olympic wrestlers while praising the exemplary young manhood of Ali, the patron of Shiis and many Sufis alike. In fact, it should be noted that some of the most successful Muslim rejuvenations in history have been orchestrated by Sufi orders (tariqas), for example, the Murabits (Almoravids) in eleventh-century Morocco and Algeria, the Safaviyyah in sixteenth-century Iran, and the Sanusiyyah in nineteenth-century Libya.

A pious Muslim, intent on enhancing the Islamic character of society, can break into the tradition at any point to which his constituency can respond, as when the leaders of the Fulani jihad in nine-teenth-century Nigeria chose Abbasid (eighth- and ninth-century Iraqi) models for their political structure; look for the principles that lie behind its historical forms; or try to escape the inhibiting and, worse, sometimes embarrassing force of tradition (compare Luther's reaction to the traditional sale of indulgences) by trying to postulate a pretradi-tional ideal, a pure normative time before decisions taken by histori-cal communities began to accumulate. For Muslims of the latter per-suasion, this has usually meant the model of the pure "primitive" or rightly guided community of Muhammad and his first four successors, and the sources of the Koran and sometimes of the Hadith. They see historical Islam as a matter of interest for anthropologists, not theo-logians, and define Islam by its fundamentals, which often run counter to the traditions of any particular community. This was as true for the Wahhabis of eighteenth-century Arabia as for other antitraditional "renewers," for example ibn-Taymiyyah in the twelfth century and al-Shafii in the eighth and ninth centuries.

So it is necessary to posit, from the very beginnings of the two most recent surges of public Islam, the existence of a set of alterna-tives: maintaining, renewing, and reforming, each of which assumed a different stance toward tradition. The outcome of the dialogue and conflict among these three and the degree to which any can gain power will probably determine the future of Islam as a twentieth-century social force, and the conflict will reflect social and economic cleav-ages as well.

The foregoing analysis also cautions us to remain skeptical about the appropriateness of our blanket term "fundamentalism" in any of the many different senses in which it might be used in the United States. In the history of Islamic cultures, fundamentalism, in the sense of a rather literalist return to or revival of narrowly defined sources and models, can call for a sharp break with tradition in part or in whole and can be politically, economically, or socially quite radical. For example, for ibn-Taymiyyah, the intellectual men-tor of the Wahhabis, even the caliphate, which most conserving tra-ditionalists of his day accepted as a fundamental of Islam, was ex-pendable. Nor can we assume that someone who talks about making society Islamic is against modernization, even though he probably is against modernization if that necessarily means Westernization, or "Westoxification," as some of Nasser's critics dubbed it.

With these points as background, we are free to proceed with our historical survey. By the end of the eighteenth century, many Muslims were beginning to make an assumption that was to be fos-tered by their nineteenth-century colonial masters soon to come. That was the assumption, which persists to this day inside and out-side the Muslim world, of an inherent causal connection between Islam, as it had accumulated over time, and Muslim peoples' backwardness vis-à-vis Europe. For most subsequent Muslim leaders, that assumption has provoked a wholesale rejection of the Islamic past and a more-or-less ambivalent acceptance of Western ways. [8]

In the nineteenth century, the nature of the Islamic dialogue began to shift in response to this assumption. Although none of the three eighteenth-century trends was lost, they were supplemented and redirected by a serious rethinking of Islam, generally associated with a movement known as the Salafiyyah. This movement, which originated in Egypt, was inspired by the social activism of the remarkable nineteenth-century figure al-Afghani and was represented most systematically by the figures of Muhammad Abduh and Rashid Rida.

One of the most succinct descriptions of the movement has been given by a scholar of the history of Islam in North Africa, where the ideas of the movement continue to resonate.

> [The Salafiyyah movement] . . . argued that the Islam of
> the early forefathers (salaf) was a religion of progress
> [and social welfare], that the backwardness and super-
> stition of Muslims was due to the corruption of Islam,
> and consequently the faith in its pristine purity was com-
> patible with the adoption of modern (European) technology
> and methods of political organization. [9]

In fact, in the leaders of the Salafiyyah, renewal combined with reform, since

> they thought it worthwhile to recast them [classical doc-
> trines of law and political theory] in a more modern
> image (emphasizing, for example, their flexibility,
> utilitarianism, and compatibility with certain institu-
> tional landmarks of European liberalism), proclaiming
> at the same time that they were returning to the original
> purity of what Islamic teaching had been before it was
> corrupted by tyranny and ignorance. [10]

The focus of this movement, whose impact has spread far to the east and west of its original home, was the reform of the Sharia —an attempt to discover in the positive principles that governed past Muslim societies the flexibility to adapt them to modern circumstances while, at the same time, developing truly Islamic forms out of the adaptation. Such principles were seen to include an "ethic of communal solidarity in the struggle against enemies, and a sense of destiny in the world. . . a concept of brotherhood and equal dignity among citizens, of the importance of justice, and of the benefits of strong and virtuous leadership."[11] In particular, one detects in these reformers a "revived interest . . . in the classical theory of the caliphate and the effort to find in it the basis for 'progressive' institutions of democracy, popular sovereignty, and the like . . . not only stimulated by the political encounter between East and West over the past century but also encouraged [ironically, one might add] by the publications of Orientalists [in the area of political theory]."[12]

Numerous intellectual and political obstacles conspired to stymie their efforts, but not to erase their impact. Put simply, their case for reform hinged on emphasizing, perhaps overemphasizing, the original flexibility and permissiveness of Islamic law. Their key demands were to abandon taqlid ("imitation"), to "reopen the gates of ijtihad" ("independent judgment in legal decision," gates that were theoretically closed by the tenth century), to revive the authority of ijma ("consensus," usually of scholars), and to reestablish the priority of the principal of maslahah ("welfare of the community"), all of which had allowed the classical schools of law to develop in the first place.

One should add, however, that even after the theoretical closing of the gates of ijtihad, jurisprudential decision making continued to be a major source of flexibility. Nevertheless, the nineteenth-century reformers still had the problems of setting acceptable limits on an officially reopened independent judgment in the context of the new modernizing purpose to which it was to be put, and of pinning down actual forms (for example, alternative leadership roles) in reanalyzed classical theory.

Their ultimate inability to do so derived not only from the fact that they had to restrict themselves to only a small part of classical theory, but that the original developers of classical theory had purposely not pinned down certain things (for example, procedures for selecting the caliph) lest an unacceptable one be selected by the "right" procedures. Also, the caliphate de facto had come much closer to an absolute monarchy than the theory allowed. Furthermore, there has always been a strong tendency for finding the modern in the classical easily, and that quickly turns into projecting the modern into the classical to the point where the latter becomes merely the hollow vessel of the former and a reflection of the personal tastes of the investigator.

In a very recent work, the contemporary Iranian scholar Seyyed Hossein Nasr, a Sufi-influenced reformist who finds deep spirituality in traditional forms, attacks the problem of form and substance in a defense of the maintenance of the Muslim lunar (hijri) calendar.

It is high time for modern Muslims to seek to understand and apply Islam rather than seek to change the Divinely

given tenets of Islam only to placate the fashions of the
times. If we penetrate the meaning of Islamic tenets, we
realize that they are all placed there for a purpose and
have a profound meaning. It is for us to understand this
purpose and to apply and defend these tenets, not to try to
change them through the excuse of rediscovering a "pure
Islam" which is usually no more than our own individualis-
tic whim and fancy moulded by various deviations of modernism. The modern history of Christianity should be a
good lesson for all Muslims on the effect of religious innovation and a defensive attitude vis-à-vis various forms of
modern thought. Only that religion survives which remains
faithful to both the spirit and form ordained for it by God. [13]

In its own time, the Salafiyyah movement failed to develop new
Islamic political and social forms, also because it could not, unlike
earlier internal Islamic reform movements, make its case solely
within an intellectual context dominated by Islam. Kerr's analysis
of the problem continues to be telling today.

The evolution of Islamic modernism from a program of
radical reform to simply a set of vague ideological attitudes has been due in large measure to the apologetic
mentality among Muslims vis-à-vis Western civilization.
Europe in the Reformation and Enlightenment could dispute theological and philosophical questions within itself,
without reference to ulterior standards; Muslims in the
nineteenth and twentieth centuries have been in no position to do so, for the ulterior standard has been there for
all to see. Hence doctrinal issues could not be disputed
solely on their own intellectual or social merits. One
must show that one's principles are no less advanced than
those of Europe, but no less Islamic than those of established indigenous conservative tradition. This dual standard has scarcely been conducive to rigorous systematic
thought; instead it has opened the market to superficial
slogans and angry polemics. [14]

The Salafiyyah did, however, introduce attitudes and ideas—spiritual
as well as material survival, self-defense, social activism, the
limits of nationalism—that continue to affect widely disparate actors,
especially in the Mediterranean parts of the Islamic world, such as
Nasser, Hasan al-Banna (founder of the Egyptian Muslim Brotherhood), Ahmad Lutfi al-Sayyid (Egyptial liberal constitutionalist), and
Ali Shariati (one of the mentors of the Iranian revolution). It also

played a pivotal role, for example, in the nationalist movements of all the North African countries.

In the twentieth century, Islamic thought has continued to be heavily reactive, sometimes even reactionary. The challenges posed by the ideology and culture of secular nationalism (to which Kerr referred) and the changes it has brought about in social and economic structures (changing, for example, the access to power of religious specialists) have had a direct impact on the ongoing Islamic dialogue. In its extreme cases, secular nationalism has displaced Islam in all three of the possible functions discussed earlier, most strikingly in the area of identification. Islam was found to be unhelpful in more cases than not as a primary definer of national loyalty.

Looking at three Mediterranean Muslim nations—Iran, Turkey, and Egypt—one finds in early nationalist thought a common search for national origins that predate the coming of Islam. For a figure like Ziya Gökalp, one of the ideological inspirations of Atatürk, that source was the pre-Islamic past of the Turkish peoples of central Asia. After all, Atatürk based his consideration of the Cyrillic alphabet as a substitute for the Arabic on the possibility of a pan-Turanian unity in which Turkey could play the central role and be reinforced in its national integration at the same time. 15

In Egypt, Mustafa Kamil, and after him Taha Hussein, argued that the Egyptian people's real roots were in Pharaonic times, that the Pharoah's Mediterranean origins meant that Egyptians were essentially Westerners or Europeans. 16

In Iran, Reza Shah named his dynasty-to-be Pahlavi, thereby symbolizing his similar claim that Islam was simply an unfortunate, perhaps retarding, interlude in the ongoing national identity of Iran, which began when Cyrus the Great founded the first Iranian dynasty, Achaemenid, more than 2,500 years ago. In the 1970s, under Muhammad Reza, there was even talk of converting the entire population to Zoroastrianism, the official Achaemenid faith, to aid in national integration, just as Iran's conversion to Shii Islam in the sixteenth century helped set it off from its Sunni neighbors. When the problem was raised that one can no longer convert to Zoroastrianism, the answer was easy—no need to; since Islam was imposed from without, all an Iranian Muslim need do was renounce Islam to become a Zoroastrian "again."

Note, too, as a symbol of the issue, the degree to which pre-Islamic archeology and linguistics have been officially encouraged to all three states. Not surprisingly, one of the first acts of the Khomeini regime was to close down all pre-Islamic archeological sites (also a source of European presence, of course) but to encourage the study of the Islamic period, entirely out of vogue before the revolution. Eventually the regime even considered razing Persepolis, a pre- and, unfortunately, non-Islamic treasure. In all of this, Islam's insistence on the primacy of a presumed international community of all Muslims has made it difficult for Islam to be used as a basis for nationalism.

If we now try to summarize the twentieth-century phase of the ongoing dialogue as a spectrum of trends, the spectrum could be broken into several sections.

Secularism: with more or less use of things visibly Islamic, often when bending to pressure for a particular dissatisfied group;

Total or partial traditionalism: breaking into the tradition at various points and often connected with a reemphasis on personal pietism; and

Rethinking: by (1) adapting Islam to the needs of a modern state, with heavy emphasis on legal reform; (2) adjusting the concept of the modern state to fit with Islam; (3) aggressive intellectual rediscovering of the importance of the Islamic past, as contributor to the world and as a source of ideas and values for modern Muslims, often connected with the reform of education; or (4) aggressive formulating of a critique of the study of Islam by non-Muslims, often connected with proselytizing.

These trends cut across national lines; no one nation's activities can be described as having only one dimension. They also engage a variety of actors and vehicles, and the most noticeable may not be the most significant. Actors can be political or military leaders, journalists, educators, scholars, religious specialists, or businessmen. Vehicles can be voluntary associations, political parties, revolutionary or military movements, governments or parts of governments, or educational systems. So far, much support for a return to public Islam has come from urbanized lower middle classes, but rapid socioeconomic change could broaden or alter that base. Although we will concentrate on examples of each type of activity from the Mediterranean parts of the Islamic world, it should be noted that in the twentieth century the geographical locuses for these trends have shifted from the Mediterranean to Asia, where countries like Pakistan and Indonesia provide stimuli for the Western parts of the Islamic world from time to time.

Secularism, with more or less use of some of the most striking elements of Islam, often taken out of context, is rather common. Very frequently this involves the reintroduction of certain Koranic and/or Shari requirements, particularly criminal punishments, for example, public flogging for use of alcohol, as introduced in 1977 in the United Arab Emirates, a country that does not present itself as an Islamic state.[17] Unfortunately, this sometimes rather casual and

selective resort to punishments as symbols of Islam misleads the
non-Islamic world into reducing Islam to them; moreover, they often
seem to be applied in a much less careful and conservative manner
than they might have been in a traditional Islamic state or than they
might be in a contemporary one.

To understand the issues surrounding the specified Islamic criminal punishments, or hudud, one must understand their place within
Islamic criminal law and Islamic law as a whole. Taken together,
Islamic criminal law (only one of four parts of Sharia law) was rather
mild for its time, conditioned by severe procedural and evidential
limitations and a strong tendency to restrict its application. The only
five specified criminal penalties became prescriptive rights of God
because of their mention in the Koran; but they were hedged around
in ways that were rather protective unless particular judges (qadis)
wished to circumvent the restrictions in the name of some excessive
personal tendency to punitiveness.

The hudud crimes are unlawful intercourse, false accusation
thereof, drinking wine, theft, and highway robbery. Penalties range
from flogging and mutilation (quite common in the premodern world)
to death. [18] But, to take one example, if four reputable eyewitnesses
were not present at the accusation of adulterers or would not throw
the first stones were they convicted, the punishment would lapse. Or,
if a thief returned the stolen object before accusation was brought,
the penalty would lapse; the amount and circumstances of the theft
were also taken into account.

Moreover, this system emerged and functioned in societies
with tight kinship organization where such crimes were much rarer
than in modern societies, where crimes we would take to court were
settled out of it, and where municipal prisons to effect long-term
punishments were virtually nonexistent. In the two hudud crimes that
involved offenses against man as well as God—theft and false accusation of adultery—the aggrieved was not even obliged to demand a public trial; and even if one did occur, the obligation to enforce was the
community's since there was usually no regular police force at the
judge's disposal.

Despite these qualifications, however, the hudud have posed
problems for modern Muslim apologists who wish to maintain the
viability of the Sharia and make it appear acceptable in the eyes of
the Western world and even among some Muslims. Such an individual
is likely to respond to a critique of the hudud by eschewing the kind of
society that prides itself on not having such penalties and then pays
the high economic and social costs of alcoholism and crime in the
streets. [19]

Like secularism, total or partial traditionalism can take various
forms and could, of course, also involve applications of the Sharia. [20]

When one views any particular case of traditionalism, it is essential to know how any given figure or group defines its traditions. The Sharia is the law of the land in Libya and Saudi Arabia—two countries with very small populations; but oddly enough, if one lived in the seventeenth-century Ottoman Empire, it would have been legitimately traditional to have the Sharia as the law of the land together with Ottoman dynastic law. Saudi Arabia, which would appear to most outsiders to be a traditional state and which presents itself as such, relies on a form of government—dynastic hereditary monarchy (hereditary in an Arab sense, not in a European one)—that may have become traditional in Saudi Arabia in recent centuries but that many Muslims would view as nontraditional in the Muslim world as a whole. In any case, it is hardly a fundamental of Islam, despite the fact that outsiders call Saudi Arabia an example of Muslim fundamentalism.

In the same sense, Khomeini's Iran, once again to many outsiders unquestionably traditional (in the most derogatory sense of the word possible), relies on a form of government, rule by religious specialists, almost unprecedented in Islamic societies, including Iran. In Islamic states, nomocracy, or rule according to the sacred law, has generally been preferred to theocracy, in the sense of rule by religious specialists. On the other hand, Saudi Arabia also preserves tradition, as in the majlis, or ruler's regular session to hear grievances and complaints, which derives perhaps from pre-Islamic Arab society and was emphasized by the Umayyad dynasty.

The changing meaning of the traditional veil shows how lively tradition can really be. The Koran urges Muslim women to dress modestly, and some of the Hadith interpret modesty to involve covering arms and hair. Some traditional men have also tended to cover hair and arms; village and nomadic women rarely do more. Nevertheless, the full body and face covering of various sorts has become customary for some Muslim women, generally upper-status urban women whose husbands could afford to keep them in seclusion and for whom the full covering became a kind of portable purdah. This custom was probably already in place among pre-Islamic upper classes in the Fertile Crescent areas taken over by Muslims, and was continued by them.

In some communities, a particular type of full body covering became such a symbol of status that lower-status women aspired to wear it.[21] For those women who do wear some kind of body and/or face covering, it can mean all sorts of things. In a certain area of North Africa, young men and women gather for an annual bridal fair, the women completely covered save hands and eyes; men and women alike testify that this fashion gives women power, since men cannot choose them for their appearance alone.[22] The hajj is said to be a

popular time among Saudi men for making matches, since women who
are normally veiled must go unveiled then, whereas at other times
only other women know what eligible girls look like. 23 In Iran all
sorts of reasons are given for wearing the chador—protection from
the ubiquitous dust, not having to put on street clothes after being in
one's own courtyard, remaining anonymous for trysts (though recognition when desired is possible through rings and shoes), going into
neighborhoods where one is not known and feels uncomfortable, and
feeling more Islamic. Flirtatious use of the chador can even be noted,
very much as chaperoned Spanish women learned to use their fans
and mantillas.

In the recent revolution, however, the chador took on a new
symbolism—revolutionary rejection of the neocolonial dominance of
Western styles and their corrupting influence. At a certain point,
to wear Western clothes was to be reactionary and unliberated. But
after Khomeini began to insist on the chador for all Muslim women,
not wearing it became a revolutionary symbol, as it had been under
Reza Shah. In Egypt, many young women influenced by the Muslim
Brotherhood and related organizations have adopted a kind of wimple
as a way of entering the job force without inviting the kind of sexual
harrassment to which U.S. women are subjected even if they dress
modestly. Yet an emphasis on Islamic manners, personal piety, and
observance of Islamic ritual demands need not be associated with activism; nor can Islamic activism without personal piety be viewed
automatically as hypocritical.

Just as tradition itself can be very lively, maintaining tradition
in a country like Saudi Arabia, which is undergoing rapid economic
growth and a unique form of modernization, can have remarkably un-
traditional consequences. A particularly good example is women's
banking. The emergence of women's banks in Saudi Arabia rests on
a tradition, decreed by the Koran and reinforced by the Sharia, that
women should inherit and hold money in their own right. Women now
hold an estimated 40 percent of the hundreds of billions of private
wealth in Saudi Arabia, and they began transferring it to women's
banks as soon as they began to open. Such banks assist women in investing and going into business for themselves, as well as offer banking careers to Saudi women. Perhaps more important, the lobbies
have apparently become the first semipublic (men cannot enter) gather-
ing places away from home for women since the bathhouse, providing
refreshments and comfortable surroundings in which to socialize. 24

Like secularism and traditionalism, rethinking of Islam, sometimes led by religious specialists and sometimes not, continues to
appear in various forms. It is possible to express the difference between traditionalism and rethinking as the difference between an Islamic state and an Islamic order. In the former, traditional forms,

however defined, define the state; in the latter, the forms of state are derived from Islamic principles. The strongest theme of nineteenth-century reformism, adapting Islam to the needs of the modern state through reinterpretation, lives on. Ali Shariati, one of the intellectual sources of the Mujahidin-i Khalq, a leftist movement now opposing Khomeini, is representative.

> His object was to reconstruct Shiah [Shii] Islamic thought in the light of Marxism, existentialism and phenomenol-ogy. . . . His politics were radical and he called for a revolutionary overthrow of the secular state and the regime of the Shah, and also what he called Safavid [six-teenth-eighteenth century] Islam, the religious and po-litical views of the established ulema [mullahs] and ayatollahs [many of whom had supported constitutional-ism in earlier stages of the Iranian revolution and some of whom still do]. Against this he set Alavid Islam . . . in its original state of purity. He was very Persian in his idealization of Ali and Hussain [first and third Shii imams].[25]

Khomeini was also very Persian in his idealization of Ali and Husayn, but used his identification with them in defense of the authority of the mullahs as representatives of the imams. Here we have a fine ex-ample of the flexibility of Islam as a language for talking about the world. Shariati and Khomeini have used the same symbols to sup-port diametrically opposed, traditional and antitraditional, views of the world. And the "Islamic Marxism" that Shariati has in part in-spired sees in Islam the communalism of Marxism without feeling a need to accept its atheism.

Rethinking can also be the converse of adapting Islam; it can instead recast the modern state and its institutions in terms of Islamic tradition. One of the best-known examples is the Muslim Brotherhood of Egypt, founded by Hasan al-Banna in 1925, but variations can be found in other movements and parties: for example, the Pakistani Jamaat-i Islami of Maududi, the Indonesian Masjumi, or the Moroc-can Istiqlal as conceived by al-Fasi. Such movements can attract all classes of people and be led by all sorts of individuals. They tend to aim for modernization without Westernization. They focus not only on the more obvious issues of the need for personal and public morality and piety, but especially on a typically Islamic concern—the organiza-tion of society.

However they derive their Islamic principles, they tend to agree that there is no place for kingship or hereditary succession and that the community has the right to choose its leaders and the right to re-

move them (a position much more akin to the Khariji movement of
early Islamic history than to standard Shii or Sunni theory). Therefore, they lean toward some kind of republic, perhaps even a constitutional monarchy, and away from dictatorship or one-man rule. For
them, Islamic principles demand that the government be elected by
the people or their representatives, obliged to consult somehow or
other, and provided with an independent judiciary (another very deep-
seated traditional Islamic principle). Theocracy, in the literal sense
of rule by religious specialists, is not popular. For many, but not
all, the state must not be excessively nationalistic, but somehow part
of a larger Islamic order that is not merely international, that is,
where the separate nations do not provide the essential building blocks.
The economic order on which such a government is to rest should provide for an equitable distribution of wealth and a turning of Islamic
principles of charity to providing for the needs of society.

Beyond these minimally shared talking points, there is wide
variation, from those who desire a very authoritarian state (Maududi)
to those who desire no state at all (Qaddafi, who a few years ago actually declared the state no longer to exist).[26] The status of women
has been one of the issues most resistant to this kind of rethinking.
One can easily demonstrate that Islam was respectful and protective
of women in its original Koranic intent, and that within certain constraints, women in traditional Islamic societies acquired remarkable
power beyond that ascribed to them, not in spite of so much as because of sex segregation. But such traditional assets assume different valences in a modernizing setting, and the sources of Islam, like
those of all patriarchal religions, can be used to validate, or be con-
fused with, customary practices that in modern terms constitute unacceptable discrimination. It will be difficult to disentangle Islam
from custom and to generate new impulses toward women without a
radical reinterpretation, probably figurative, of the Koran.

Rethinking need not, however, focus on questions of morality,
law, and social organization. It can also take the form of an assertive recapturing of the breadth and global significance of historical
Islamic civilization (one of the key aspects of al-Afghani's chauvinism
a century ago). In fact, this activity may be central to the future of
Islam as a public social force, since one of the obstacles to change
in any tradition is the tendency of particular communities to cultivate
and commit themselves to only a tiny fraction of their heritage, thus
narrowing their options considerably. This form of rethinking frequently involves demands for a reform of the educational system and
a reintroduction of non-European subjects and languages. A very
recent example of it is the publication by Seyyed Hossein Nasr of Islamic Life and Thought.[27] Here Nasr attempts to scan the panorama
of Islamic cultural and intellectual life, dwelling particularly on

philosophy and science down through the seventeenth century. He thereby demonstrates, as did al-Afghani, that Islam does not have to be incompatible with scientific advance just because its scientific tradition, which had long led the world, was not being cultivated actively when Europeans began to dominate Muslim lands.

Connected with insiders' rediscovery of their own civilization's history is their discovery of the pitfalls of depending on Western Orientalists' understanding of it. That discovery has led to an aggressive critique of Western Orientalism as politically and intellectually insidious. Although biting critiques of Western distortions of Islam, scholarly and otherwise, have been written by Westerners before, they have not had near the impact of the publication of one insider's book (and related books and articles), Orientalism.[28] Although its author, Edward Said, could be faulted for having defined the Orientalist tradition in such a way as to support his critique of it, the content of his book is much less important than the event of its publication, for the book provides important scholarly underpinnings for the politics and rhetoric of anticolonialism.

Finally, this kind of rethinking is reflected in another phenomenon—increased proselytizing and efforts to put explanations of Islam outside Muslim lands into Muslin hands wherever possible.[29] In 1960, for example, the Egyptian Ministry of Religious Affairs set up a missionary corps. Saudi Arabia helps fund the Muslim World League, an international educational mission. The publications of the Islamic Teaching Center in the United States constantly demonstrate the ways in which a growing number of U.S. Muslims are finding relevance in Islam.[30]

Ironically, some of the most energetic foreign missionary and educational activity has been undertaken by a syncretistic movement that has been declared un-Islamic in its home base (Pakistan) and abroad (Saudi Arabia). The Ahmadiyyah was founded in the 1890s by Mirza Ghulam Ahmad, a man who claimed to be the Mahdi (Messiah of Islam), the Messiah of Christianity, and an incarnation of Krishna. The movement's encouragement of numerous translations of the Koran was a major cause of opposition. There had been translations of the Koran before, but Muslims had always insisted on viewing them as interpretations or commentaries, and on having the Arabic text appear together with the translation. Furthermore, most Muslims certainly felt that one could not study scripture through them, just as Jews cannot study Torah except in Hebrew.

Paradoxically, Muslims' general reluctance to make serious use of vernacular Korans has helped Islam retain the unity Christianity lost by doing so, but has also kept it from reforming in ways made possible for Christianity by vernacular translation. What has not emerged, however, from all this activity, is a really effective pan-

Islamic congress or society. Muslim leaders do assemble from time to time in various formats, but activity has remained local or regional. It does not help, either, that there are no formally ordained clergy among Muslims to be constituted as official representatives to such meetings.

CONCLUSION

When one has set the many contemporary Islamic trends in a variety of contexts, one is led to wonder whether there really is a resurgence of Islam and whether we are dealing with one phenomenon or many. If we are ever to learn the answers to these and other questions, we must stop preferring occasions of conflict as vehicles for learning about Islam; and we must avoid at all costs the cliché of an atavistic monolithic fanaticism that is conveyed in the following headline: "A fundamentalist fever is rising among 600 [actually 750] million Muslims. The prognosis for the Western world as well as the countries directly involved? Danger. Islam on the March!"[31]

We cannot even estimate the number of Muslims for whom Islam is serving as a major source of symbols, norms, and identities today or begin to describe all the forms that service might be taking. Although Islam seems to be vital as a private religion, there is no necessary correlation between the degree to which it flourishes as a personal faith and the degree to which it will serve as a source of public action, particularly as the ideological impetus for the kind of mass movement that has occurred in Iran. Furthermore, we must be very careful not to assume that everything a Muslim individual or group does springs from Islam, even though very old traditions for reviving Islam as an active social force seem to be at work on a widespread basis.

Whether the whole order of various Muslim societies can be given, as at times in the past, some kind of Islamic spiritual cast without communal intolerance is impossible to say. Perhaps the full possibilities of the Islamic past will be uncovered and conveyed in such a way as to unify and motivate persons with very different backgrounds and interests to press for the Islamization of their societies. Or perhaps Islam will simply be available, as is Christianity, for separate individuals and groups to draw on privately, and occasionally publicly, in states that remain or become self-consciously secular as well as in ones that do not.

Whatever role Islam plays in the future, one can envisage the development of a more complementary relationship between the spiritual and the mundane than has obtained in modernized Western societies. One must observe in passing, however, that the modernization of Western countries has not in fact produced total secularization; in the United States, where Christmas is a national holiday and Sunday the official day of rest, the degree to which religion plays a role in public life is still open to question. Yet the separation of church and state, on which the partial secularization of Western countries rests, is not necessary or possible in Muslim countries because no Islamic "church" exists. Nevertheless, many say that what Islam needs is a reformation like Christianity's. The comparison is as instructive as it is inappropriate. The ideologies of the Protestant Reformation played into the hands of the emerging secularizing national rulers of Western Europe, who, buoyed by a surge of material prosperity and vitality, wished to break the unity of the Christian Church. An Islamic reformation would have to reconstruct a politically relevant lost spiritual unity in times of material uncertainty, fighting or accommodating to the secularizing tendencies of those generally in control.

One cannot help but feel the impatience of the Islamic folk-hero Nasrudin (also known as Jeha and the Hoja) as he waited for a new shirt.

> Nasrudin had saved up to buy a new shirt. He went to a tailor's shop, full of excitement. The tailor measured him and said: "Come back in a week, and—if Allah wills—your shirt will be ready." The Mulla contained himself for a week and then went back to the shop. "There has been a delay. But—if Allah wills—your shirt will be ready tomorrow." The following day Nasrudin returned. "I am sorry," said the tailor, "but it is not quite finished. Try tomorrow, and—if Allah wills—it will be ready." "How long will it take," asked the exasperated Nasrudin, "if you leave Allah out of it?"[32]

NOTES

1. Mircea Eliade, The Sacred and the Profane, trans. Willar R. Trask (New York: Harcourt, Brace & World, 1959).

2. The classification that made this possible was ahl al-kitab ("people of the book"); it conveyed a degree of respect for groups whom God had previously blessed with scriptural guidance. Originally it meant Jews and Christians, but its eventual extension to Zoroastrians and Hindus shows the limits to which Muslim religious tolerance could go. Through this analysis, the multiple connotations of the difficult term jihad can be approached. Jihad means "any striving in the way of the Lord" and has been used to characterize an effort

at inner purity (banishing the demons within the human soul) as well as purification of the rule of territory controlled by lukewarm Muslims and forceful acquisition of territory ruled by non-Muslims. The latter usage, though best known in the West, is probably not the most common. The rules of military jihads are strictly regulated by the Sharia, which prefers peaceful surrender to battle. Yet it also appears that strains of true communal intolerance have crept into the Islamic tradition and found their reinforcement in this concept. All time references throughout are to the Christian calendar unless noted otherwise.

3. Most notable is the new political saliency of the Sunni-Shii split in places like Syria, Iraq, Turkey, and Bahrain. For some of the Islamic world, the split is unknown; for much, irrelevant. Even for the areas named, the political utility of the labels has been enormously enhanced by its significance in Khomeini's Iran. Notable, too, is the way in which "Muslim-Jew" for some persons has recently come to stand for "Arab-Israeli" and therefore for inevitable conflict in ways it rarely did before.

4. S. N. Eisenstadt, "Social Institutions," International Encyclopedia of the Social Sciences (New York: Macmillan and Free Press, 1968), 14: 409-29.

5. The discussion that follows owes much of its formulation to Hao Chang, Wyant Professor of Chinese History, Ohio State University.

6. Marshall G. S. Hodgson, The Venture of Islam: Conscience and History in a World Civilization (Chicago: University of Chicago Press, 1974), 1: 80-81.

7. Antoine de Saint-Exupéry, The Little Prince, trans. Katherine Woods (New York: Harcourt, Brace & World, 1943), p. 17. It is interesting that the major traditional area of button sellers in Istanbul is called European Row as well as Button Row, and that the Doghrib Eskimo of Alaska referred to French trappers as the men with buttons down their coats.

8. The same argument has been made for Catholic areas, but there too the connection is not unidirectional. Traditions reflect as well as produce circumstances.

9. Jamil M. Abun Nasr, A History of the Maghrib (Cambridge: At the University Press, 1971), p. 320.

10. Malcolm H. Kerr, Islamic Reform (Berkeley: University of California Press, 1966), p. 12.

11. Ibid., p. 222.

12. Ibid., p. 19.

13. Seyyed Hossein Nasr, Islamic Life and Thought (Albany: University of New York Press, 1981), p. 217.

14. Kerr, Islamic Reform, p. 16. In this context, the excruciating xenophobia of certain Iranian Muslim leaders and followers in recent years may make a little more sense.

15. See Robert G. Landen, The Emergence of the Modern Middle East: Selected Readings (New York: Van Nostrand Reinhold, 1970), pp. 119-25.

16. Ibid., pp. 110-16; William H. McNeill and Marilyn Robinson Waldman, The Islamic World, Readings in World History, vol. 6 (New York: Oxford University Press, 1973), pp. 412-22.

17. G. H. Jansen, Militant Islam (New York: Harper & Row, 1979), p. 142. (This is a much better book than its title and hasty composition would lead one to expect.) Sometimes this kind of use of the symbols of Islam involves official encouragement of a particular act of piety, as when Kuwait banned the sale of pork in the 1970s; or making a claim to superior spiritual descent, as the kings of Morocco and Jordan do; or forcing women to wear traditional garb.

18. See Joseph Schacht, An Introduction to Islamic Law (Oxford: Clarendon Press, 1964), p. 176, for a good introduction.

19. Such a response was made by Ismail Faruqi in a lecture on Islamic law at Ohio State University in 1979.

20. In fact, human societies often distinguish as well between legitimate and illegitimate tradition; the latter they may demote to custom or attribute to a poor understanding of religion. Or, tradition can have more force in some areas of life than in others.

21. Dawn Chatty, "Changing Sex Roles in Bedouin Society," in Women in the Muslim World, ed. Lois Beck and Nikki Keddie (Cambridge, Mass.: Harvard University Press, 1978), p. 403. It is amusing to discover how the black veil, hardly a fundamental of Islam, is said to have become fashionable.

> A merchant from Kufa came to Medina with veils. He
> sold all but the black ones, which were left on his hands.
> He was a friend of Al-Darimi (a poet and jurist who died
> in 869) and complained to him about this. At that time
> Al-Darimi had become an ascetic and had given up music
> and poetry. He said to the merchant, "Don't worry, I
> shall get rid of them for you; you will sell the whole lot!"
> Then he composed these verses:
>
> "Go ask the lovely one in the black veil
> What have you done to a devout monk?
> He had already girded up his garments for prayer
> Until you appeared to him by the door of the Mosque."
>
> He set it to music and Sinan, the scribe, also set it to
> music and it became popular. People said, "Al-Darimi

is at it again and has given up his asceticism." And there
was not a lady of refinement in Medina who did not buy a
black veil and the Iraqi merchant sold all he had. When
Al-Darimi heard this he returned to his asceticism and
again spent his time in the mosque.

Taken from Bernard Lewis, ed. , Islam (New York: Harper & Row,
1974), 2: 147-48; and quoting Al-Isfahani's Kitab al-Aghani.
  22. Carla Hunt, "Berber Brides' Fair," National Geographic
157, no. 1 (January 1980): 118-29.
  23. Jansen, Militant Islam, p. 35.
  24. Douglas Martin, "Saudi Banks for Women Thriving," New
York Times, January 27, 1982, sec. 4, pp. 1, 33.
  25. Jansen, Militant Islam, p. 158.
  26. Ibid., p. 177. In fact, Qaddafi's notion of participatory
democracy run by committees reminds one of early Khariji doctrine,
which, after all, had its greatest impact around that part of North
Africa now occupied by Libya. Other leadership models have also
been worked with. Part of Khomeini's original legitimacy came from
his being one of very few ayatollahs to be considered marja-i taqlid,
a reference point for imitation in legal matters. In the constitution
of the Islamic Republic of Iran, he is the faqih, legal arbiter.
  27. Nasr, Islamic Life. Here he makes much more pious and
less secular use of the material than he did when he was developing
educational programs for the shah. Pakistan's exploration of the
value of traditional Islamic medicine is also a notable example of this
trend. Many traditional fields must be rediscovered, since their continuity was interrupted during the colonial period.
  28. Edward Said, Orientalism (New York: Pantheon Books,
1978); see also idem, The Question of Palestine (New York: Times
Books, 1979); and idem, Covering Islam (New York: Pantheon Books,
1981). For Western critiques see Jean-Jacques Waardenburg, L'islam
dans le miroir de l'occident (The Hague: Mouton, 1963); Norman
Daniel, Islam, Europe, and Empire (Edinburgh: Edinburgh University
Press, 1966); and Hodgson, Venture of Islam, vol. 1.
  29. See Jansen, Militant Islam, p. 90. Contrary to popular
opinion, Islamic missionizing was not predominantly self-conscious
and organized prior to the twentieth century.
  30. For example, see ITC News, published monthly by the Islamic Teaching Center, P.O. Box 38, Plainfield, IN 46168.
  31. William E. Griffith, "Islam on the March!" Reader's Digest,
June 1979, p. 81.
  32. Idries Shah, The Pleasantries of the Incredible Mulla Nasrudin (New York: E. P. Dutton, 1971), p. 29.

8

EGYPT'S REGIONAL POLICY
FROM MUHAMMAD ALI TO
MUHAMMAD ANWAR AL-SADAT

GABRIEL R. WARBURG

THE NINETEENTH CENTURY

After Muhammad Ali assumed power in 1805, Egypt reemerged as a dominant factor in regional politics both in the Eastern Mediterranean and in East Africa. Though Egypt remained de jure an Ottoman province until World War I, its foreign policy during most of the nineteenth century was formulated and executed from Cairo and not from Istanbul. Hence, any attempt to examine Egypt's foreign policy following the European penetration into this area will have to start with the turbulent period of Muhammad Ali's reign in the first half of the nineteenth century. Such an examination will yield the following conclusions:

1. Modern Egypt, based on its strategic position, was set on playing a leading role in regional politics.
2. Geopolitical factors suggest two major trends: first, the Nile Valley as a natural area for Egyptian expansion; and second, the recognition of the northeast as an area from which potential enemies could penetrate into Egypt and that, consequently, should be under Egyptian influence.
3. Based on the Napoleonic experience in Egypt (1798-1802), Muhammad Ali and his successors realized their limitations as semi-independent rulers. In other words, they had to limit their ambitions

in accordance with the constraints of the "Eastern question," namely, to British predominance in the Eastern Mediterranean. Britain ruled the seas, as had been adequately demonstrated both by the annihilation of Bonaparte's navy by Nelson at Abukir (1798) and in the battle of Navarino (1827), when the combined Turco-Egyptian navy was destroyed. Hence, any new venture in Egyptian regional politics could only be viable if it received at least the tacit blessings of Whitehall.

The highlights of Egypt's regional politics in the nineteenth century can therefore be illustrated in this manner. The centrality of the African hinterland, and in particular of the Nile Valley, is a dominant and constant factor throughout the century. The conquest of the Sudan in 1820-21 and the expansion of Egyptian rule into equatorial Africa, as well as in Bahr al-Ghazal and Dar Fur in the 1870s, are clear indications of the importance of these regions for Egypt's new rulers. The discovery of the sources of the Nile in the 1860s and the opening of the Suez Canal in 1869 add another dimension to this area and shed light on the Khedive Ismail's abortive venture into Ethiopia (1875-76). There is an additional reason for focusing on Africa rather than on the Middle East in the second half of the nineteenth century. Egypt had been rebuffed in Syria and, hence, had to relinquish its conquests in the northeast. Thus, following the Treaty of London (1841), Africa became the only area for legitimate expansion, especially since Egypt became a reluctant, though active, partner in the Anglo-Egyptian treaty for the suppression of the slave trade (1877). When in 1881-85 Egypt was forced by Britain to evacuate the Sudan following the Mahdist revolt, it did so under protest. The unity of the Nile Valley remained a dominant feature both in Egypt's evolving nationalist ideology and in its regional politics throughout the nineteenth and twentieth centuries.

Egyptian conquests in the east and the northeast in the first half of the nineteenth century illustrate Egypt's determination to play a leading role in the region, on the one hand, and its perception of the northeast as an area of potential danger, on the other. First came the destruction of the Wahhabis and the conquest of Najd and the Hejaz, including Mecca and Medina, in 1811-18. From Egypt's point of view, this venture had the following advantages. It proved Egypt's military power vis-à-vis the Ottoman sultan, who had not been able to overcome the Wahhabi challenge with his forces stationed in Syria and Iraq. Second, by bringing the holy cities under Egyptian suzerainty, Muhammad Ali's prestige within the Islamic world had been considerably enhanced. Third, Egypt's pacifying role in the Red Sea and its ability to renew the pilgrimage to Mecca were also clear demonstrations of its power and goodwill as far as the Muslims under British rule were concerned. Hence, Britain's first attempt to coop-

erate with Ibrahim Pasha goes back to this period. Last, the conquest
of the Hejaz and later of the Sudan brought both the eastern and the
western coasts of the Red Sea under Egyptian control. Even after its
expulsion from Syria in 1841, Egypt retained at least part of its influence in the Arabian peninsula, especially along the Red Sea coast,
where several harbors in northern Hejaz remained under Egyptian administration until the 1890s.

The Syrian campaigns (1831-40) of Ibrahim Pasha provide the
sole example of Egyptian expansion in the northeastern direction in
the nineteenth century. In contrast to the campaigns in the Arabian
peninsula and the Sudan, undertaken under Ottoman auspices, this was
a direct challenge to the sultan and an abortive attempt to link Egypt
with British interests, as interpreted in Cairo. Muhammad Ali tried
to convince the British authorities that the shaky Ottoman sultan could
no longer provide for the safety of British imperial communications
and, hence, a strong Egyptian government in the Fertile Crescent
would be an asset as far as Her Majesty's Government (HMG) was
concerned. Muhammad Ali's own ambitions went far beyond those
linked to British interests, for there is little doubt that Egyptian
forces could and would have conquered Anatolia itself if the Russians
had not intervened with the Treaty of Hünkar Iskelesi in 1833.

Egyptian rule in Syria lasted for an additional seven years.
However, when, following the London Convention for the Pacification
of the Levant (July-September 1840), the Egyptian army was forced
to retreat, the sole gain of the Muhammad Ali's dynasty in the northeastern region was a right granted in the Ferman of 1841 to administer
a narrow section of the Sinai west of a line running from Suez toward
Rafa. This brought an end to Egyptian expansion in that region during
the nineteenth century. The opening of the Suez Canal (1869) and the
British occupation of Egypt (1882) created a new situation with regard
to the Sinai peninsula. However, when Great Britain forced the Ottomans to accept Egyptian administration of the Sinai in 1892 and 1906,
it did so out of concern for its own imperial interests, while the
Egyptian Nationalist party of Mustafa Kamil defended Ottoman sovereignty over the peninsula. [1]

THE SUDAN IN ANGLO-EGYPTIAN RELATIONS

Egypt's dependency on England remained a dominant factor in
its foreign policy throughout the first half of the twentieth century.
Despite its de jure independence since February 1922, Egypt was in
no position to determine its own foreign relations as long as the
British army occupied Egypt and the so-called advice of HMG,
granted through the British high commissioners (later ambassadors)
in Cairo, was binding.

Egypt's active venture into regional politics until World War II was in relation to the Sudan. No Egyptian king, government, or political party ever recognized the legitimacy of the Anglo-Egyptian condominium leading to British predominance in the Sudan since 1899. Hence, the unity of the Nile Valley remained a constant and central issue of controversy throughout the period. In fact, a close look at the numerous attempts to reach an agreement undertaken by the British and Egyptian governments, both before the signing of the Anglo-Egyptian treaty of 1936 and after World War II, proves that they foundered on the Sudan question. The two governments succeeded in reaching tentative agreements with regard to the evacuation of British forces from Egyptian territory. But the gap in their respective positions regarding the Sudan was so great that no compromise could be found. England regarded its trusteeship over the Sudan as one that would ultimately lead to Sudanese independence. This desire was based primarily on two assumptions: first, that the Sudanese revolted against Egyptian rule in 1881-85 for good reasons and had no desire to come once again under Egyptian domination; and second, that Egypt's nationalist anti-British fervor, as demonstrated in the revolt of 1919, would sooner or later force Britain to relinquish its direct hold over that country. Therefore, British presence south of Wadi Halfa, leading to an independent, pro-British Sudan, would facilitate the safeguarding of HMG's interests in Egypt and the Suez even after the evacuation.

Egypt's position could be summarized under similar headings: first, that the Sudan was part of Egypt and any attempt to grant the Sudanese independence was an imperialist plot aimed at creating artificial divisions against the natural aspirations of the people of the Nile Valley; and second, while Britain was bound to relinquish its hold over Egypt sooner or later, its presence in the Sudan was of much greater concern, as its separatist policy was irreversible, and every additional year of British administration and education in the Sudan would further diminish the chances of eventual unity.

This gap was finally bridged in 1953, when following the "free officers" coup and the deposal of King Faruq, the Revolutionary Command Council, under President Muhammad Neguib, decided to grant the Sudanese the right of self-determination. This was based on a better understanding of the Sudanese, on the one hand, and on the belief that Ismail al-Azhari and his National Unionist party would ultimately lead the Sudan to unity with Egypt, on the other. However, the Sudanese opted for independence, and since January 1956, the unity of the Nile Valley ceased to be a realistic political aim.[2]

ACTIVE PAN-ARABISM

World War II and its aftermath was a turning point in Egypt's foreign and regional policy. First, there was a beginning of renewed interest in the northeast. Second, the gradual exit of the United Kingdom from the Middle Eastern scene and the emergence of the superpowers as the main actors created a new situation.

The two are, of course, interrelated, for even before World War II the United Kingdom had attempted to involve Egypt in its broader Arab policy, both in order to counteract Jewish ambitions in mandatory Palestine and to withstand the growing tide of Arab nationalism. The claim that Anglo-Arab interests are in harmony and that by acting in concert with the United Kingdom and with each other the Arabs would serve their own interests was first aired by Sir Anthony Eden in 1943. Egypt's leading role in the founding of the Arab League in 1945 was, in large measure, the result of British initiatives and the fear of Hashimite ambition. The unity of Greater Syria and a Hashimite federation of the Fertile Crescent were regarded as a threat to both Egypt and Saudi Arabia. The Arab League thus provided Egypt with a convenient vehicle with which to withstand prospective challengers for supremacy in the Arab Middle East. While there have been strong ideological currents since the 1930s in quest of greater Egyptian involvement in the Arab and Islamic arenas, politically there have been no serious moves in that direction. The Muslim Brothers had been involved in the Arab revolts in Palestine in 1936-39, while King Faruq's close associates Ali Mahir and Abd al-Rahman Azzam took part in the Palestine Round Table Conference, convened in London in 1939. However, the major political parties, including Wafd, paid little more than lip service to these efforts. Only after the war, when it became clear that the United Kingdom and France were on their way out and, hence, the vacuum left in their wake might be filled by undesirable opponents, did Egypt venture once again into the northeastern region, with the 1948 war in Palestine as its first real manifestation. But while the Muslim Brothers and their Jawwala ("Rovers") were enthusiastic fighters and martyrs for the Palestinian cause, the Egyptian government was reluctant to commit the army to full-scale fighting.[3] Even when Faruq gave the order for the Egyptian army to invade Palestine on May 15, 1948, he did so both in order to fight the new Jewish state and to stop the Hashimites from annexing Arab Palestine.

The next phase in Egypt's regional and international relations started in 1955 and came to its end following the 1967 debacle. It was characterized by an aggressive leadership role, played by Nasser, that ultimately led Egypt to a fiasco not dissimilar to the one suffered by Muhammad Ali following his Syrian venture. Four major events, all of which occurred in 1955, symbolize the opening of this new chapter: the decision of the Sudan to opt for independence rather than unity with Egypt, the Baghdad Pact, the Bandung Conference, and the Soviet-inspired arms deal with Czechoslovakia. The fact that the Sudanese opted for independence rather than unity was, as mentioned, a great disappointment for the Egyptians. However, it enabled Nasser to shift the focus of his regional politics from the south, where it had been centered for over a century, to the northeast. The Baghdad Pact symbolized the United Kingdom's last attempt to retain a foothold in the region following its reluctant agreement to evacuate Egypt and Palestine. Moreover, it proved to Nasser that, in a divided Arab world, Iraq, his main rival for Arab leadership, could with imperialist backing, still play an important role and thereby frustrate Egypt's regional ambitions. The Bandung Conference, Nasser's first venture into the international arena, proved that with the support of prominent leaders such as Tito and Nehru, Nasser could assume the mantle of leadership of the Arab Middle East and independent Africa. To do so he had to create conditions under which he could challenge the West openly and force his fellow leaders of the Arab world to fall into step. The announcement of the arms deal with the Soviet bloc in September 1955 was such an event, and the nationalization of the Suez Canal in 1956 was yet another. Here was an aggressive, independent Arab leader, at the head of the biggest and strongest state in the Middle East, commanding a well-equipped army trained in modern warfare, who had achieved all this without submitting to the dictates of the superpowers. Through positive neutrality, as formulated in Delhi, Belgrade, and Bandung, Nasser hoped to succeed where Muhammad Ali had failed, namely, to bring the Arabs under Egyptian control without firing a single shot and without outside interference.

His success lasted for less than six years. Its major manifestations were the Suez War and the final humiliation of the United Kingdom and France at the hands of the United States and the Soviet Union. The fact that, at the height of the cold war and despite the Soviet arms deal and Soviet financing of the high dam at Aswan, Nasser also succeeded to maintain, for most of that period, cordial relations with the United States and to receive from both superpowers massive aid for his five-year development plans and his armed forces, convinced him that his regional supremacy and his so-called positive neutralism were his major political and economic assets. Arabism thus became the dominant feature of Egypt's regional politics. Through it Nasser aborted the Eisenhower Doctrine and, in February 1958, came to the rescue of Syria in creating the United Arab Republic (UAR). When in July of that year the Hashimite Kingdom of Iraq was toppled by what appeared to be a Nasser-type military coup, it seemed that Egypt was well on the way to achieving the supremacy that had eluded it a hundred years earlier.

However, within a year the dream of Arab unity and Nasser's aggressive Arabism were both challenged. Opposition stiffened within the traditional monarchies of Saudi Arabia and Jordan. In the latter, as well as in Lebanon, Nasserist attempts to topple the regime had, with Western aid, been successfully resisted. But even more discouraging was the attitude of fellow revolutionaries in Iraq, who upon having assumed power, had no intention of succumbing to the Egyptian embrace. In March 1959 an Egyptian-inspired and -backed coup in Mosul was crushed and all known pro-Nasserists were either annihilated or put behind bars. Even in Syria the Bath, which had been the main instigator of unity, was seeking a greater role in running the state and had to be brought under strict control. With Field Marshal Abd-al-Hakim Amer in charge of Syria and Egyptian officers replacing Syrians in the army, relations within the UAR were under considerable strain long before the unity finally broke down in 1961. Thus, the only modest manifestation of Nasser-style Arab unity and his attempt to dominate the Fertile Crescent came to its untimely end exactly 120 years after Muhammad Ali failed in a similar venture.

Nasser's involvement in the war in Yemen (1962-67) and his growing commitment to the conflict with Israel, culminating in the June 1967 war, are probably the two major elements in the final failure of his regional ambitions. When, under the banner of "Unity of Purpose," Nasser decided to back the Yemenite revolutionaries against the traditional imamate and its Saudi Arabian ally, he entered into direct confrontation with one of the most important clients of the United States in the region. This was connected with the decline of the Afro-Asian positive-neutralist bloc through internal strife and the inherent weakness of its members. The failure to convene a second Bandung conference in 1965 combined with the Sino-Soviet conflict forced many of the weaker members of this bloc to become clients of one or the other of the superpowers. Nasser opted for Soviet patronage, and by 1966 President Johnson came to the conclusion that there was no longer any logical reason to pour U.S. aid into a Soviet-dominated economy.

By 1966 Egypt's power in the region had also declined considerably. The revolutionary regimes in Syria and Iraq were openly challenging Nasser's supremacy, using his own revolutionary jargon refined by Bathist ideology. The stalemate in Yemen was proving Egypt's military weakness in the face of growing resistance from the traditional, conservative camp. Nasser's attempt to revive his regional leadership in the Arab summit conferences in 1964-65 ended in a fiasco. It only proved that even the Palestinian issue, as symbolized in the creation of the Palestine Liberation Organization (PLO), was not enough to create unity in the Arab camp, where suspicions and open hostility were the order of the day.

It was under these circumstances that Nasser decided to take the gamble that ultimately led to the June 1967 war. The instigators were the Syrians, who had again and again challenged his leadership and questioned his willingness to commit the Egyptian army against the main foe, namely, Israel. Ever since the Suez War of 1956, Nasser had claimed that he was preparing his armed forces, with Soviet aid, for the final battle against Israel. But when, in May 1967, he ordered the UN emergency forces out of the Sinai and closed the Straits of Tiran to Israeli shipping, he believed that Israel would submit without battle. This was an Israel that did not have the backing of the United Kingdom and France as in 1956, without which, as Ben Gurion had declared, he would not have dared to challenge Egypt. Now Israel was standing alone against three well-equipped and Soviet-trained armies and with a prospective fourth ally in King Hussein. Moreover, Israel itself was in the midst of a severe economic crisis that seemed to have weakened its political leadership and undermined its ability to reach critical decisions. Hence, Nasser's gamble was based on the assumption that even if Israel would decide to fight it could be overcome by the combined Arab forces. Had the gamble succeeded, Nasser's leadership role in the Arab circle would have gained a new lease on life. Having failed and been humbled by Israel's complete and fast victory in the Six Day War, Nasser had lost his last shred of credibility as the regional leader. It was also the end of Nasserism, the creed that had been cultivated by Nasser's followers in the belief that it would lead the Arab world to modernism and enable the Arabs to assume their rightful position among the advanced nations of the world. All that remained of this messianic belief, which Nasser had successfully evoked, was the promise made by Nasser that "what had been taken by force shall be returned by force." However, following Egypt's failure in the War of Attrition of 1969-70, even that announcement seemed to have become an empty slogan.

It may be an irony of history that Nasser's last attempt at Arab politics and unity was in northeast Africa. Having been humiliated and crushed in the Fertile Crescent, first by his Arab compatriots and then by Israel, and having been forced out of the Arabian peninsula after a long and protracted war in Yemen, the military coups in Libya and the Sudan in 1969 suddenly provided Nasser's Arabism with a new ray of hope. It was symbolical because of the fact that this new prospect of Arab unity followed Nasser's final humiliation at the Rabat summit conference in December 1969, where all his requests had been bluntly turned down by his fellow Arab leaders and he had walked out empty-handed.

In Tripoli, whence he traveled from Rabat, Nasser was welcomed by his new protégés Qaddafi of Libya and Numeiri of the Sudan, and signed with them an agreement for the coordination of political, economic, and military action. Here was what seemed to be a new area for action: A possible revival of the unity of the Nile Valley, providing a prospective solution for Egypt's population explosion, combined with unity with Libya, where recent oil discoveries had boosted the economy and, hence, could provide considerable aid for Egypt's development. It seemed a much more realistic approach to Arab unity than the failures of the past, and it shifted the focus of Egypt's regional politics in the direction from which it had been diverted by Nasser since 1955. Nasser's last act in this direction was to sign an agreement of economic unity with Libya and the Sudan in April 1970. But when he died in September of that year, having just accepted the Rogers Plan in July, a clear admission of Egypt's failure in the War of Attrition, Nasser's standing as a regional leader was in shambles.[4]

PREPARING THE GROUND FOR A NEW POLICY

President Sadat's approach to regional politics cannot be separated from his overall view of Egypt's internal problems, on the one hand, and its crucial involvement in international relations, on the other. To put it more bluntly, Egypt's economy was in shambles, with no solution in sight. The continued conflict with Israel consumed all of Egypt's energy and resources; moreover, its dependency on the Soviet bloc and on the complexities of inter-Arab relations created the magic circle within which Nasser had gambled and lost.

Several steps were required in order to break out of what seemed a hopeless situation. In order to arrive at a settlement with Israel, on terms acceptable to Egypt, the post-1967 status quo had to be broken. This in turn would serve to convince both Israel and the United States, its major protector, that a compromise was essential. The steps leading to an Egyptian solution of its conflict with Israel started with Sadat's hesitant initiative in February 1971, through the October 1973 war, which brought about the Sinai disengagement agreements of 1974-75, and ultimately to the peace initiative in November 1977. By 1981 Egypt had regained most of its territories, improved its image both internally and internationally, and been able to move economically from near stagnation to economic growth. This growth was made possible by a considerable growth in capital imports, primarily from the United States (nearly $1 billion per annum in 1977-79) and other Western sources. The unilateral transfer of over $2 billion per annum from Egyptian workers in the Arab oil states (primarily Saudi Arabia and Libya) was not interrupted, despite the economic sanctions decided upon at the Arab summit in Baghdad in 1979. If one adds the foreign loans received by Egypt from other sources, such as the International Development Association of the World Bank (in 1980, $215 million for infrastructural projects) and the Gulf Organization for the Development of Egypt, the pressing economic hardships of the pre-1977 economy had indeed been eased. However, none of the serious economic problems have yet been solved. These include a large foreign debt that must be serviced, growing inflation, heavy expenditure on armaments, and an enormous bureaucracy. When Robert McNamara, president of the World Bank, stated in February 1981 that owing to its improved economic performance Egypt was no longer eligible for soft loans, it certainly signified an improvement in Egypt's economy, but not a solution to its problems. In the long term, Egyptian economists hope to transform the economy with increased foreign investments, lured by the open-door policy; fast-growing revenues from the Suez Canal and from crude oil; and the 1978-82 development plan, which in its first year envisaged an investment of nearly $4.5 billion.[5] But while the economic base has thus been broadened and the sources of revenue diversified, the benefits accruing from this development have not been spread evenly, and relatively little has trickled down to the lowest and most populous strata of the Egyptian population. The widespread discontent within these strata, aggravated by the fact that the peace treaty with Israel had not brought about the expected economic salvation, was a major factor in the open criticism leveled against Sadat and his entourage immediately after his assassination.

One can discern several stages in the process that led Egypt from complete dependence on the Soviet Union to a pro-United States orientation. When Sadat was forced to sign the Treaty of Friendship with the Soviet Union in May 1971, he still hoped that Soviet aid on a massive scale would enable him both to boost Egypt's stagnant economy and to reequip his armed forces. Neither of these hopes were fully realized, and while the Soviets fulfilled some of their promises as far as equipment was concerned, they did so, according to Sadat, on a selective and humiliating basis. Moreover, the Soviet-inspired abortive coup in the Sudan in July 1971 as well as growing Soviet involvement in other parts of Africa and the Middle East convinced Sadat that there was a Soviet master plan to dominate the region.

Furthermore, the Soviet Union, despite its military and economic aid since 1955, had neither been able to modernize Egypt's economy nor to help it in regaining its lost territories in the Sinai. The answer clearly lay in Washington and not in Moscow, as the Americans had the money, the technology, and the leverage over Israel required to satisfy Egyptian needs. The steps leading to the shift in

Egypt's superpower alliance started with the expulsion of Soviet military personnel from Egypt in July 1972 and culminated in the exclusion of the Soviets from the post-1973 settlements and in Sadat's open objection to a future Soviet role in the negotiations.

Hostility reached a further peak following the Soviet Union's open rejection of Sadat's peace initiative and the subsequent Camp David accords. Sadat accused the Soviets of trying to topple his regime through aiding fanatic opposition groups to commit acts of sabotage in Egypt and of attempting to undermine his regime through the good offices of Soviet puppet states in Libya, Ethiopia, and the People's Democratic Republic of Yemen. Even in the Sudan, Sadat suspected the Soviets of attempting once again to topple his closest ally in the Libyan-backed Ansar revolt of July 1976. Attempts at rapprochement, undertaken during 1978, failed completely and only strengthened Sadat's conviction regarding Soviet intrigues and hostility.

The shift toward the United States was therefore based both on the belief that the United States would be able and willing to deliver and on anti-Soviet sentiments and suspicions. Economically, as mentioned above, Egypt's reliance on the United States and its allies has already born considerable fruit, far beyond what the Soviet bloc would have undertaken. Politically, Sadat set out to prove that, while he appreciated the U.S. special relationship with Israel, Egypt, the Horn of Africa, and the Persian Gulf were also cardinal Western interests and had to be treated accordingly. In other words, while Egypt needed U.S. economic aid and technology, Egypt could guarantee the stability of the region and its security as well as provide a barrier against the Soviets once the Middle East conflict was brought to a satisfactory end. Sadat's success in conveying this message both to the administration and to the U.S. public, especially during the hostage crisis in Iran, clearly indicates that the bilateral relations between Egypt and the United States have moved according to his original plans. [6]

INTER-ARAB RELATIONS

In examining the shift in Sadat's regional policy, which is our major topic, it is advisable to examine this policy in relation to the Arab Middle East and to assess its implications in Africa, especially in relation to Libya, the Sudan, and the Horn of Africa.

When Sadat assumed power in September 1970, Egypt's standing in the Arab world was, as mentioned above, at a very low ebb. Even the two youngest and weakest so-called revolutionary regimes in Libya and the Sudan, who had sought protection under Nasser's umbrella, were not so sure whether Sadat could provide them with similar services. Sadat's own preferences and personal friendships were among the "reactionary" rulers such as King Hasan of Morocco, King Faysal of Saudi Arabia, the Sabah ruling family of Kuwait, President Franjieh of Lebanon, and others. Although he did pursue the planned federation with Libya, the Sudan, and later also Syria, he did so primarily for reasons connected with internal Egyptian politics, and especially as part of his fight for survival against the Nasserist power centers. In May 1971, with the internal battle won, the quest for unity could be discarded. [7] Thus, by 1972 Sadat's earlier adherence to Arab unity had become subordinate to what he regarded as the values of true Egyptian patriotism. Two observers of the Middle Eastern scene have described this shift. Fuad Ajami has stated that "the idea that has dominated the political consciousness of modern Arabs is nearing its end. . . . It is the myth of pan-Arabism. . . . Slowly and grimly, with a great deal of anguish and of outright violence, a 'normal' state system is becoming a fact of life."[8] Burrell and Kelidar have phrased it somewhat more critically. "Sadat has sought . . . to extricate Egypt as far as possible from the labyrinth of inter-Arab politics and rivalries; but in doing so he has rendered Egypt dependent upon the goodwill of the conservative states of the Arab world. The result of this has been to reduce the number of policy options open to the Egyptian leadership, a position Nasser would have striven to avoid."[9] In fact, Sadat had gambled his future on this policy, and his survival was thus questionable.

Mohamed Hassanein Heikal, one of the main exponents of Nasserism, went even further when he stated that since Egypt was culturally, linguistically, and religiously part of the Arab nation and could never lose its identity, Sadat's mistaken foreign policy had alienated the Arabs. This led, according to Heikal, to Sadat's losing his real constituency and with it his regional and internal standing. [10]

Sadat's premise when determining his new regional policy was based on a number of considerations. Egyptian interests made a political solution of the Middle East conflict essential, and such a solution could be arrived at, on Egyptian terms, only with U.S. backing. That backing could only be marshaled if the United States was convinced that a shift away from Israel, leading to a more balanced Middle East policy, would serve its own interests. To convince the United States, Sadat needed the backing of the conservative pro-U.S. states, primarily Saudi Arabia and Iran. And finally, there was an acute danger of a Soviet pincer movement starting from Afghanistan and moving through Iran or Iraq to the Persian Gulf and from there through Oman to South Yemen, the Horn of Africa, Ethiopia, and Libya. To Sadat it seemed clear that only a pro-United States alliance, headed by Egypt and including all possible anti-Soviet regimes, could with U.S. aid, stop this danger. Hence, his choice of allies was quite obvious. Iran, Saudi Arabia, and Morocco seemed more reliable allies against the Soviets than the Fertile Crescent, let alone the Palestinians. Therefore, after the October 1973 war, when Asad and Arafat became obstacles to Sadat's political program, he all but discarded them and concentrated on the traditional rulers, who continued to back him, at least tacitly, until the Camp David accords. Sadat's regional policy was therefore based on areas of possible cooperation that would benefit Egypt, rather than on ideological principles or pan-Arab considerations.

By 1975-76 Syria was accusing Sadat of betraying the Arab world, and although a temporary reconciliation was reached under Saudi auspices, it lasted for less than a year and disappeared with Sadat's trip to Jerusalem. The fate of Egypt's relations with other radical Arab states and with the PLO was similar. In the case of Libya, relations deteriorated even earlier, and in the summer of 1977 military conflict broke out along Egypt's western border. Sadat branded Qaddafi a fanatic lunatic and regarded him as a menace not only to his own people but to the whole continent. Libyans and Libyan-paid Egyptian agents were accused of numerous acts of sabotage in Egypt, and Qaddafi's regime was accused of financing the anti-Sadat opposition and its propaganda machine both in Egypt itself and in the diaspora. [11] More dangerous were Libya's continued attempts to overthrow the regime in the Sudan, both in an abortive coup in July 1976 and later. In March 1979 Qaddafi declared his willingness to sign a treaty of defense with the Marxist regime in Ethiopia aimed against the Cairo-Khartoum axis. [12] Later in that month, Libyan and Syrian forces were again accused of massing on the Egyptian border. [13] In June 1979 Qaddafi asked, through Moscow, for Iraqi troops to come to his aid against Egypt. According to Iraqi sources, there were at that time some 5,000 Cuban and 2,000 Hungarian troops in Libya, while there was no proof of Egyptian military presence on the Libyan border. [14] In June 1980 Egypt imposed martial law on its border regions with Libya in response to Qaddafi's threats against Egypt and the Sudan. [15] In November 1980 Libyan troops invaded Chad and "liberated" its northern province. Tripoli accused Egypt and the Sudan of aiding the separatist forces in Chad under Hisseine Habre. [16] In December the Libyan attack reached a new peak with the conquest of Ndjamena, the capital of Chad. [17] While the Libyan forces were fighting in Chad, Qaddafi and Asad had agreed to form a "revolutionary leadership," which would formulate the plans for uniting Libya and Syria. From a Libyan point of view, it was an advantage to have at least one ally in face of the general outcry in Africa against its involvement in Chad, especially since there were constant rumors about the concentration of Egyptian and Sudanese forces. [18] Egyptian-Libyan relations had thus reached a new low ebb as a result of these developments. In February 1981 Syria and Libya agreed upon a plan to spread the Islamic socialist revolution in Africa through military and subversive means. If one takes into account that during the same period Colonel Hisseine Habre was preparing his forces with Sudanese and Egyptian aid for a counterattack against his antagonists in Chad, it is clear that the Libyan-Syrian axis might be heading for a confrontation with Cairo and Khartoum. [19]

EGYPTIAN RELATIONS IN THE NILE VALLEY AND THE RED SEA

Sadat's special attitude toward the Sudan was clearly demonstrated in July 1971, when Egyptian intervention helped Numeiri to regain power following the abortive procommunist coup. But this special relationship began to flourish only after the October 1973 war, when the two countries actually signed, in February 1974, an agreement for their political and economic integration. They started to coordinate their development projects as well as plan combined cultural and religious institutions. This development was part of a common regional and international outlook. Both Numeiri and Sadat regarded Qaddafi as a dangerous threat to their regimes. They had similar feelings regarding the new regime in Ethiopia (discussed below), and hence cooperation was a natural conclusion. In September 1975 Numeiri was the only Arab leader who openly supported Sadat over the second Sinai disengagement agreement. But if one has to point to a single event that brought the two leaders even closer, it was the Egyptian intervention in the July 1976 coup, which saved Numeiri's regime once again. Moreover, it clearly proved that Qaddafi was openly involved in attempts to overthrow Numeiri and in helping the Sudanese opposition parties organize in the National Front. In July 1976 Egypt and the Sudan signed a joint defense agreement that provided for closer cooperation in all matters of security and defense than existed between Egypt and any other country. Matters of mutual concern, such as the security of the Red Sea, the prevention of Soviet penetration into adjacent regions, or Libyan plots against the Sudan or other African states were hereafter dealt with by the joint defense council that was set up by the two countries. [20]

Sadat and Numeiri also agreed with regard to the Soviet threat to Africa. In 1977 they cooperated in repulsing the Angolan invasion into Zaire, while in May 1977 the Sudan announced the termination of contracts of all Soviet experts stationed in the Sudan defense force. Thus, by the end of 1977, there was complete agreement between Egypt and the Sudan and close cooperation between the two of them and Saudi Arabia, as the three countries shared a common concern for the Arabization of the Red Sea and the necessity to stop Soviet penetration into the region. This trilateral harmony was disturbed by Sadat's visit to Jerusalem. While Numeiri gave his immediate blessing to Sadat's initiative, the Saudis reacted negatively and thereby created a difficult situation for Numeiri, who was torn between his political loyalty to Sadat and his economic dependence on Saudi Arabia. But Numeiri prevailed despite Saudi and Libyan pressures and growing internal opposition, especially from the Ansar, in his support of Egypt's peace initiative. The Sudan did not join the economic sanctions imposed against Egypt in November 1978, nor did it sever diplomatic relations. Throughout the period following the signing of the Camp David accords in September 1978, Numeiri continually emphasized the Sudan's close relations with Egypt, based on their historic ties and common heritage. Hence, Numeiri claimed that the Sudan had a better understanding of Egypt's circumstances and of the enormous sacrifices made by the Egyptians for the sake of the Palestinians. An attempt to boycott and isolate Egypt was therefor uncalled for.[21] The convening of the national assemblies of Egypt and the Sudan in a joint session in January 1979 symbolized the Sudan's determination not to succumb to Arab pressures. But in March 1979, following the signing of the Egyptian-Israeli peace treaty, there were signs that Numeiri was beginning to succumb to internal and external pressure. Criticism of Sadat's policy started to appear frequently in the Sudanese press, and consequently, the relationship between Sudan, Saudi Arabia, and Libya improved. Saudi Arabia supplied all the Sudan's requirements as far as oil was concerned in order to save it from the consequences of an Iraqi-imposed embargo. Libya filled its part of the bargain and closed the Ansar's training camp in Kufra and helped in furthering the reconciliation between Numeiri and the opposition National Front.[22]

Numeiri's participation in the Tunisia summit in November 1979 and his subsequent signing of the resolution condemning the Egyptian-Israeli peace treaty was regarded as a low ebb in Sudanese-Egyptian relations. The Sudan seemed to be returning to the Arab fold. The anti-Egyptian resolution, calling for the liberation of Palestine and the realization of the full rights of the Palestinians, was supported even by Oman and Somalia, who along with the Sudan had hitherto sided with Egypt.[23] One month later the Sudan announced that if the Israeli flag would be raised in Cairo, its ambassador in Egypt would be recalled. Thus, while diplomatic relations between Egypt and the Sudan were not broken, the establishment of diplomatic relations between Egypt and Israel brought about a further decline in their relationship.[24]

But even after this setback in the relationship, Sadat's attitude toward the Sudan and Numeiri continued to be moderate. Egypt blamed

Saudi Arabia for plotting against the Sudan and putting pressure on its leaders. Sadat declared that the Sudan was free to return its ambassador to Cairo whenever it desired to do so, while he stated that this would not be granted to other Arab states.[25] However, throughout 1980, the Sudan continued to improve its relations with Saudi Arabia, the Gulf emirates, and even with Iraq. Only the Libyan intervention in Chad since December 1980 changed the course of Sudanese politics, and by March 1981 diplomatic relations with Egypt were once again at an ambassadorial level. On March 18 Numeiri announced that foreign elements were again plotting against his regime, mentioning the Syrian Bath as a possible suspect, and invited the United States to use Sudanese territory for its military facilities. On the same day, Egypt's Kamal Hasan Ali promised full Egyptian backing to the Sudan against its Libyan enemy.[26] In June 1981 the Sudan expelled all Libyan diplomats from the Sudan and suspended all flights between Khartoum and Tripoli. The new deterioration was connected with the situation in Chad, where opposition to the Libyan presence was increasing.[27] The close relationship between Egypt and the Sudan, which had characterized most of the years since Sadat and Numeiri came to power, was once again approaching its peak, and the economic, cultural, and political integration of Egypt and the Sudan was advancing smoothly.

If proof was needed regarding the depth of Numeiri's solidarity, it came after Sadat's assassination. Numeiri attended Sadat's funeral, while other Arab heads of state were either openly jubilant or suspiciously quiet. Furthermore, while in Cairo, Numeiri declared his full and unflinching support for the late president's policy and even took part in the election of Husni Mubarak as Egypt's new president, surely an unheard-of act of solidarity and close alliance.

Ethiopia and the Horn of Africa have emerged as another region of concern for Egypt. This is, of course, not a new venture; as mentioned above, Egypt attempted to conquer Ethiopia 100 years ago, while the threat of manipulating the waters of the Blue Nile had arisen many times, as early as the eleventh century. However, the developments of the 1970s were unique owing to the fact that there was a close connection between the Soviet threat, as perceived in Cairo and Khartoum, and internal developments on the Horn of Africa, on the one hand, and Egypt's increased involvement in the Red Sea region, on the other.[28]

In a way Egypt and Ethiopia moved in opposite directions in the 1970s: Egypt moved from revolution to conservatism, from being a Soviet client to reliance on the United States, while Ethiopia moved to pro-Soviet revolution and to becoming the most important Soviet base in the Red Sea. Egypt's policy toward Ethiopia and the Horn of Africa was therefore formulated in 1976-77 to try to counteract these trends. It seemed at that time that since the Soviets were entrenched in Ethiopia and Somalia on the African side of the Horn and in the People's Democratic Republic of Yemen on the eastern shores of the Red Sea, they were in a position to hold Egypt to ransom. It was under these circumstances that Egypt, Saudi Arabia, and the Sudan formulated and coordinated their policy in 1976 to preserve the Arab predominance in the Red Sea. When, in 1977, Somalia and Ethiopia went to war over the Ogaden Desert, and the Soviets, having decided to back Ethiopia, were expelled from Somalia, the Arabization of the Red Sea appeared to be on its way toward realization. Further gains for Egypt and its allies were made in Eritrea, where the various independence movements conquered most of that province, while newly independent Djibuti joined the Arab League as a junior ally of Egypt and Saudi Arabia.

However, since 1978 the Egyptian policy in the Red Sea and on the Horn of Africa has suffered considerable setbacks. First, Ethiopia has all but succeeded in overcoming Somalia as well as the Eritrean separatists, owing primarily to their internal divisions. Second, Sadat's peace initiative has undermined the Cairo-Riad axis, which was further weakened following the Islamic revolution in Iran and its repercussions in the Persian Gulf. Hence, the Egyptian initiative on the Horn of Africa rested primarily on its two African allies, the Sudan and Somalia, while Oman was its sole solid support in the Arabian peninsula. Even the Sudan, following the influx of Eritrean refugees and the realization that Ethiopia has regained the initiative, has all but mended its fences with Mangistu and the Derg and closed its borders to the continuing flow of refugees. While the Sudan remained committed to the anti-Soviet policy centered in Cairo, it had stopped playing an active role against Ethiopia, the current center of Soviet power in the Red Sea.

Egypt thus seemed to follow a policy that enjoyed little support in Africa. By aiding Somalia both politically and militarily, Egypt was undermining one of the most sacred principles of that continent, namely, the permanency of the present borders. Somalia is claiming territories from both Kenya and Ethiopia, thereby undermining the very precarious balance in the region. In following this line, Sadat seems to have alienated prospective African support for his regional policy. However, recent developments in Chad, the Sudan, and Ethiopia suggest that Egypt may not be as isolated as it seemed.

ISLAMIC MOTIVES IN EGYPT'S
REGIONAL RELATIONS

The decline of Arabism as a major ideological factor after the 1967 defeat had enhanced the position of Islam both internally, in fighting the anti-Sadat opposition in the early years, and externally, in rallying as many supporters as possible behind Sadat's new foreign policy. This, of course, was not a new venture. Islam had been exploited in Egypt's foreign relations in the nineteenth century, and the "Islamic circle" had been an important feature in Nasser's "philosophy of the Revolution." However, under Nasser the emphasis was definitely on the Arab circle, while Islam never achieved similar prominence.[29] Sadat's foreign policy was, as stated above, concerned with two major considerations: first, how to combat Soviet-inspired intrigues in the region, and second, how to enable Egypt to determine its own priorities vis-à-vis the Middle East conflict without undue concern for possible Arab reactions. The role of Islam in legitimizing these policies, both inside Egypt and within the Islamic world, should not be overrated, but neither should it be ignored.

It was fairly easy to receive full backing from both the Islamic establishment and the more popular Islamic leadership in the fight against communism, both inside Egypt and in its external relations. Islam was employed against the Soviet-backed Ethiopians in their fighting in Erithrea and the Ogaden Desert. Indeed, even after Sadat had been branded a traitor by most of the Arab governments, anti-communism continued to serve as a common denominator with other Muslim states. The ban on all communist movements, which was adopted by the League of the Islamic World in its twentieth session, held in August 1979 in Mecca, illustrated the affinity in views in this respect. Moreover, the declaration of Shaykh Muhammad Ali al-Harakan, the league's general secretary, that "there is no place in our Islamic World for the emergence of any ideology other than Islam,"[30] was a clear indication that Egypt's policy in this respect was synonymous with that of most Muslim governments. But when Sadat had attempted several months earlier to receive Muslim support for his plan for the liberation of Arab Jerusalem, he was, of course, rebuffed. The "traitor" of Camp David could not be trusted by those who condemned him. Indeed, Egypt's declaration that it would not attend the Rabat meeting of Muslim foreign ministers in May 1979 unless an Islamic summit would be convened to discuss Jerusalem[31] was probably a face-saving device. However, when the Islamic meeting decided to consider Egypt's suspension from membership, Hasan al-Tuhami, Egypt's deputy prime minister, denounced it as illegal, claiming that Egypt was not only one of the most important pillars of Islam but had also "contributed most to the preservation of Islamic heritage."[32] Sadat himself reacted almost immediately and declared that the grand imam of al-Azhar, backed by the ulama and the Islamic Research Council, had already condemned this decision as illegal. They stated, according to Sadat, that for the last thousand years al-Azhar had defended Islam, "and has protected it from those who are now exploiting it." Hence, "neither the Arab nor the Islamic mission will proceed without Egypt and without Egypt's Al-Azhar." Sadat complimented the six African states, who did not support Egypt's suspension, saying "the Africans know the truth about our cousins, the Arabs."[33]

So, while Sadat's emphasis on confronting the Soviet Union and communism has continued to receive impressive support in the Muslim world, his peace initiative with Israel was, at least on the official level, rejected. Arab pressures, and in particular those of Saudi Arabia, had succeeded to sway, at least temporarily, even Egypt's allies, like Morocco and the Sudan, whose future depends to a very large extent on continued Egyptian support against their radical enemies. It was therefore of particular importance to Sadat to have the full support of al-Azhar and of as many Muslim ulama as possible for his peace initiative since its inception. Sadat's visit to Jerusalem in November 1977 received blessings of the Islamic establishment. Those supporting this historic act included the rector of al-Azhar, the presidents of Egyptian student unions, and the Muslim Youth association. The latter conveyed its blessings through its chairman, Shaykh Ahmad Hasan al-Baquri. In thanking those who had come to bless him, Sadat stated that only faith in God had prompted his initiative and stressed that the only way to build a better Egypt was to educate the young Egyptians to be faithful to their religious principles.[34] Similar support was granted to the peace initiative and the Camp David accords by the shaykh of al-Azhar and the minister of Waqfs.[35] When the Islamic Conference in Rabat denounced Sadat's initiative, the staff of al-Azhar and all its affiliated institutions, under the chairmanship of Shaykh Rajab al-Ayidi, reaffirmed their full support of Sadat's policy and stated that "the peace treaty between Egypt and Israel is a blessed Islamic step, founded on the principles of religion."[36] Sadat himself has continuously emphasized the religious aspects of his mission. He did so when he addressed the Knesset in Jerusalem in November 1977. He stressed the historic ties between Islam and Judaism when he opened his speech at the Ben Gurion University in Beersheba, in May 1979, stating that it was the prophet Muhammad who had ordered the people of Yathrab (Medina)—Jews and Muslims—to form one nation and to practice their respective religions in peace and harmony. Following the return of Santa Catarina to Egyptian rule, Sadat emphasized his desire to build a house of prayer for Christians, Jews, and Muslims on that mountain and expressed his wish to be buried there when his day would come. He also ordered a study of the feasibility of a canal from the Nile to Jerusalem. Of the latter project he said the following:

> In the name of Egypt and its great Al-Azhar and in the name of defending peace, the Nile water will become the

new "Zamzam well" for believers in the three monotheistic religions. . . . The water will serve all pilgrims visit-
ing the holy shrines in Jerusalem.[37]

So, despite the general trend among the Muslim rulers outside
of Egypt to denounce Sadat's peace initiative as a betrayal of Islam,
the Islamic establishment of Egypt continued to stress its religious
blessings. Indeed, there was a growing tendency to stress the unre-
liability and the treason prevailing among the Arab rulers while they
praised Islam and its potential blessings for the future of the region.
In a long interview published in the October magazine on September 30,
1979, Sadat listed the crimes committed by the Arabs against the
Palestinians and against each other, not sparing even his erstwhile
allies such as Saudi Arabia and Morocco. He concluded, "I regret to
say that the Arabs have not changed, but have instead regressed.
Egypt alone has changed. We were and still are more civilized and
advanced. The Arabs have become small in our eyes because we
have grown bigger."[38] It is therefore hardly surprising that after the
Arab League summit, which was convened in Tunisia in November
1979 following the removal of its headquarters from Cairo, Sadat declared that "the latest comedy in Arab solidarity has ended. . . .
This new Arab League has ended, and it had to end. There will be no
Arab League, but there must be a wider and greater Islamic League."[39]
A week later, the editor of October, Anis Mansur, called on all Muslims to establish an "Islamic Peoples' League to confront the enemies
of Islam and to rise from the abyss of Arab policy to the glory of Islam." Only the Muslim people could achieve this, as certain Muslim
rulers, notably those of Saudi Arabia, Syria, Iran, and Afghanistan,
were distorting Islam in order to serve their own interests.[40]
But this was the accusation leveled with ever-growing vehemence
against Sadat himself. The spokesman of the more radical populist
Islamic movements in Egypt, including the very strong Muslim Brothers, denounced Sadat's peace treaty with Israel as an act of heresy,
while his internal policies in Egypt fared no better. Sadat's attempt
to crush this radical opposition, when he ordered the arrest of some
1,500 of its leaders in September 1981, was probably an admission
that he had underestimated their popularity.

CONCLUSION

This brief survey of Egypt's regional policy over the last two
centuries has shown that there are definite shifts in emphasis rather
than permanent changes in Egypt's foreign relations. If we evaluate
these changing relations and analyze them, we will discover, first,

that the Sudan and the Nile Valley played a constant and central role
in Egypt's regional considerations for most of that period in both the
nineteenth and twentieth centuries, and second, that shifts in emphasis between the Arab Middle East and the East African region were
to a large degree based on Egypt's reaction to international power
politics and their repercussions, as interpreted in Cairo. It was
Muhammad Ali's reading of the "Eastern question" that dictated his
handling—or mishandling—of the Syrian campaigns in the 1830s. Ismail's appreciation of the difficulties faced by his grandfather in the
Fertile Crescent and the potential offered by aligning Egypt with British interests in Africa was a major cause for his African campaigns
and conquests in the 1860s and 1870s. When Egypt once again turned
to the northeast, after World War II, it did so because of its interpretation of the repercussions of the Anglo-French evacuation of the
region and the consequent dangers of the vacuum being filled by a hostile Hashimite bloc. Later, in the 1950s, Nasser's concept of Arabism was based on his reading of the advantages offered by the cold
war. The exploitation of superpower antagonisms in a regional context, with the aid of so-called positive neutralism, enabled Egypt to
play a leading role in that region for a period of some ten years.
Nasser failed to realize in the early 1960s that the cold war was declining, and hence, that positive neutralism no longer offered the same
dividends. Egypt therefore found itself in a position of growing dependence on the Soviet Union. Its regional policy, leading to the June
1967 war, was based to a certain degree on Soviet desires to undermine Western influence in the Middle East to a vanishing point. With
the 1969 coups in Libya and the Sudan, this policy seemed to be paying off, but in reality, neither Nasser nor Soviet Middle Eastern interests ever recovered completely from the 1967 defeat.

Sadat's shift in regional politics was based on several assumptions. First, the Soviet connection no longer carried with it any advantages for Egypt but, rather, presented acute dangers. Second, in
the Fertile Crescent, with its ever-growing entanglements, Egypt
would be served best by certain aloofness. Sadat did neither foresee
nor plan the breakdown in Egypt's relations with the Arab world that
followed the Camp David accords. But his policy, even prior to 1977,
clearly indicated his determination to formulate Egypt's priorities in
foreign policy in Cairo according to Egyptian interests, regardless
of the repercussions in Damascus, Baghdad, or even Riad. According to these considerations, Egypt's top priority, based on its strategic position as interpreted by Sadat and his close associates, was to
forge the kind of alliances that would enable it to withstand the ever-
growing Soviet threat. This meant, in the first place, closer relations with the United States, who had to be convinced that an alliance
with Egypt was at least as important for Western interests as U.S.

commitment to Israel or to Saudi Arabia. It further indicated the
necessity of peace with Israel, both because of Egypt's economic
problems and in order to facilitate the shift in U.S. Middle East policy
as indicated above. For Sadat realized quite clearly, and especially
after his futile pro-U.S. gestures in 1971-72, that only a major breakthrough in regional politics ending the Arab-Israeli stalemate would
convince Washington that Egypt was serious in its desire to return
to the Western fold. The October 1973 war, leading to the Sinai disengagement agreements and from there to the November 1977 peace
initiative, supplied the necessary proof.

The changing pattern of Egypt's regional relations was therefore the result of several accumulated factors rather than a carefully
considered, consistent, and well-planned policy. Had Saudi Arabia,
Morocco, Jordan, or even Syria and the PLO fallen into line, Sadat's
Egypt would probably have regarded it as a blessing. Since the reverse happened, Sadat, unlike Nasser, refused to change course and
was willing to risk a confrontation with the Arabs. Sadat was convinced that Egypt's isolation among the Arabs was a passing phase and
that its superior strength and geopolitical position enabled it to ignore
the objections of those whom he defined derogatively as "dwarfs."

But the Sudan, owing primarily to geopolitical considerations,
was in a different category. The Sudan is Egypt's hinterland, and the
danger of a hostile government in the Upper-Nile regions was always
regarded in Cairo as a far greater threat than the accumulated noise
made by Qaddafi, Asad, and Arafat. True, Qaddafi could and did encourage the anti-Sadat opposition both in Egypt and in Tripoli, but he
was never more than a nuisance who could, if it became necessary,
be disposed of. Numeiri's continued support for Egypt's regional and
international policy continued to be an important feature of Sadat's
foreign relations and is likely to remain so under Egypt's new president, Husni Mubarak. Numeiri's presence in Cairo throughout the
crucial week following Sadat's assassination and his outspoken support for the new president are clear indications of his intentions.

Sadat's tragic disappearance from the Middle Eastern scene
may yet present a further shift in Egypt's regional relations. It may
enable the more conservative Arab rulers to modify their anti-Egyptian stance and to ease Egypt's reentering the scene of inter-Arab relations in order to weaken the extremist pro-Soviet front. Egypt's open
support of Iraq in its war with Iran was an indication as to the willingness of the anti-Egyptian front to overlook the Arab boycott of
Egypt in case of need. Egypt's support for Morocco against the
Libyan-Algerian front in the recent upsurge of the conflict in the western Sahara could present Egypt with an additional way out of its isolation. Mubarak's announcement, on October 21, 1981, that Egypt's
mass media would cease to attack and denounce the Arab leaders who

criticized it was an indication of a desire to forge a more balanced policy in the Arab Middle East.

NOTES

1. For details on Muhammad Ali's and Ismail's foreign policy, see Henry Dodwell, The Founder of Modern Egypt (Cambridge, 1931), pp. 39-67, 125-91; L. A. Fabunmi, The Sudan in Anglo-Egyptian Relations (London, 1964), pp. 22-51; J. C. Hurewitz, The Middle East and North Africa: A Documentary Record (New Haven and London, 1975), 1:271-78; Lord Gromer, Modern Egypt (London, 1908), 1:371-95; Rifaat Bey, The Awakening of Modern Egypt (Lahore, 1964), pp. 140-51; Gabriel Warburg, "The Sinai Peninsula Borders, 1906-47," Journal of Contemporary History 14 (1979): 677-92; and M. S. Anderson, The Eastern Question (New York, 1966), pp. 88-109.

2. Fabunmi, The Sudan, pp. 62-112, 114-202; Gabriel Warburg, The Sudan under Wingate (London, 1971), pp. 13-45; idem, Islam, Nationalism, and Communism in a Traditional Society (London, 1978), pp. 67-89; and Lord Lloyd, Egypt since Cromer (New York, 1970), 2:123-39, 285-302, 395-99.

3. On Great Britain's Middle East policy, see E. Monroe, Britain's Moment in the Middle East (London, 1963), pp. 131-77; E. Kedourie, The Chatham House Version and Other Middle Eastern Studies (London, 1970), pp. 213-35, 351-94; P. J. Vatikiotis, Nasser and His Generation (London, 1978), pp. 85-96; and Y. Gershuni, Egypt between Distinctiveness and Unity: The Search for National Identity, 1919-1948 (Tel Aviv, 1980), pp. 200-82.

4. Vatikiotis, Nasser, pp. 225-62; S. Shamir, ed., The Decline of Nasserism, 1965-1970 (Tel Aviv, 1978), pp. 1-38, 208-309 (in Hebrew); M. Kerr, The Arab Cold War 1958-1970 (London, New York, Toronto, 1967); and O. M. Smolansky, The Soviet Union and the Arab East under Khrushchev (Lewisburg, 1974).

5. For details, see G. G. Gilbar, "Egypt's Economy: The Challenge of Peace," The Opening of the Peace Process (Jerusalem: Levi Eshkol Institute, 1981), pp. v-xxii; see also E. Kanovsky, "Major Trends in Middle East Economic Developments," Middle East Contemporary Survey (hereafter cited as MECS), ed. C. Legum, vol. 1, 1976-1977 (New York and London, 1978), pp. 227-34; MECS, vol. 2, 1977-1978 (New York and London, 1979), pp. 398-406.

6. A. Z. Rubinstein, Red Star on the Nile (Princeton, N.J., 1977), especially pp. 129-345; Anwar al-Sadat, In Search of Identity ((New York and London, 1978), pp. 210-313; see also MECS, 1:35-36, 305-10; MECS, 2:19-39, 387-90; and MECS, vol. 3, 1978-1979 (New York and London, 1980), pp. 409-12.

7. Sadat, In Search of Identity, pp. 204–31; and A. I. Dawisha, Egypt in the Arab World (London, 1976), pp. 197–98.

8. F. Ajami, "The End of Pan-Arabism," Foreign Affairs, Winter 1979.

9. R. M. Burrell and A. R. Kelidar, Egypt: The Dilemmas of a Nation 1970–1977, Washington Papers, vol. 5, no. 48 (Beverly Hills and London, 1977), pp. 59, 69.

10. M. H. Heikal, "Egyptian Foreign Policy," Foreign Affairs, Summer 1978.

11. For details, see article in Al-Sharq al-awsat, September 17, 1979, quoted in FBIS-MEA, September 19, 1979. Iraq was also aiding the anti-Sadat opposition; see also MECS, 1:314–16; and MECS, 2:391.

12. Al-Dustur, March 2, 1979. The treaty was, in effect, signed in the summer of 1981 and included South Yemen, too.

13. New York Times, March 29, 1979.

14. Afro-Asian Affairs (London), June 1979.

15. New York Times, June 17, 1980.

16. Quoted from Tripoli Radio, November 5–6, 1980, by BBC-ME, November 7–8, 1980.

17. Quoted from Paris home service, December 15, 1980, by BBC-ME, December 17, 1980.

18. Quoted from Tripoli Radio, December 17, 1980, by BBC-ME, December 19, 1980.

19. C. Legum, "The Syrian Connection in Africa," Jerusalem Post, March 1, 1981.

20. MECS, 1:316–17, 595–98; MECS, 2:213–15, 237–43, 390–95; and MECS, 3:75–78, 416–17.

21. Y. Ronen, "Sudan's Position towards the Egyptian Peace Policy," Shiloah Center Occasional Paper no. 72 (Tel Aviv: Shiloah Center, 1980) (in Hebrew).

22. G. R. Warburg, "Islam in Sudanese Politics," Jerusalem Quarterly 13 (Fall 1979): 47–61.

23. FBIS-MEA, November 23, 1979, quoting QNA, November 22, 1979.

24. Herald Tribune, December 20, 1979; October, December 31, 1979.

25. Al-sharq al-awsat, June 11, 1980.

26. BBC, March 18, 1981, quoting MENA, March 16, 1981, and SUNA, March 16, 1981.

27. UPA, quoted by Davar, June 26, 1981.

28. The following is based on MECS, 1:58–73; and MECS, 2:56–60, 237–42. I am also grateful to H. Erlich of the Tel Aviv University for letting me use his notes of a lecture on this topic, delivered to the Israel Oriental Society in March 1981.

29. P. J. Vatikiotis, "Islam and the Foreign Policy of Egypt," in Islam and International Relations, ed. J. H. Proctor (London, 1965), pp. 120–57.

30. FBIS–MEA, August 8, 1979; for details on the Muslim Brothers' contribution, see below.

31. FBIS–MEA, May 4, 1979, quoting MENA, May 3, 1979.

32. FBIS–MEA, May 8, 1979, quoting MENA, May 7, 1979.

33. FBIS–MEA, May 11, 1979, quoting Sadat's speech to the people of Kafr al–Shaykh, from Radio Cairo, May 10, 1979.

34. FBIS–MEA, December 1, 1977, quoting Radio Cairo, November 30, 1977.

35. I. Altman, "Recent Radical Trends in the Positions of the Moslem Brothers," Shiloah Center Occasional Papers (Tel Aviv: Shiloah Center, 1979) (in Hebrew), quoting from Ruz al–Yusuf, December 12, 1977, and Al–Ahram, September 28, 1978.

36. FBIS–MEA, May 18, 1979, quoting MENA.

37. FBIS–MEA, December 21, 1979, quoting October, December 16, 1979; the Zamzam is the holy well in Mecca. See also FBIS–MEA, May 29, 1979, quoting Radio Cairo.

38. Quoted from FBIS–MEA, October 2, 1979.

39. Quoted from October, December 30, 1979, by FBIS–MEA, December 31, 1979.

40. FBIS–MEA, January 8, 1980.

PART II

CONSEQUENCES OF INTERDEPENDENCE

9

THE EUROPEAN COMMUNITY IN SEARCH OF A NEW MEDITERRANEAN POLICY: A CHANCE FOR A MORE SYMMETRICAL INTERDEPENDENCE?

STEFAN A. MUSTO

> The capacity to create wealth is infinitely more important than the wealth itself. It guarantees not only possession of that created but also replacement of that which is lost.
>
> Friedrich List

INTERDEPENDENCE AND SELF-RELIANCE

Collective self-reliance is, according to Galtung, an "open-ended" concept.[1] It does not mean autosufficiency, but a somewhat vaguely defined development approach, "implying (a) the severance of existing links of dependence operated through the international system by the dominant countries, (b) a full mobilization of domestic capabilities and resources, (c) the strengthening of links and collaboration with other underdeveloped countries, and (d) the reorientation of development efforts in order to satisfy the basic social needs . . . of the peoples involved."[2] If this definition by Oteiza is accepted, then there is no necessity to postulate a contradiction, but rather a complementary relationship between both concepts, self-reliance and interdependence.

The subject of this chapter is the Mediterranean policy of the European Community (EC) as a response of the Western European industrialized nations to the increasing interdependence between them and the countries situated on the southern and eastern shores of the Mediterranean. This is an interdependence between economically unequal partners. So, the question can be raised whether the Mediterranean policy of the EC is to be considered as a contribution toward reinforcing the collective self-reliance capabilities of the structurally weaker partners or rather as an instrument for strengthening the existing links of dependence operated through the international system by the dominant countries.

The topicality and practical significance of this question derive from the fact that the EC, subsequent to its southward enlargement, will have to review the Mediterranean policy it has pursued to date and redesign its relations with the southern Mediterranean countries in line with this new point of departure. The Community has already begun work on formulating a new policy concept for the Mediterranean area. That this is so is an indication of the special significance that the Mediterranean area has always held in the foreign relations of the West European industrial nations. The special relationship between the northern and southern littoral countries derives from a basis of strong mutual interests, even if the individual concrete interests of the various partners differ considerably on account of their divergent political and economic structures.

It is true, as Vaitsos points out, that "there is no single and uniform European position vis-à-vis the developing nations. Instead, there is a set of complex considerations with conflicting objectives and vested interests depending on the circumstances and specific issues involved."3 This applies equally to the position of the EC vis-à-vis the Mediterranean countries. Nevertheless, points of common interest do exist, and it is these that dictate the European interest in close cooperation.

Historico-cultural factors: The Mediterranean has always effected not a division but a connection. It has been an area of cultural encounter, but also of conquest and colonization. It was by no means coincidental "that from Alexander the Great to the Euro-Arab dialogue we have both experienced so much of one and the same life."4

Economic factors: The Arab region provides 70 percent of the Community's energy imports. The Mediterranean is a transport route for Western Europe's supplies of crude oil and other raw materials. The independence and security of these supply routes constitute an issue of vital interest to the industrial nations of Europe.

Trade factors: The southern Mediterranean countries are particularly important markets for European exports and are, furthermore, capable of expansion in the long term. The 385 percent increase in EC exports to the Arab region between 1970 and 1976—de-

spite and because of oil price increases—was still substantially higher
than the corresponding 246 percent increase in EC imports from the
Arab region. Economic recession and current account deficits are
increasingly obliging the Community to pursue a policy of activating
trade relations and securing market outlets.

Security factors: The motivation for increased cooperation derives from the strategic importance of the Mediterranean area at the
open southern flank of the North Atlantic Treaty Organization Alliance
and from Western Europe's endeavors to exert a stabilizing influence
on developments in the areas of conflict and tension in the southeast
Mediterranean (Middle East, Cyprus, and Turkey).

From the viewpoint of the Mediterranean countries, the common
interest in closer cooperation derives from a threefold motivation.

In historico-cultural terms, Western Europe is a frame of reference for the Mediterranean countries. That this is so is partly
the result of the colonial past and partly the consequence of an ancient tradition based on a common perception of humanity and civilization.

In economic and commercial terms, the EC is a vital sales
market for the traditional exports of the Mediterranean countries.
On average, one half of their trading volume is accounted for by the
European Community.

In terms of economic development, cooperation with the EC in
the fields of industry, technology, and finance is an essential condition for continued economic and social progress in the majority of
the Mediterranean countries.

Because of these mutual interests, the Community sought from
the outset to accord the countries of the Mediterranean area a privileged status in its foreign-relations policy.[5] The first steps undertaken by the EC toward a common Mediterranean policy were the association agreements concluded with Greece in 1962 and with Turkey
in 1964. The first approaches toward a global Mediterranean concept
were subsequently initiated at the 1972 Paris summit; since that date
a complex network of trade and cooperation agreements between the
European Community and the Mediterranean countries (with the exception of Libya and Albania) has been established.[6] However, a
uniform political concept has proved not to be realizable because of
numerous problems relating to the production structure of the Mediterranean countries and the bilateral relations between these and the
Community. The existing network of contractual agreements has not
proved to be fully satisfactory for either the EC or the Mediterranean
partner countries. The second enlargement of the Community to in-

clude Greece, Spain, and Portugal will undermine, in a number of fields, the trade preferences accorded to the southern Mediterranean countries and thereby deprive the existing agreements of at least part of their substance. [7] This circumstance implies the necessity for the EC to reformulate its overall Mediterranean policy concept.

The problem posed by this undertaking resides in the fact that the Community's need for action and its possibilities for action are in inverse proportion to one another: increased cooperation with the countries of the Mediterranean region necessitates new concessions on the part of the Community at a point in time in which its scope for compromise is at its most limited. In a situation of economic stagnation and political tension, additional concessions or new instruments of cooperation could operate only at the expense of relations with fourth countries. Thus, any new EC Mediterranean concept can ultimately be reduced to a question of evaluating political priorities.

HIERARCHY OF PREFERENCES

The European Economic Community (EEC), predecessor to the EC, was founded with a political mandate; its first president, Walter Hallstein, remarked that the EEC was in politics, not business. With the passage of time, however, it has been precisely trade and business that have proved to be the most readily conducive to integration; it is thus primarily these areas that shape the EC's external relations.

The system of these relations is a hierarchy of trade preferences (see Table 9.1) that the EC has conceded, in differing measure, to partner countries both individually and collectively. "It is a shifting pyramid. The countries at the top have preference not only over developed countries, but also over other members of the pyramid. Hence a change in the EEC's terms with one group may affect the value of its terms with another." [8] Accordingly, the Mediterranean countries enjoying association status complain that their preferences are devalued by the EC concessions to the African, Caribbean, and Pacific (ACP) states; the ACP states complain that their contractually guaranteed free access to the EC market will be devalued by the extension of the generalized system of preferences.

In periods of worldwide economic expansion, it was justified to assume that a relative deterioration in the preferential terms conceded to one group of countries resulting from concessions granted to fourth countries could be compensated by a generalized expansion of markets. Rising energy prices and the economic crisis, however, have caused demand to stagnate and, in some instances, even to decline, and this alone has already eroded the value of some preferences.

TABLE 9.1

A Hierarchy of Preferences

Status	Preferred Countries
Full EC membership	EC Ten, acceding countries, Spain and Portugal
Association agreement	Turkey, Malta, Cyprus
Preferential trade and co-operation agreement	Algeria, Tunisia, Morocco, Egypt, Syria, Jordan, Lebanon, Israel, Yugoslavia
Lomé Convention	ACP countries
Generalized system	Least-developed countries
Industrial free-trade agreement	EFTA countries
Nonpreferential trade agreement	Bangladesh, India, Pakistan, Sri Lanka, ASEAN countries, Argentina, Brazil, Mexico, Uruguay, China, Romania

EC   = European Community
ACP   = African, Caribbean, and Pacific
EFTA = European Free Trade Association
ASEAN = Association of Southeast Asian Nations

A much stronger influence on the practical value of the trade preferences enjoyed by the EC partner countries will be exerted by the Community's southward enlargement: the import demand of EC member states for a number of agricultural and industrial products for which many Mediterranean countries have developed a specialized export production will decrease substantially.

The advantages that the EC have conceded to the countries of the Mediterranean region and that are now jeopardized by the prospect of erosion take the form of free or preferential access for the products of the favored country to the EC market, access to the European Development Fund (EDF) and the credit facilities of the European Investment Bank (EIB), closer economic-technological cooperation with the Community, and in some instances (the Maghreb countries, Turkey, Yugoslavia), special arrangements regarding migrant labor. Since the original (1972) plan for an industrial free-trade zone to cover the EC and the Mediterranean region proved impossible to implement (and has since been deprived of its attractiveness for the

TABLE 9.2

Network of EC Agreements with Mediterranean Countries

Country	Year	Agricultural Concessions	Industrial Concessions	Financial Aid, 1977-81 (in millions of European Currency Units)			Social Measures	Economic-Technical Cooperation
				European Investment Bank Loans	Special Loans	Grants		
Algeria	1977	Tariff preferences of 30 percent to 100 percent for agricultural products; special arrangements for vegetables and olive oil	Free access to EC market for industrial goods (exceptions: cork and refinery products)	70	19	25	x	x
Tunisia				41	39	15	x	x
Morocco				56	58	16	x	x
Egypt	1976	Tariff reductions of 40 percent to 100 percent for various agricultural products	Free access to EC market (exceptions: limitations and tariff ceilings for some sensitive products)	93	14	63		x
Jordan				18	4	18		x
Syria				34	7	19		x
Lebanon				20	2	8		x
Turkey	1964	Tariff reductions for a number of products	Free access to EC market (exceptions: oil and textile products)	total of 745*			x	x
Cyprus	1973	Tariff reductions for a number of products	Tariff reductions of 70 percent	20	4	6		x
Malta	1971	Tariff reductions of 40 percent to 75 percent for certain products (calendar limitations)	Tariff preference of 70 percent (exception: textile products)	16	5	5		x
Israel	1975	Tariff reduction of 20 percent to 80 percent; special arrangements for a number of goods	Free access to EC market	30				x
Yugoslavia	1980	Tariff reductions in connection with tariff ceilings for wine, tobacco, and meat	Free access to EC market, tariff ceiling for 29 products	total of 200 for a period until 1985*			x	x

*Specific information not available.

156

Community on account of the structural crisis in certain industrial
sectors), the subsequent surrogate agreements were each designed
individually (see Table 9.2) to take into account the structure of the
partner country's or countries' exports to the Community. In the agricultural sector, EC concessions are subject to a number of qualifications: the value of the customs-tariff reductions for agricultural
products is diminished by market organizations, calendar limitations,
and quantity restrictions whenever the import product is in competition with a Community product.[9] In the industrial sector, goods exported by Mediterranean countries to the EC enjoy customs preferences ranging from 70 percent to 100 percent, although the value of
these is restricted when the product in question is one that is classified within the EC as "sensitive" but is of central importance to the
export structure of a number of Mediterranean countries (cork, refined petroleum products, cotton, textiles, and so forth). Each agreement contains a safeguard clause that permits the Community to can-
cel trade preferences in the event of problems emerging within the
EC. Accordingly, France and the United Kingdom imposed an import ban on early potatoes from the Mediterranean region in 1980;
Tunisia and Morocco were induced to subscribe to a self-limitation
agreement limiting their textile exports to the EC. Community preferences governing trade in industrial goods (abolition or reduction
of customs duties) are generally conceded on a nonreciprocal basis;
exceptions to this arrangement are the agreements with Turkey, Cyprus, Malta, and Israel, which countries grant reciprocity of preferences for industrial exports from EC member states.

   The motivation for close cooperation between the Community
and the countries of the Mediterranean region rests on interests that
are different in each case: for the Community, a broad-based political interest predominates (security policy, oil supplies from the
Arab region), whereas trade interests are attributed relatively subordinate importance; for the developing countries of the Mediterranean
region, in contrast, it is uncontestedly the economic, that is to say,
trade interest that predominates, whereas foreign relations and
security considerations play only a very limited role in their links
with Western Europe.

   This inverse motivation is discernible in the asymmetry in the
design of the groups' mutual trade relations. The seven Maghreb
and Mashrek countries account for only 4.8 percent of the Community's export volume and only 2.7 percent of its import volume. In
contrast, the share of total exports absorbed by the Community lies
at 42.7 percent for Egypt, 59.0 percent for Morocco, 47.8 percent
for Syria, 51.8 percent for Tunisia, and 37.3 percent for Algeria.
The share of Community products among total imports lies at 38.8
percent for Egypt, 51.4 percent for Morocco, 35.4 percent for Syria,

TABLE 9.3

Trade of the Mediterranean Countries with the EC, 1978

Country	Exports to EC Nine		Imports from EC Nine		Deficits
	Value in U.S. Dollars	Percentage of Total Exports	Value in U.S. Dollars	Percentage of Total Imports	
Egypt	1,135.6	42.7	2,559.6	38.8	1,424.0
Algeria	2,328.5	37.7	5,103.5	65.1	2,775.0
Morocco	991.7	59.0	1,878.9	51.4	887.2
Syria	518.9	47.8	901.3	35.4	382.4
Lebanon	41.3	5.1	847.1	39.1	805.8
Jordan	8.4	3.0	529.0	34.6	520.6
Israel	1,338.2	34.2	2,434.3	34.2	1,096.1
Tunisia	762.8	51.8*	1,633.5	61.2*	870.7
Cyprus	168.3	33.2*	422.3	47.4*	254.0
Malta	258.8	69.0*	438.2	65.7*	179.5

*Data from 1975/76, taken from United Nations, Trade by Country, Yearbook of International Trade Statistics, vol. 1 (New York, 1977), pp. 315, 635, 918.

Source: United Nations, International Monetary Fund, Direction of Trade 1979 (Washington, D.C.: IMF, 1979).

61.2 percent for Tunisia, and 65.1 percent for Algeria[10] (see also Table 9.3). From the outset, this imbalanced interdependence has been the source of problems and conflicts of interest that, under the prevailing critical economic and political circumstances, will become still more acute in the wake of the Community's southward enlargement. However, before proceeding to examine the implications of this southward enlargement, it is expedient to undertake a brief historical analysis of the development of the contractual relations between the EC and the Mediterranean countries.

THE ASYMMETRY OF RELATIONS

The relevant available data clearly reveal that the volume of trade transacted between the EC and the Mediterranean countries increased considerably between 1970 and 1978 (with isolated fall-backs such as in the case of Lebanon). However, it also reveals that the Community's exports to the Mediterranean have increased considerably more rapidly than its imports from the Mediterranean.[11] Taylor refers to these trade surpluses on the part of the EC as reassuring, since increased export sales to the countries of the southern Mediterranean have enabled the Community to partly compensate its oil bill deficits.[12] For the countries thus affected, however, this situation is extremely disconcerting, since their trade balance deficits with the Community (see Table 9.3) are likely to worsen in the future for structural reasons alone, and even without the consequences of southward enlargement.

An analysis of this historical development shows that the share of Mediterranean exports absorbed by the Community (excepting those from Egypt and Syria) during the period investigated (1970-1978) stagnated in the case of Jordan, Israel, Tunisia, and Malta and diminished in the case of Morocco, Lebanon, Cyprus, and Yugoslavia.[13] Only in isolated cases (Algeria, Morocco) is this situation attributable to a deliberate policy to diversify foreign-trade relations; in all other instances it is attributable to de facto import restrictions imposed by the EC (on agricultural products and sensitive products) and the low competitiveness of the other industrial products of Mediterranean supplier countries. It is thus not without justification that critics of the Community's present Mediterranean policy have observed that "the trade concessions which are being offered by the EEC to these countries are granted either on those products, an insignificant quantity of which will be marketed or even if there occurs a gradual rise in their exports to the Community market, their imports from the Community market will always outweigh their exports. Indeed, the imbalance in the structure of trade between the EEC and the Mediterranean countries is particularly striking."[14]

This imbalance is of particularly grave consequence because the production structures of the majority of Mediterranean countries have traditionally been oriented toward the West European market. A fall in demand on this market or recourse by the Community to protectionist policies can have negative implications for the economies of the Mediterranean countries affected. In the final analysis, this traditional division of labor benefits neither the Mediterranean countries, which are vulnerable to every fluctuation in the EC market, nor the Community, which is increasingly obliged to act to protect its own agricultural products and sensitive industries and thereby to accept the further limitation of its scope for cooperation with its partner countries. Roberto Albioni views the one-sided orientation of some Mediterranean countries toward the needs of Western Europe as a sign of conservative, structure-preserving economic policy:

> It is rather the lack of a progressive will that has induced Morocco and Tunisia to choose association with the EEC, but above all, to utilize as if it were the instrument for preservation of their agricultural structures. Beyond this is the lack of political decision required to set in motion the development process. In the case of Algeria, which has made these political decisions, the possibility of an alternative use of relations with the EEC is significant. [15]

The characteristics of the traditional division of labor can be seen from the composition of exports from Mediterranean countries into the Community (see Table 9.4). These figures show that the share of agricultural products and commodities in the exports of these countries to the Community is, in general, tending to decrease, although it still remained very high during the period in question in some countries (Morocco, Egypt, Israel, Cyprus, Turkey). The export of combustibles is increasing, and the share of industrial goods also reveals an upward trend, although it should be borne in mind that a large proportion of these exports consist of products that are classified as "sensitive" within the Community (textile and clothing, footwear, some chemical products, and so forth). A further part of industrial exports consists of relatively unprocessed raw materials or commodities; examples in this connection are phosphates from Jordan, crude animal materials from Lebanon, and certain cotton products from Egypt. It can thus be stated that, in general, the Mediterranean countries export to the Community products with only a small degree of net value added, whereas they import from the Community goods, services, and technologies with a high degree of net value added.

TABLE 9.4

Structure of Exports of the Mediterranean Countries to the EC, 1973-77
(in percent)

Country	Food, Tobacco, Beverage			Combustibles			Primary Goods, Oils and Fats			Industrial Products		
	1973	1977	T	1973	1977	T	1973	1977	T	1973	1977	T
Egypt	24.9	16.9	-	26.7	49.0	+	33.4	20.1	-	14.8	13.9	o
Algeria	8.0	2.1	-	87.9	95.1	+	1.0	1.2	o	2.8	1.5	-
Morocco	54.6	34.8	-	0.5	0.3	o	33.5	37.1	o	2.4	27.7	+
Tunisia	22.6	8.1	-	11.5	25.8	+	38.0	15.8	-	27.6	50.3	+
Syria	10.4	0.7*	-	53.5	89.9*	+	32.1	8.4	-	3.8	0.8	-
Lebanon	-	-	-	-	-		-	-	-	33.0	45.0*	
Jordan	9.7	0.1	-	-	-		18.8	99.0	+	65.6	0.8	-
Israel	39.5	25.9	-	4.0	-		6.0	8.2	+	50.5	65.8	+
Cyprus	72.0	78.7	o	-	-		16.0	9.0	-	5.0	11.6	+
Malta	-	5.7		-	-		-	-		98.0	93.0	o
Turkey	42.0	45.0	o	3.0	0.0	-	33.0	22.8	-	20.0	31.3	+
Yugoslavia	20.0	18.0	o	-	-		12.0	18.7	+	68.0	55.4	-

*1976.

Note: T = Tendency, + = growing, - = diminishing, o = constant.

Source: Based on E. Guth and H. O. Aeikens, Consequences of the Southward Enlargement for the Mediterranean Policy and the ACP Policy of the Community (Brussels, 1981), p. 9, table 4.

Accordingly, trade between the Community and the Mediterranean countries has made virtually no contribution to the latters' industrial development. In view of the evolution in the price of oil, several Arab countries launched a number of projects to establish oil refineries and petrochemical industries. The criticism leveled by these countries, in particular Algeria, at EC policy refers to the import restrictions it imposes on refined oil products and petrochemical products, while (and because) the Community, despite dwindling demand in the wake of rising oil prices, created excess capacities in these production areas during the period in question.[16] In the view of the countries thus affected, it is uncontested that points of departure for a reasonable division of labor between the northern and southern Mediterranean states existed in these sectors of basic industries.

The EC proceeded in a more generous manner in the field of financial cooperation, the main objective of which was to create incentives for potential investors in Mediterranean countries. However, the future prospects of a further intensification of this instrument of cooperation, which is in no way insignificant for industrial development, are not particularly favorable because of the Community's strained financial situation and the increasing costs of its agricultural sector. Correspondingly, it is not very realistic to expect that the trade disadvantages befalling the southern Mediterranean countries as a consequence of the Community's southward enlargement might be compensated by additional capital transfers from the Community budget.[17]

THE CONSEQUENCES OF ENLARGEMENT

The implications of the accession of Greece, Spain, and Portugal into the EC have already largely been explored from a static viewpoint and with respect to the ceteris paribus clause.[18] More difficult, however, is the assessment of dynamic effects of their accession, for these will depend in the medium and long term on the development of the world economy, any possible reform in the structure of West European integration (in particular regarding the common agricultural policy), and, not least, on structural changes in the acceding countries. Generally speaking, however, it can be forecast that the enlargement will be reflected in more acute and, for the Mediterranean countries, more demanding competition with the acceding countries for trade, financial, and social advantages.

In the agricultural sector, the immediate effect of enlargement will be an increase in the degree of the Community's self-sufficiency. This will refer in particular to those products that to date, have primarily been imported from the Mediterranean countries (see Table

TABLE 9.5

Agricultural Self-sufficiency in the EC

Product	Calculation, 1975-77		Projection, 1990	
	EC Nine	EC Twelve	EC Nine	EC Twelve
Northern				
Wheat	103	102	117	116
Grain (aggregate)	87	85	93	92
Rice	75	82	91	94
Milk	108	106	118	116
Beef	99	97	103	101
Pork	99	99	100	100
Mediterranean				
Wine	116	122	118	123
Tobacco	28	42	32	48
Oranges	32	64	40	70
Lemons	85	113	70	112
Citrus fruits (aggregate)	51	89	—	—
Olive oil	88	109	—	—
Tomatoes	94	99	—	—
Potatoes	99	100	—	—

Sources: United Nations, Food and Agriculture Organization, Commodity Review and Outlook 1979-1980 (Rome: FAO); Eurostat; and Agrarstatistisches Jahrbuch 1974-77.

9.5). A marginal relaxation of the EC market will be incurred only in the case of products that, like grain and animal products, do not belong to the traditional exports of the Mediterranean countries concerned. The consequence of a rising degree of self-sufficiency in the Community will manifest itself in the form of declining demand for imported goods. In the case of wine, olive oil, and certain types of fruit and vegetables, the Community will have surpassed the state of self-sufficiency and proceeded into that of structural surpluses, a situation that will inflict additional financial burdens on the agriculture budget. For the majority of the Mediterannean export products concerned, there exist hardly any alternatives on the world market: olive oil, for example, is impossible to market outside the Community.[19] Other countries that will be hard-hit are those that have hitherto exported citrus fruits, early potatoes, tomatoes, and fish conserves into the Community.

The gravest consequences will affect Morocco, Tunisia, and Cyprus, although Turkey and Israel will also suffer. One-third of the Moroccan, approximately one-quarter of the Tunisian, and more than half of the Cypriot agricultural exports into the Community will be directly affected by the enlargement. If one examines, per product group, the areas in which the Community's import demand will radically diminish, the following picture emerges: wine (Cyprus, Morocco, Algeria, and Tunisia), fish conserves (Morocco), olive oil (Tunisia and, to some extent, Morocco and Turkey), tomatoes (Morocco), citrus fruits (Morocco and Israel), and early potatoes (Cyprus, Egypt, and Morocco). As far as processed fruit and vegetable products are concerned, Turkey in particular will have to anticipate negative consequences.[20]

But the culculated static degree of self-sufficiency does not illustrate the whole impact of enlargement on the surplus production of the EC Twelve. "In addition, at least two aspects of further dynamic effects have to be taken into consideration. One is the fact that the taking over of the Common Agricultural Policy will cause price effects in the acceding countries which could result in production increases, especially in Spain, where production capacity can be largely extended. These price effects could become still stronger through the raising of the intervention prices for Mediterranean products, which constitutes a common interest of France, Italy, Greece and Spain."[21] The other possibly dynamic effect is that the Community could seek to sell at least some of its agricultural surplus products on the world market by way of subsidizing prices, thereby entering into competition with similar products from the Mediterranean countries in third-country markets. All in all, the competition handicaps anticipated for the Mediterranean countries will oblige both these countries and the EC to reconsider the terms and structure of its future trade relations.

In the industrial sector, too, enlargement will lead to a downward trend in the EC import demand. Although it is true that the Mediterranean countries will in the future have easy access to the markets of the acceding countries as well (since these will have to adopt the EC preference agreements in full), the significance of these potential new markets will be very much moderated by the similarity of their production structures. Since the industries of the acceding countries will add weight above all to those production areas in the Community that already count as sensitive (textiles, clothing, footwear, steel, ship construction), it is probable that the Community's predisposition to protectionism vis-à-vis third countries will increase considerably.

This is particularly true of the textile and clothing industries, a field in which the countries of the southern

Mediterranean have considerably expanded their production capacities, often with European capital aid, via
joint ventures with EEC enterprises, or via a processing scheme whereby Community undertakings sent them
semi-processed goods for further processing and
thereby took advantage of the labour available in these
countries cheaper than in the Community. They have
also expanded their capacities in the footwear industry.
Nevertheless, both textiles and footwear are sectors
which, in the Community, are in a very critical
state. [22]

This has particular consequences for some Mediterranean countries:
textiles and clothing alone represent, respectively, 33 percent, 11
percent, and 13 percent of Tunisian, Israeli, and Moroccan exports
into the EC. In the case of Egypt, although its share is smaller, the
industries concerned have a special strategic value because of their
close interrelation with the agricultural sector. Other industrial
sectors will also be affected, although to a lesser extent than textiles
and clothing: cork products (Morocco and Tunisia), chemicals and
fertilizers (Morocco, Tunisia, and Israel), iron and steel (Algeria),
and aluminum products (Egypt).[23] Regarding steel production, the
accession of Spain will increase the Community's existing surplus
capacities to the effect that Algeria's and Egypt's medium-term prospects of being able to build up an export-oriented steel industry will
be markedly reduced.[24]
　　It appears, moreover, that the Community is ever less prepared, because of the present economic crisis and the strained employment situation, to embark on an effective structural adaptation
policy in those industrial branches that are no longer able to produce
competitively, and that, instead, it is moving further in the direction
of administrated markets in these areas of production. Gsänger has
stated in this connection, "The first Multi-Fibre Arrangement and
the thereby resulting EEC textile agreements with different exporting
countries in Asia, Europe, and Latin America are starting points
for preserving crisis-management."[25] Such a development, which
might display certain similarities with the existing Community agricultural market organization, would put a definitive end to the original
concept of an industrial free-trade zone covering the EC and the Mediterranean countries. However, it would also imply the end of the
Mediterranean policy pursued to date by the Community.

PROSPECTS AND ALTERNATIVES

The problems pointed out thus far will undoubtedly severely strain Community relations in the future. Since, although for different reasons, both sides are interested in close cooperation, the task for the coming years is to elaborate a new, viable concept to form the basis of the future Mediterranean policy of the Community. This will in no way be facilitated by the inflexibility of the structures within the EC; realists are anticipating neither a fundamental reform of the Common Agricultural Policy nor a readiness "to simply leave entire branches of intermediate industrial technology to countries willing to accede and the developing countries."[26] Yet, on the other hand, one cannot fail to acknowledge that whenever the Community has found itself under severe political or economic pressure, it has always been able to react with flexibility. Examples of such flexibility include not only the ideas set in motion regarding the reform of the Common Agricultural Policy as a response to U.K. demands and the strained budgetary situation but also the effective concessions that the EC has been willing to grant on the strength of political opportunism to, for example, Yugoslavia (immediately after the death of Tito) and Turkey and Cyprus (to stabilize the eastern Mediterranean). Such a disposition toward matters of principle indicates that the question depends partly on the extent to which the Mediterranean countries can be skillful in playing off their political potential and influence within the Arab world vis-à-vis the Community.

All of this notwithstanding, what shape could the contours of a Mediterranean concept that guarantees compensation for the disadvantages incurred by the Community's southward enlargement and its increasing recourse to protectionism take? Three basic solution approaches can be outlined in response to this question.

1. Upgrading of EC preferences in favor of the Mediterranean countries by shifting, in full, effects to fourth countries;

2. A policy of small solutions consisting of compensating, at least in the most important, sensitive areas, the trade disadvantages incurred by the Mediterranean countries by means of specific individual measures taken in connection with a more intensively pursued policy of financial cooperation; and

3. The initiation of a policy that would seek to effect a new division of labor among the countries situated on the northern and southern shores of the Mediterranean.

Each of these solution approaches has its own specific plausibility, but likewise its own specific problems. The first possibility is uncomplicated, consisting merely of granting the Mediterranean countries greater trade and cooperation privileges at the expense of the ACP states and the remaining developing world. However, by adopting such a proposal, the Community would not only simply procrastinate its existing problems, it would also reduce its scope for political and economic action in the Third World. Viewed in the long term, such a policy would not only endanger the West European industrial states' supply of raw materials, it could provoke retaliatory measures on the part of the remaining countries of the Third World. [27] In the final analysis, this approach would solve no problems, indeed, on the contrary, it would undoubtedly create new ones.

The second possibility, which is at present being examined by an EC commission, could be effected in the form of a package of individual measures. [28] The most important would be an increase in the compensatory payments made for loss of earnings. It is also conceivable that a system to stabilize export earnings could be set up along the lines of the Stabex system introduced in the Lomé Agreement concluded with the ACP states. Other measures would affect the individual production areas concerned: the levying of customs duties on imported, cheap vegetable oils and fats could stimulate the consumption of olive oil (although at the expense of the consumer); quality controls and the limitation of new, additional viticulture could contribute toward reducing the wine surpluses; a reduction in the aids granted for the cultivation of fruit and vegetables in Dutch glass-houses would not only benefit the Community budget but also reduce energy consumption. Such proposals are by no means unrealistic, yet they too harbor difficulties. The introduction of a system to stabilize agricultural earnings analagous to the Stabex system would imply that the agricultural export products of the Mediterranean countries would at least partially be integrated into the Community's agricultural market, a development that would have dangerous consequences for both sides: additional structural surpluses, new financial burdens for the Community, and the consolidation of agricultural production structures that are oriented toward the EC market for the Mediterranean countries. Thus, in the final analysis, this approach would hinder the structural reforms that are required on both sides. At the same time, such a policy of individual measures could hardly contribute toward solving the problems in the industrial sector. However, it is undisputed that further industrial development in the Mediterranean countries and more intensive trade with the Community are important prerequisites for any beneficial cooperation in the future.

Accordingly, the search for a new concept for an EC Mediterranean policy would have to focus on the circumstances required for steps to be taken to redesign the division of labor between the EC and the Mediterranean countries and the form that such steps might take.

This third approach places emphasis on finding a solution that permits further industrial development in the Mediterranean countries,
on the one hand, and takes account of the long-term interests of the
Community, on the other.[29] It implies a structural change in both
the agricultural sector and the industrial sector and on both sides of
the Mediterranean.

This concept would require of the West European industrial
countries an accelerated implementation of structural adaptation
measures in those branches of industry that have become sensitive,
the modernization of the production structure of the three partly industrialized acceding countries, and the long-overdue reform of the
Common Agricultural Policy.[30]

For their part, the Mediterranean countries would have to intensify their endeavors to diversify their production structures and
trade relations, develop their own domestic markets, and in accordance with their respective interests, become integrated in associations providing larger regional markets. The initiation of such
steps would simplify relations between these countries and the EC in
the various problem areas without damaging trade in the respective
competitive areas and without restricting the scope of action for continued industrial, technological, and financial cooperation. Any special needs for protection on either side could be accommodated by
negotiating the terms of an agreed specialization.[31]

TOWARD A MORE SYMMETRICAL INTERDEPENDENCE

It would appear that this proposal would require the almost-
impossible from both sides: from the European industrial countries,
a comprehensive structural adaptation and adjustment process; and
from the Mediterranean countries, a broad reorientation of the structural and economic components of their development models. That
these requirements are not completely unrealistic is already being
documented by initial steps now being taken on both sides that witness
the emergence of a certain consciousness of the problem at hand. In
the north, the Community has been endeavoring for many years—although as yet without any great success—to formulate a common structural adjustment policy; in the south, at least some of the littoral
states are making serious efforts to place both their trade relations
and their domestically oriented development policies on a broader
and more effective basis. This structural change, which in Western
Europe presupposes an adaptation to changed conditions in the international division of labor and, on the other side of the Mediterranean,
a more domestically oriented strategy of economic and social development, admittedly finds itself confronted with three grave obstacles:

(1) the difficulty in these latter countries of implementing a development policy based on the domestic market and greater self-reliance, (2) the problem of subregional cooperation in the southern Mediterranean region, and (3) the fear of the West European industrial nations of "the commercial threats created by Third World manufactures exports."[32] These obstacles are in fact surmountable only on a long-term basis and with the aid of a package of specially designed policies.

The first obstacle is the result of a variety of circumstances that are mainly related to the structural imbalances in the various economic systems concerned and a shortage of political initiatives for structural reform. One major symptom of such imbalances is the fact that, despite favorable conditions, the majority of the Mediterranean countries are still not in a position to satisfy the domestic demand for basic foodstuffs. Tunisia is now further away than at any time previously from fulfilling the objective of self-sufficiency in basic foodstuffs envisaged in its 1977-81 Five-Year Plan. Morocco has had to watch foodstuffs, including 1.2 million tons of grain, account for as much as 30 percent of its total imports in recent years. Algeria, despite a domestically oriented development policy, yet precisely because of its massive industrialization, is still dependent on imported foodstuffs. Egypt, too, uses a considerable part of its foreign-currency earnings for the importation and subsidization of basic foodstuffs. Furthermore, in some of these countries, for example Morocco, there exists an ever-widening gap between the modern export-oriented agricultural sector and the rather ailing domestic-market-oriented agricultural sector that is compounded by a political preference for promoting the export sector. Grave imbalances are also manifest in the industrial processing sector. In some countries, such as Egypt and Morocco, industrial development was neglected for decades. The bottleneck here resides less in the dimensions of the potential domestic market or the scarcity of investment capital than in the absence of social reforms and policies to raise the general standard of living and the production level. Algeria pursued a one-sided, capital-intensive and technology-intensive industrialization policy based on crude oil production that is manifesting its incompatibilities in the form of a growing discrepancy between the group of highly qualified skilled workers and the broad masses of the population. Although the industrial "big push" was driven on, "the dynamism for [their] own initiative expected from it has not materialized";[33] accordingly, up to 50 percent of industrial capacity has not been fully exploited. The continued industrial development of Tunisia, in particular the planned extension of the electrical and mechanical engineering industries, would require close cooperation with the other Maghreb states, because the minimum capacity at the microeconomic level would otherwise exceed the absorption capacity of the Tunisian market at the macroeconomic level.[34]

This leads to the second obstacle, namely, the difficulty of arriving at effective subregional cooperation in the southern Mediterranean region. Bourrinet is unequivocal in his assertion that "all attempts at unification within the Mashrek region and the Maghreb region have failed; this is as true of attempts to create political unity as it is of those to create economic unity."[35] The Arab Economic Union (AEU) founded in 1957 remained a paper proclamation; the Arab Common Market established in Cairo in 1964 has failed on account of the heterogeneity of the economic structures and political interests of the participating countries. Intraregional trade among the member countries of this common market accounts for only 8 percent of their total export trade, while the trade among the Maghreb countries is less still, accounting for only 3 percent.[36] Even the Organization of Petroleum Exporting Countries (OPEC) members' massive capital investment in the other Arab states since 1974 has not been able to prevent a considerable widening of the disparities within the Arab region. Yet, although at the present time it is difficult to perceive any concrete approaches toward intraregional cooperation other than in the petroleum sector, the possibilities of coordinating production structures and harmonizing economic interests in the region are by no means fully exhausted. Better mobilization of local financial and human resources and cooperation in industrial development could reformulate and consolidate relations not only within the southern and eastern Mediterranean region but also between this area and Western Europe. Bourrinet is undoubtedly correct when he ascertains that "vertical cooperation (between countries with different levels of development) and horizontal cooperation (at subregional level between countries with comparable structures) are not contradictory concepts: they are mutually determinant and mutually reinforcing. Progress in European cohesiveness and in cooperation with the Arab states seems to be the precondition for organizing a European–Arab development area."[37]

Finally, the third obstacle is the lethargy of some industrial production structures of Western Europe or, more precisely, in those areas of production that are no longer competitive on the world market. Theoretically, a whole range of products could be relocated away from Europe to the developing countries, thereby relieving the West European countries of protectionist constraints and endless subsidies and at the same time activating the economic structures of countries in the Third World. Vaitsos lists three such product groups: (1) unwrought nonferrous metals, (2) primary and secondary petrochemical products, and (3) labor-intensive manufacturing in the areas of consumer final products and component subcontracting.[38] However, such a relocation of industrial activities encounters difficulties that are attributable partly to political and social considerations (loss of jobs) and partly to the monopolistic control of markets and technologies as exercized by the large transnational companies. The erosion of the situational advantages of the established industrial nations would not necessarily imply a weakening of their competence in the fields of marketing, finance, and technology.[39] But it is precisely the fact that large companies have these competencies and high innovative capacities that implies that the industrial countries are predestined to concentrate primarily on innovative, high-technology production areas with promising prospects for the future and not hinder the development of their future market outlets by mummifying a number of obsolete fields in their own production systems. What is implied is not the transplantation of entire production areas to developing countries, but a more effective structuring of the division of labor between the areas north and south of the Mediterranean, with the aid of a forceful structural adjustment policy and an intensification of industrial and technological cooperation.

The opening-up of the domestic markets of the southern Mediterranean countries, a greater measure of self-sufficiency in the agricultural sector, progressive industrialization, and intensive cooperation among these countries would be advantageous for countries on both sides of the Mediterranean region: for the Mediterranean countries, which could thereby achieve a greater degree of collective self-reliance without renouncing the advantages of their relations with Western Europe; and for the European Community, which would thereby not only gain expanding market outlets, in particular for investment goods and other high-value industrial exports, but would also find itself able to examine a broader range of options for its future cooperation policy. A reorientation of the EC Mediterranean policy in this direction could release forces on both sides that, as Friedrich List pointed out, would guarantee not only possession of that created but also replacement of that which is lost.

NOTES

1. J. Galtung, Self-reliance: Concept, Practice and Rationale, International Peace Research Association, Paper no. 35 (Oslo: University of Oslo).

2. E. Oteiza and F. Sercovich, "Collective Self-reliance: Selected Issues," International Social Science Journal 28, no. 4 (1976): 665.

3. C. V. Vaitsos, "From a Colonial Past to Asymmetrical Interdependences: The Role of Europe in North-South Relations," in Europe's Role in World Development, Papers of the EADI Conference in Milan, September 1978 (Milan: EADI, 1981), p. 40.

4. Quoted in European Economic Community Commission, "Die europäische Gemeinschaft und die arabische Welt," EEC Paper 169, 1978, p. 4.

5. Concerning European interests in the Mediterranean area, see among others, W. Hager, "Das Mittelmeer—'Mare Nostrum' Europas?" in M. Kohnstamm and W. Hager, Zivilmacht Europa—Supermacht oder Partner? (Frankfurt, 1973); R. Regul, ed., Die Europäischen Gemeinschaften und die Mittelmeerländer (Baden-Baden, 1977); A. Shlaim, "The Community and the Mediterranean Basin," in Europe and the World, ed. K. J. Twitchett (London, 1976); G. van Well, "Mittelmeerpolitik und Süderweiterung," Integration, no. 1 (1978); A. J. Kononassis, The European Economic Community in the Mediterranean: Developments and Prospects on a Mediterranean Policy (Athens, Greece, 1976); and W. Wallace and I. Herreman, eds., A Community of Twelve? The Impact of Further Enlargement on the European Communities (Bruges, 1978).

6. Compare European Economic Community Commission, Sechster Gesamtbericht über die Tätigkeit der Gemeinschaft 1972 (Brussels, 1973); and idem, "Globalansatz für die Mittelmeerpolitik der Europäischen Gemeinschaft—Vorschläge der Kommission vom 22.9.1972," Europa-Archiv, no. 21, 1972.

7. B. Claus, K. Esser, C. Heimpel, and W. Hummen, Zur Erweiterung der Europäischen Gemeinschaft in Südeuropa (Berlin: German Development Institute, 1977); K. Esser, H. Gsänger, C. Heimpel, W. Hillebrand, and W. Hummen, European Community and Acceding Countries of Southern Europe (Berlin: German Development Institute, 1979); S. A. Musto, Spanien und die Europäische Gemeinschaft (Bonn, 1977); K. Esser, Portugal—Industrie und Industriepolitik vor dem Beitritt zur Europäischen Gemeinschaft (Berlin: German Development Institute, 1977); W. Hummen, Greek Industry in the European Community: Prospects and Problems (Berlin: German Development Institute, 1977); G. Ashoff, "The Southward Enlargement of the EC," Intereconomics, November-December 1980, pp. 299-307; and S. A. Musto, "Die Süderweiterung der Europäischen Gemeinschaft," Kyklos 34, no. 2 (1981): 242-73.

8. P. Mishalani, A. Robert, C. Stevens, and A. Weston, "The Pyramid of Privilege," in EEC and the Third World: A Survey, ed. C. Stevens (London: Hodder, 1981), p. 60.

9. For further analysis, see E. Guth and H. O. Aeikens, Konsequenzen der Süderweiterung für die Mittelmeerpolitik und die AKP-Politik der Gemeinschaft, European Economic Community Commission, Directorate General of Information (Brussels: EEC, 1980); and R. Taylor, Auswirkungen der zweiten Erweiterung der Europäischen Gemeinschaft auf die Länder des südlichen Mittelmeerraumes, European Economic Community Commission, Directorate General of Information (Brussels: EEC, 1980).

10. Data and further analysis in Guth and Aeikens, Konsequenzen der Süderweiterung; Taylor, Auswirkungen der zweiten Erweiterung; and Mishalani, Robert, Stevens, and Weston, "The Pyramid."

11. Analysis of data in Mishalani, Robert, Stevens, and Weston, "The Pyramid."

12. Taylor, Auswirkungen der zweiten Erweiterung, p. 6.

13. United Nations, UN Yearbook of International Trade Statistics, 1977 (New York, 1978), 1:315, 527, 558, 588, 635, 663, 918, 1002.

14. S. K. Chatterjee, "The impact of Enlargement of the EEC upon the Mediterranean Basis Policy through Trade Agreements" (Paper presented to the Conference on the EEC and the Mediterranean Countries, Istituto Universitario Orientale, Naples, March 1980).

15. R. Albioni, "The Maghreb's Development and Its Relations with the EEC," in The EEC and the Mediterranean Countries, ed. A. Shlaim and G. Yannopoulos (Cambridge: At the University Press, 1976), pp. 179-98.

16. Compare Issam El-Zaim, "La politique Mediterranéenne de la Communauté economique européeanne" (Paper presented to the Conference on the EEC and the Mediterranean Countries, Istituto Universitario Orientale, Naples, March 1980).

17. Compare H. Gsäanger, "The Southward Enlargement of the European Community and the Future of the Common Mediterranean Policy," in European Community and Acceding Countries of Southern Europe, ed. K. Esser, H. Gsänger, C. Heimpel, W. Hillebrand, W. Hummen, and J. Wiemann (Berlin: German Development Institute, 1979). See also C. Heimpel, W. Hummen, and S. A. Musto, Studie zur Erweiterung der Europäischen Gemeinschaft um Griechenland, Spanien und Portugal (Berlin: German Development Institute, 1976).

18. In addition to the studies already quoted in notes 4, 6, 11, and 14, see also H. Kramer, Konsequenzen der Süderweiterung für die Stellung der EG im Nord-Süd-Konflikt (Ebenhausen: Stiftung Wissenschaft und Politik, 1980); R. Morawitz, "Die Auswirkungen der Süderweiterung der Europäischen Gemeinschaft auf das Mittelmeerbecken," Europa-Archiv, no. 6 (1980); and H. v.d. Groeben and H. Müller, eds., Die Erweiterung der Europäischen Gemeinschaft durch Beitritt der Länder Griechenland, Spanien und Portugal (Baden-Baden, 1979).

19. Compare Taylor, Auswirkungen der zweiten Erweiterung, p. 10.

20. See the data of the European Economic Community Commission reproduced in ibid., pp. 26-27, tables 4 and 7. See also the data analysis in Mishalani, Robert, Stevens, and Weston, "The Pyramid," p. 78, table 4.6.

21. Gsänger, "The Southward Enlargement," p. 151.

22. Taylor, Auswirkungen der zweiten Erweiterung, p. 12.

23. Data presented by Mishalani, Robert, Stevens, and Weston, "The Pyramid," p. 81.

24. See Kramer, Konsequenzen der Süderweiterung, p. 51.

25. Gsänger, "The Southward Enlargement," p. 153.

26. Taylor, Auswirkungen der zweiten Erweiterung, p. 16.

27. See Gsänger, "The Southward Enlargement," p. 157.

28. Compare European Economic Community Commission, Wirtschaftliche und sektorielle Aspekte: Analysen der Kommission als Ergänzung zu den Betrachtungen über die Probleme der Erweiterung (Brussels: EEC, 1978); and idem, Verhandlungen über den Beitritt Spaniens—Vorschlag betreffend den landwirtschaftlichen Bereich (Brussels: EEC, 1980).

29. It is impossible to present here detailed proposals related to a new division of labor between the EC and the Mediterranean countries. Some concepts and proposals concerning the division of labor between the EC Nine and the three acceding countries have been elaborated on by G. Ashoff, K. Esser, A. Eussner, and W. Hummen, Überlegungen zur portugiesischen Industriepolitik angesichts des Beitritts (Berlin: German Development Institute, 1980); see also S. A. Musto, "Regional Disparities in an Enlarged EEC and the Problem of an Intra-Community Division of Labour" (Paper presented to the Conference of Integration and Unequal Development, Madrid, November 1979).

30. Some summarized proposals for the reform of the Common Agricultural Policy in C. Heimpel, "Die Struktur der mediterranen Landwirtschaft als agrarpolitisches Problem der erweiterten Gemeinschaft—Ansätze für eine Reform der Gemeinsamen Agrarpolitik in einem Europa der 12," in Zur Erweiterung, ed. Claus, Esser, Heimpel, and Hummen. See also J. March, The Impact of Enlargement on the Common Agricultural Policy (Bruges, 1978).

31. Some remarks on the problem of agreed specialization can be found in Gsänger, "The Southward Enlargement," p. 157; and Musto, "Regional Disparities."

32. Vaitsos, "From a Colonial Past," p. 65.

33. Handelsblatt, April 29, 1981.

34. C. Eikenberg, "Rück- und Ausblick auf die soziale und wirtschaftliche Entwicklung Tunesiens," Technisch-wirtschaftlicher Dienst der Ländervereine, November 1980.

35. J. Bourrinet, "Interdependance et developpement Euro-Arabes," in Le dialogue Euro-Arabe, ed. J. Bourrinet (Paris, 1979), p. 270.

36. Ibid., p. 271.

37. Ibid., p. 276.

38. Vaitsos, "The Southward Expansion," p. 65.

39. Compare S. Borner, "Die Internationalisierung der Industrie," Kyklos 34 (1981): 23.

10

POLITICAL CHANGES IN THE EUROPEAN MEDITERRANEAN ARENA: AN OVERVIEW OF CONSIDERATIONS FOR POLICY MAKERS

SUSAN L. WOODWARD

A striking characteristic in the 1980s of the European countries that share Mediterranean borders is the similarity in their economic difficulties and in the political responses to those difficulties despite the vast cultural and governmental differences among them. This common fate, though it remains for the most part unrecognized, can be traced to major policy decisions made independently in each country except Albania during the period between 1947 and 1960 (1953 appears particularly important). [1] In each case, political leaders aimed at the same goals—economic development and national independence (on the condition, of course, that their own power be reinforced). More significantly, they chose the same strategy. As a result, these policies came to have fundamentally the same consequences for domestic politics and for the lives of the people, whether in Portugal, Spain, Greece, Yugoslavia, or Turkey.

These similarities did not arise only in the late 1950s. Their origins can be traced in particular to the nineteenth century, and one

I wish to acknowledge here the invaluable assistance and patience of Raghbendra Jha, in writing this, and the generous financial support from Dean Francis Oakley of Williams College to attend the Conference on Stability and Change in the Contemporary Mediterranean World.

175

could begin the story as easily perhaps in the 1920s or 1930s. By the
late 1940s, however, these earlier developments had become only
conditions that, although in each case they revealed the failure of interwar policy and the need to begin anew in the postwar era, did not
predetermine the road to be taken.

By the standards in the rest of Europe, per capita income was
low. Furthermore, living conditions were even worse for the majority than such income averages would suggest, because the distribution
of income was highly unequal and because governments neglected investment in public services under the heavy military and debt service
obligations of interwar budgets. Industrialization had begun in all
cases before World War II—mining, shipbuilding, textiles, and commerce had survived the changing fortunes of several centuries, and
some regions of Spain could be called industrial by the 1880s. Every
country was ruled by groups committed to economic development and
to state-assisted industrialization. The pace of industrial development and to state-assisted industrialization. The pace of industrial
development still suffered, however, from the conditions of the previous decades: extensive foreign investment during the 1920s and
1930s in extractive industries, finance that expatriated both resources
and profits, stagnation during world depression and the war, massive
destruction during the war (in the case of Greece and Yugoslavia),
and the understandable, but hastily and thus crudely applied, defensive
protectionism of state policy during the depression.

The social structure of these European Mediterranean countries
combined elements of the new industrial forces and of the former
economy, based on agriculture and some commerce. The majority
of the population still lived in the countryside, either working in traditional subsistence agriculture or looking for opportunities to emigrate.
The industrial working class was small, slightly more than 20 percent of the major force in Spain, although it was regionally concentrated, and under 15 percent elsewhere. Except in Yugoslavia, a
few domestic industrialists, landlords, shipowners, or financiers
controlled quite extraordinary wealth in relation to the rest of the
population. In contrast, a relatively large group of public employees,
produced by the expansion of the state bureaucracy to deal with some
of the interwar unemployment, had middle-class status but meager
incomes.

Although the government was relatively large, acting as employer
and coercive presence and having firm alliances with the army and a
significant portion of the leaders in the economy—particularly in the
state-assisted industrial and financial sectors—and with landlords
where landholdings were still concentrated, as in Portugal, it was
weak in its legitimate hold over the population. In each case, military
rule and police suppression of significant opposition groups were a

part of the immediate past, if not also the present. Government alternatives seemed to be either authoritarian rule with labor-repres-
sive policies or parliamentary regimes that were highly unstable as
a result of wrangling at the top—within parties and coalitions, among
parties, among king, army, and parliament, and so forth. Revolution had changed these conditions in Yugoslavia by 1945. In Greece,
however, a revolution had been prevented by British and U.S. intervention, and they continued to undermine national sovereignty. In
the remaining three states, prewar political forces were unaffected
by the war.

This description neither accounts for these conditions, nor reveals the variety within the area. One could cite the differences
among regions within Spain and Yugoslavia, for example, or the contrast in industrial development and living conditions in Portugal and
Turkey, on the one hand, and Spain and parts of Greece on the other.
There is further disparity in the separate fates of civil war in each
country, in the imperial past of Spain and Turkey, in the continuing
empire of Portugal, and in the colonial past of Yugoslavia and Greece.
From the perspective of national goals and governmental policies in
pursuit of those goals, however, the conditions defining alternatives
in the 1950s were remarkably similar. Fiercely nationalist indus-
trializers with an eye toward the United States and northern Europe
as models and potential allies (in this Yugoslavia was again an exception), these ruling groups all chose to abandon their protectionist
(even autarkic in some cases) mode of industrial development and
to seek external sources of finance during the 1950s. The timing and
perhaps the best explanation for this choice probably has as much to
do with external events as with internal politics—the Truman Doctrine
in 1947, the Cominform Resolution in 1948, the recognition of Franco's Spain by the United States in 1953, and the recovery in Europe.
Finally, although the policies were initiated at different moments—
Yugoslavia and Greece began somewhat earlier, Turkey somewhat
later than the others—by the end of the 1950s all were actively promoting an export-led strategy for economic development.

According to this strategy, economic development at home was
limited by the size of the local market, the lack of capital equipment,
and an insufficient domestic surplus for industrial investment. But
by turning outward to a world market, firms already producing capital goods in more developed countries, and foreign sources of capital, these barriers to economic growth at home will disappear. Revenue from export sales of those goods with which the country has a
comparative advantage in the world market will be invested in the domestic economy both by profit-seeking entrepreneurs producing goods
for local demand and by the state guiding resources into infrastructural development and strategic industries. The pressures of inter-

national competition will make domestic firms more productive. The growth of the economy and specific inducements such as cheap labor, an unexplored market, and governmental fiscal concessions will encourage foreign investment. In the early stages, some foreign borrowing may be necessary, but a growing economy will have no difficulty paying those obligations. At first, some foreign-exchange revenue will be diverted to the import of capital goods for new industries and processed goods for the growing internal market, but soon, again with state guidance and subsidy, local industries will respond to this market by producing such goods themselves. This substitution of domestic production for formerly imported goods will contribute further to growth and keep the country's balance of payments sound. As the economy grows, per capita income will also increase, providing the increased demand for products that, through multiplier effects of various sorts, will make domestic production self-sustaining. Furthermore, all of this can be accomplished without disturbing the internal structure of economic and political power.

It is not surprising that this choice was made. The political risk of alternatives was high. It is difficult to imagine who might even have proposed significant redistribution of income and wealth to create larger internal markets, for example, although it is easy to predict resistance to it within the ruling coalitions. [2] The cold war foreign policies of creditor and major trading nations would have scarcely been sanguine about serious planning, or even about trade to the East. Even where these policies had actually been adopted, after the necessary political revolution as in Yugoslavia, their abandonment in the face of external aid from Western sources suggests that the pressures against this route in the international arena must have been powerful indeed.

The stages of this integration into the world economy have proceeded slowly. There were some political hindrances: political disagreements over the policy itself or the form of integration into world trade; governmental instability preventing consistent economic policy of any kind; the usual problems of policy in a market economy where decision makers are autonomous even when the state has an accepted tradition of intervention and assistance; the colonial policy (in Portugal); and the time necessary to negotiate agreements with trading partners, not to speak of the political costs of some of the terms of such agreements. As time went on, however, economic difficulties became more significant.

The number of false assumptions and unlikely conditions built into this strategy are so great as to make the probability of its success extremely small and dependent, furthermore, on special conditions in the international economy, over which there is no control. One assumption is that internal growth will be spurred by export earnings. The structure of the world economy, and in particular the monopoly nature of world prices, makes it very difficult in fact for these countries to earn more from their exports than they pay for imports. Those processed goods that they are likely to be able to produce for export meet stiff competition from firms in other countries, either because the goods are almost indistinguishable, because they come from countries where costs are lower owing mainly to even lower wage levels, or because they come from more developed countries where start-up costs have long been paid and economies of scale, vertical integration of production, and near-monopoly control of merchandising give advantages on both costs and pricing. Also, these goods exchange at extremely unfavorable terms with the manufactured capital goods imported for new domestic industry.

To adjust to this terms-of-trade effect, balance-of-payments theory argues that either the costs of export goods should be reduced to increase their competitiveness—in other words, wages should be reduced—or, a switch should take place to exports for which there are more willing buyers and for which a country may have more of a comparative advantage, such as raw materials—minerals and agricultural products—necessary to industry in the developed countries. These latter goods trade on even less favorable terms, however, and their price is highly vulnerable to economic conditions in developed countries. Revenue from such exports, therefore, is highly variable and subject to sometimes sharp, often unpredictable, decrease during world recession. Simply to maintain some balance between import expenditures and export earnings, it becomes necessary to sell ever more for less, to reorient an even larger part of production to export needs, to strip the country of its natural wealth (thus jeopardizing its own industrial future), and to shift the agricultural sector increasingly toward cash crops and away from food production for the local population. In addition, since imports must be paid for in acceptable foreign currency, activities that directly earn foreign exchange must be encouraged—trade with convertible-currency countries of North America and Western Europe and, increasingly, the "invisible" exports of human services (tourism and the export of labor) in construction crews at sites in Tunisia, Iraq, and Saudi Arabia, and especially as "guestworkers" in northern Europe.

To enter such trade at all, agreements must be made. To take one example, the United States placed great pressure on the European Community (EC) to make all trade agreements between the EC and other countries (especially Mediterranean countries) consistent with the provisions of the General Agreement on Tariffs and Trade. As the United Nations Conference on Trade and Development (UNCTAD) has persistently demonstrated, these terms put all countries outside the Group of Ten at a distinct disadvantage. Since the terms require minimizing restrictions on imports, agricultural protection, and other trade controls, they also eliminate one of the main measures available to governments, according to both balance-of-payments theory and the export-led growth strategy, to keep the balance of payments in line and to guide domestic industry toward self-sustaining development.

Finally, when the world economy is in crisis, these are the countries that suffer the most. Their trade is dependent on conditions in the economies of their trading partners, and they have the fewest alternatives with which to respond, given the absence of self-sufficient domestic development. When recession appears to threaten industry in the developed countries, protectionist tariffs interrupt the free trade to which they have agreed (at a price), and migrant workers are laid off and sent home. Export earnings then fall, as does the growth on which it depends, and unemployment climbs. Such crises, furthermore, are not unusual, as the southern European experience shows. In the postwar period, integration into world trade during the late 1950s and early 1960s encountered the developed economies in boom times, encouraging the freer movement of labor, capital, and goods by a range of liberalizing treaties. These apparently advantageous conditions were interrupted by recession in 1961, again in 1965-67, and, when the extent of integration made these countries even more vulnerable, by the deep recession of the 1970s. One might even say that protectionist policies in the developed countries are a normal part of the environment, except that they vary enough to be unpredictable and are completely outside the control of the southern countries.

A second assumption of this strategy for development is that state policy and economic growth in these countries will be able to attract foreign investment that assists domestic development. Unlike the massive foreign investment of the nineteenth century that built the industrial infrastructure of the United States, Canada, Australia, New Zealand, and South Africa, foreign investment has not amounted to very much in the mid-twentieth century.[3] To the extent that is available, foreign investment comes increasingly only from multinational companies aiming to reduce wage costs for certain stages of their production, thus bypassing southern Europe for poorer countries, or to produce and sell goods whose growth potential, and thus profit margin, in richer countries is declining. For both purposes, as in the past, foreign investment continues to avoid southern Europe as it travels among countries within the Group of Ten with large markets or between the headquarters of multinational companies based in the most-developed countries and their subsidiaries in parts of the Third World. The small portion that reaches southern Europe in search of markets is very costly to the local economy in terms of licensing and management payments for the multinational company, profit expatriation, and lost revenues from the tax exemptions and subsidies granted by governments to attract the foreign investment. This foreign-controlled industry also often competes with the very goods these countries are trying to encourage domestically, and it does so at an advantage over domestic firms that do not have the same organizational, technological, capital, and market resources, or even the privileges offered to foreign capital by their own governments. Thus, local investment is driven out of that sector into less-dynamic sectors or even less-risky areas, such as real estate and tourism. Foreign capital also tends to create industries that are highly dependent on imports—machinery, petroleum, personnel, and technology—thus increasing the trade imbalance in a sector over which the government has already conceded much control. Finally, where such industry produces for the local market, it reaches the aggregate demand limits soon, and the initial investment is not increased as the multinational company pulls out and searches for another market. In the meantime, less easily abandoned structural changes have taken place for the recipient country.[4]

A third assumption is that domestic investment will grow in response to the larger aggregate demand created by increased employment and growth in the national income. There are two difficulties here. The first is that this assumption says nothing about how that national income will be distributed. Yet we already know that the distribution of income in these countries is highly unequal, both among persons and households and among regions. Effective consumer demand is present for luxuries, but not for basic consumer goods, and since the marginal propensity to consume declines as incomes rise, demand reaches its limit fairly quickly. To the extent that local industry does take hold, either from domestic or foreign investment, it will soon encounter insufficient demand owing to the low incomes of the majority of the population. Overcapacity will bring on reduced investment, some of which will be directed into nonproductive forms of investment or even go abroad, as in Spain, in search of larger markets, and the increased unemployment that follows will reduce demand further. Poorer regions will be unlikely to attract the initial industry.

The second difficulty is that the progressive effects of the export strategy on employment, inflation, wages, and income distribution reduce and in some cases even negate the contribution that is made by growing wage-sector employment and national income: For example, the export of labor—northern European employers recruit the more highly skilled workers already in cities who are then replaced by less-skilled workers from the countryside, depressing the skill levels of industrial workers within domestic industry, probably lowering productivity, and certainly reducing the aggregate growth in wages that would expand the local market.[5] As those migrants return, their tastes for the consumption goods of northern Europe and their higher incomes become translated into a higher demand for imports and even luxury goods, not for local mass products. As agriculture is transformed from subsistence to cash crops for exports, the concentration of landownership and the capital-intensive techniques of agribusiness both reduce employment and increase income inequality in the countryside, again reducing effective demand (in 1971, only 4.8 percent of the active agricultural population in Greece—40.2 percent of the labor force—were wage earners).[6] The diversion of agriculture away from domestic food production also leads to increasing imports of food (by 1970, still-agricultural Portugal imported half of its food, and that dependence increased throughout the 1970s). The burden of policies designed to adjust the balance of payments and to stabilize currency is always borne by labor, as real wages, food subsidies, and welfare benefits are cut, inflation raises the cost of food and other necessities, and unemployment rises, once again reducing the redistribution of income necessary for growth. The consequences of world recession and crisis on the domestic economy are felt in aggravated terms because of the fragility of their domestic industrial structures and the few means available to these countries to counteract the effects, limits that only magnify the effects on unemployment and inflation and, thus, on demand.

The consequences in fact of the export-led strategy have been chronic trade imbalances and persistent balance-of-payments deficits. Instead of reducing imports and expanding exports, the import needs of these countries rose faster than export production. Insofar as exports did grow, export revenue declined as a result of declining terms of trade and international competition. A deficit in itself might not be a problem; indeed, governments might try to finance industrial development at home with such deficits. A far more severe problem than an accounts deficit is the distortion of domestic production to the needs of export. Yet prevailing powers in the international economy will pressure for reduction of these deficits. If they wish to continue trading, they must prove themselves tradeworthy. As long as the export-led strategy is retained, balance-of-payments adjustment becomes the precondition for further trade, aid, foreign investment, and in effect, for development. The result is that state economic and foreign policy increasingly becomes preoccupied with reducing balance-of-payments deficits rather than with development.

Balance-of-payments theory prescribes that if a deficit cannot be easily financed by a country's foreign reserves or by outright grants from international financial institutions (such as the United Kingdom occasionally receives), and especially if the deficit is not temporary, then policy to reduce the deficit, whatever its cause, must effect a reallocation of resources either to reduce foreign expenditures or to increase foreign-revenue-producing output. These policy measures fall into two categories: (1) deflationary fiscal and monetary policies to reduce expenditures by reducing aggregate demand (for example, tight monetary policy to restrict the money supply and credit, reductions in government expenditures, and incomes policies to restrict wage and price increases) and (2) policies to reduce purchases of foreign goods and increase sales of domestically produced goods (for example, selective trade controls such as tariffs, import restrictions, and export subsidies, restrictions on the outflow of capital and inducements to its inflow, rations of foreign exchange, and either variations in the foreign-exchange rate or devaluation). [7]

The effectiveness of any particular measure in reducing a country's balance-of-payments deficit depends on many factors: the specific cause of the deficit itself; the elasticity of demand in the international market for a country's exports; the levels of inflation and employment in the domestic economy; the "assignment problem," that is, the difficulty of coordinating policies that are, in a capitalist economy, the jurisdiction of separate economic decision makers; the ability of domestic output to increase as needed and the limits that can be placed on the inflationary side effects of growth; the internal price ratio for exports; and the policies of other countries and of international trade and financial institutions. A mixture of policies is usually recommended, but the probability of success, given the lack of control over so many necessary influences on the outcome, is not particularly high.

Perhaps more important for the countries of southern Europe are the consequences these policies must have for their domestic economies. If at all successful, policies reducing trade deficits must effect a permanent decrease in the goods a country imports in relation to the goods it exports, affecting both productive inputs in industry and living standards and redirecting the country's orientation away from internal development toward export-oriented economic activities. Such an adjustment takes time, however. In the meantime, invisible foreign-exchange earnings become relied upon to make up the difference. Resources crucial to internal development are then lost in massive investment, to transform the country's most beautiful areas for the luxury leisure of foreigners, and in the temporary migration of millions of the youngest and most-skilled workers to employment outside the country.

All these policies lower rather than increase the growth of the country's output. Policies that are then implemented in order to switch aggregate demand to domestic goods will tend to be inflationary—the opposite of what is needed to reduce the deficit and stimulate productive investment. As the rate of inflation rises, it becomes more variable, thus injecting more uncertainty into an economy that can little afford it. Business reacts to the uncertainty by further reducing productive investment, productive capacity does not grow, and unemployment, particularly of the young, rises. When deflationary policies are added to this situation, as export-led strategy prescribes, unemployment is further increased since wages and prices tend to be downwardly rigid. Living standards also decline, and aggregate demand is further reduced. Contrary to solving the balance-of-payments difficulties, it is obvious that these policies have a tendency to exacerbate their causes and at the same time to depress further the conditions necessary for domestic development.

Since these policies reduce growth, may reduce inflation in the short run but exacerbate it wildly soon after, depress the living standards and purchasing power of the majority of the population, and increase unemployment, yet do not relieve the deficit, further steps to reduce the balance of payments become necessary almost at once. These decisions are rarely made without a political battle, since the consequences for most citizens are the opposite of those promised, but the argument is always made successfully that too many steps in the direction of international trade and alliances have been taken to reverse them. Thus, the government begins to seek loans to finance the deficit.

The terms for such credit are either tied to the purchase of goods produced elsewhere or are more of the same stabilization policies. They not only do not reduce dependence on imports, but the government also must commit an ever-increasing portion of its budget to guarantee repayment; that is, an ever-larger debt service becomes a portion of revenues that more appropriately should be committed to investment in development projects such as roads, schools, and irrigation facilities, initial investment in major capital industries, and stimulation of employment-generating activities. Since the loans do not in themselves assist domestic growth and structural changes in an economy vulnerable to all the changes in the international economy take longer than the period over which most credit is available, the government finds itself trying to renegotiate the loan payment schedule. At each step, the fundamental goal of national political and economic sovereignty is compromised further as external sources of financing dictate domestic-policy changes in exchange for loans. Nonetheless, there usually comes a point when external sources of finance are no longer available without the assurance of a credit rating that comes from an International Monetary Fund (IMF) loan.

The terms of the IMF loan are based on the same logic as balance-of-payments adjustment theory, but they tend to be more nar-

rowly addressed to stabilization alone, resulting in policies that are
even more counterdevelopmental than those designed for adjustment.
The IMF usually begins by insisting on devaluation of the domestic
currency, automatically making all exports cheaper and imports more
expensive, ostensibly to make a country's goods more competitive in
the world economy and to force reductions in imports. The immediate consequence is dramatic domestic inflation, withdrawal of business from investment, reduced living standards, and soon more unemployment. To counter this inflation, the IMF insists on orthodox
monetary policies—the contraction of credit through tight monetary
policy, restriction of wage increases, and reduction of domestic
budgetary deficits by raising taxes, cutting public expenditures, reducing the share of public enterprise in the economy or increasing
its rates, and removing the public subsidies that are designed to assist domestic industry and to protect the population from the effects
of inflation on basic, life-sustaining commodities.

Two examples of the inappropriateness of these policies and
their political consequences come from Yugoslavia and Portugal. In
the Yugoslav case, balance-of-payments deficits and dangerously de-
pleted foreign-exchange reserves were the consequence of an external
shock—the economic blockade by Cominform countries in 1948-49—
not of domestic prodigality. The decision to seek Western aid within
four years of the socialist revolution was not made easily, but once
taken, it led to a systematic decentralization of the Yugoslav economy,
abandonment of development planning in favor of the market, and integration into the world economy in response to the policies dictated
by the IMF in exchange for credit over the next 20 years (1951, 1960,
and 1965 are particularly important). This dismantling of socialism
has been accompanied by persistent balance-of-payments deficits,
high unemployment, high inflation, and increasing internal inequality.
In Portugal, an export-led growth strategy, beginning formally with
membership in the European Free Trade Association (EFTA) in 1960,
led to a fourfold increase in foreign-trade deficits between 1964 and
1973, but large foreign-currency reserves from the African colonies
were able to make up the difference until 1974. These deficits were
compounded between 1974 and 1976, however, by political retaliation
against the revolution: businessmen abandoned their failing firms and
lobbied with former foreign customers and creditors to withdraw association from the now worker-run enterprises; trading partners un-
derinvoiced exports to the estimated amount of U.S.$120 million,
canceled orders, and shortened credit repayment periods; many
larger employers fled; transnational companies closed their factories
in Portugal; and a foreign media campaign reduced tourism and worker
remittances. Nonetheless, the government responded to the resulting
jump in the balance-of-payments deficits with orthodox deflationary

policies. By 1978 the deficit had worsened and aid from the United States and Germany was available only if IMF approval was obtained. The IMF program aroused major opposition within the ruling Socialist party, but it was implemented nonetheless. Within a year—by 1979 gross national product was declining, several critical industries had stagnated, inflation was at 25 percent, and the real-wage gains of the 1975 revolutionary phase had been reversed—real wages were 30 percent lower in industry than in 1975, and a transfer of income from urban and rural workers to export-import trading groups and to certain sections of the prerevolutionary ruling class had occurred. In December 1979, the Socialist party was defeated in national elections, and less than six years after the revolution, government economic policy was growing progressively more conservative. [8]

Chronic trade imbalances, growing balance-of-payments deficits, increasing inflation and unemployment, and declining growth and productive investment have characterized the economies of Portugal, Spain, Greece, Yugoslavia, and Turkey since the early 1960s. Their vulnerability to the external sector is vividly demonstrated by the sharp deterioration of their domestic economies during the world recession of 1974-76 and, as Tables 10.1-10.6 illustrate, increasingly since then. None are in as dire straits at the end of 1981 as Turkey, which has been declared no longer a good place for foreign investment by the most influential investment advisory council and which lost untold millions in a run on several major banks in December 1981 by their directors, who then fled abroad.

Yet, the conditions underlying the mutual hardships of these countries show no sign of reversal. Inflation rates average 20 percent to 30 percent (although they have jumped as high as 100 percent, as in Turkey in 1980, and 40 percent, as in Portugal in 1974 and 1976). Official unemployment rates (the percent of the active labor force registered at unemployment bureaus) reached 16 percent to 20 percent by 1981, but unofficial estimates put this number at 25 percent and higher everywhere. Furthermore, this reckoning does not take into account workers employed abroad or the increasing shift to part-time work among the employed. Growth in national production between 1979 and 1980 was 0.6 percent in Greece, 1.3 percent in Turkey, 1.7 percent in Spain, 3.0 percent in Yugoslavia, and 4.7 percent in Portugal. Portugal resumed negotiations with the IMF at the end of 1981, and the Yugoslav financial consortium to attract European capital was doing poorly.

The terms of trade in the international market for these countries' exports declined further in 1981, and workers' remittances, which are the largest source of foreign exchange (usually paying more than half the balance-of-payments deficit), suffered a sharp setback for at least three years following 1974. [9] The hopes of many that the

TABLE 10.1

Balance of Trade, 1969-80
(in millions of SDRs)

Year	Greece	Portugal	Spain	Turkey	Yugoslavia
1969	-904	-352	-1,871	-189	-660
1970	-1,093	-491	-1,874	-262	-1,195
1971	-1,320	-639	-1,599	-378	-1,436
1972	-1,480	-683	-2,133	-481	-917
1973	-2,367	-763	-2,939	-470	-1,391
1974	-1,955	-1,659	-5,788	-1,522	-2,592
1975	-1,944	-1,319	-6,079	-2,334	-2,458
1976	-2,333	-1,880	-6,331	-2,256	-1,614
1977	-2,707	-1,711	-5,314	-2,854	-3,243
1978	-2,796	-1,632	-3,287	-1,398	-3,006
1979	-3,881	-2,037	-4,409	-1,687	-4,691
1980	-4,274	-3,096	-8,914	-2,945	n.a.

n.a. = not available

Note: SDRs are the official unit of account for the IMF; the value of a Special Drawing Right is based on a basket of currencies. Its value in 1980 was about U.S. $1.3.

Source: United Nations, International Monetary Fund, Balance of Payments Yearbook/Statistics (Washington, D.C.: IMF, 1981).

TABLE 10.2

Balance of Payments: Current Account, 1969-80
(in millions of SDRs)

Year	Greece	Portugal	Spain	Turkey	Yugoslavia
1969	-349	131	-392	-174	-64
1970	-405	53	79	-93	-348
1971	-341	141	856	-49	-358
1972	-367	285	527	98	382
1973	-998	286	461	497	407
1974	-950	-690	-2,616	-527	-791
1975	-722	-624	-3,513	-1,522	-515
1976	-802	-1,109	-3,710	-1,701	156
1977	-923	-822	-2,114	-2,765	-1,152
1978	-774	-379	1,398	-975	-1,022
1979	-1,461	-44	1,154	-778	-2,834
1980	-1,704	-827	-3,561	-2,122	n.a.

n.a. = not available

Note: SDRs are the official unit of account for the IMF; the value of a Special Drawing Right is based on a basket of currencies. Its value in 1980 was about U.S. $1.3.

Source: United Nations, International Monetary Fund, Balance of Payments Yearbook/Statistics (Washington, D.C.: IMF, 1981).

TABLE 10.3

Outstanding External Public Debt, 1970-79, Disbursed Only
(in millions of U.S. dollars)

Year	Greece	Portugal	Spain	Turkey	Yugoslavia
1970	905	473	1,209	1,840	1,198
1971	1,031	472	1,362	2,227	1,340
1972	1,339	496	1,452	2,450	1,615
1973	1,541	609	1,704	2,869	1,877
1974	2,037	723	2,514	3,120	2,093
1975	2,552	825	3,409	3,159	2,327
1976	2,377	989	4,775	3,583	2,806
1977	2,612	1,458	6,970	4,287	3,104
1978	3,123	2,806	7,606	6,228	3,477
1979	3,532	3,708	8,656	10,972	3,700

Note: Does not include undisbursed public debt or external private non-guaranteed debt, which is very substantial for these countries.

Source: United Nations, International Bank for Reconstruction and Development, World Debt Tables (Washington, D.C.: IBRD, 1981).

TABLE 10.4

Debt-Service Payments, 1970-79
(in millions of U.S. dollars)

Year	Greece	Portugal	Spain	Turkey	Yugoslavia
1970	102	90	195	171	241
1971	146	92	332	163	183
1972	172	97	264	224	218
1973	266	98	373	207	307
1974	330	90	325	241	394
1975	444	113	390	262	447
1976	487	100	522	310	305
1977	525	152	844	362	379
1978	559	226	2,427	448	445
1979	740	411	1,703	640	563

Note: Does not include undisbursed public debt or external private non-guaranteed debt, which is very substantial for these countries.

Source: United Nations, International Bank for Reconstruction and Development, World Debt Tables (Washington, D.C.: IBRD, 1981).

TABLE 10.5

External Public Debt as Percentage of GNP, 1973-79

Year	Greece	Portugal	Spain	Turkey	Yugoslavia
1973	12.7	8.6	5.0	13.6	19.9
1974	10.5	9.2	2.8	10.6	7.4
1975	11.9	8.6	3.3	8.9	7.2
1976	10.3	10.8	4.4	9.2	7.3
1977	9.7	13.5	5.8	9.8	6.6
1978	9.6	22.0	5.2	12.9	6.3
1979	8.9	25.0	4.6	20.0	5.2

Source: United Nations, International Bank for Reconstruction and Development, World Debt Tables (Washington, D.C.: IBRD, 1981).

TABLE 10.6

External Public Debt as Percentage of Exports, 1973-79

Year	Greece	Portugal	Spain	Turkey	Yugoslavia
1973	68.2	25.5	31.4	97.8	75.8
1974	53.8	27.9	18.3	91.2	28.2
1975	60.0	33.2	23.4	99.7	28.5
1976	47.8	49.7	30.0	100.0	30.2
1977	45.5	50.1	37.6	133.8	30.8
1978	45.7	74.1	31.8	170.5	29.8
1979	40.1	68.5	n.a.	233.2	n.a.

n.a. = not available

Source: United Nations, International Bank for Reconstruction and Development, World Debt Tables (Washington, D.C.: IBRD, 1981).

structural dislocations caused by the return of migrant workers from northern Europe, both from the lost foreign exchange and the increased domestic unemployment, would be compensated for by the export of capital from recession-plagued northern Europe and the United States to southern Europe are not being fulfilled. The old pattern that leads foreign investment to bypass the northern Mediterranean persists, except in small amounts in the kind of manufacturing industries that have the fewest internal repercussions.[10] For Greece, Spain, and Portugal, adaptation to the membership terms and the more competitive environment of the European Community will reinforce these conditions further; for Yugoslavia and Turkey, along with Malta, Cyprus, Morocco, and Tunisia, these conditions will seriously worsen as the self-sufficiency in agricultural products of an expanded European Community will eliminate most of the current demand for their agricultural exports.[11]

For most observers, current developments in the countries lying on the northern shores of the Mediterranean begin with the fundamental regime changes of the mid-1970s, not with inchoate economic-policy changes in the 1950s. Yet the transformation of political institutions—the end of fascist dictatorship in Portugal and in Spain, the return of parliamentary democracy after military rule in Greece, the simultaneous return to Leninist principles in the Yugoslav League of Communists and further decentralization of governmental institutions, and the precarious vacillation between military coup and unstable parliamentary regime in Turkey—is the consequence of these earlier economic decisions and of the social and political changes that they required. Furthermore, it would be difficult to understand the politics of these new regimes (after 1974-75) without considering the conditions of growing deficits, indebtedness, external vulnerability, unemployment, inflation, and inequalities facing political leaders.

The reorientation of economic activities toward international trade and an open economy could not have occurred without facilitating changes in local conditions. No later than the mid-1960s, liberalizing reforms made their appearance: abandonment of the most repressive and corporatist aspects of economic control; legislation to free up the movement of labor and of capital among regions and across national borders; greater autonomy from state regulation for private enterprise; growth in private financial institutions; rationalization and expansion of the educational system, especially at the secondary and higher levels; emphasis on technocratic values in the ruling ideology; and greater contact with trading partners, including greater receptivity to their cultural and political influences through tourism, advertising, international conferences, and television.

In keeping with the liberal assumptions of the new strategy that the market would bring these freed agents and forces into equilibrium periodically, the social changes unleashed by these reforms were not regulated. Some of the consequences were those desired, some were not, but the social upheaval in any case was of significant proportions. People moved into cities in search of jobs, often selling their land to agribusiness, thus changing property relations in the countryside and foreclosing the possibility of returning to agriculture and self-sufficiency in food, and leaving the aged, the women, and children behind to continue cultivation. Industrial jobs do not expand as fast as people move, so that unemployment, expanding service-sector jobs (which mostly do not contribute to increases in productivity and are often largely public-sector jobs), and severely strained services and infrastructure in the cities increased the demands on public finances faster than revenue grew. Regional inequalities increased as capital moved through the banking system toward the more industrial regions in the country (for example, in Spain, from Andalusia and Galicia to the Basque country and Catalonia; in Yugoslavia, remaining in Croatia and Slovenia rather than "trickling down" to Kosovo, Macedonia, and Bosnia).

In some cases, social mobility even changed the recruitment base, usually downward, of such politically significant institutions as the police and the army. The personalistic bonds of the patronage-based political system weakened as technical skills and urban networks became more important, and traditional institutions, such as the church, came under pressure to retreat or to reform. Urbanization brings many more individuals into the orbit of urban organizations such as political parties and trade unions, thus increasing the potential for effective protest against unemployment and increasing inequalities. At the same time, many others—the more successful—become less easily mobilized for political action, hoping thereby to preserve their improved status. The changes also create new opportunities for conflict over government policy as traditional groups of influence are joined by professional groups and leaders of new industries and as foreign alliances are adapted to the needs of trade. By the second half of the 1960s, student demonstrations in the expanded universities (exacerbated by the probability of unemployment after graduation), regionalist movements for autonomy from the central government, and labor protests against growing unemployment and declining real-wage rates disturbed the calm of the repressive years, though in most cases the organizational expression of these grievances was met with government suppression.

Between 1953 and the early 1960s governments in all five countries enacted legislation to encourage foreign investment, with privileges or relaxed controls to redirect domestic investment to export-oriented firms, to join European trade associations, and to facilitate and regulate labor migration to northern Europe. In Spain ruling groups began to abandon Francoism more than a decade before his death in 1975. [12] The resources of Portugal's colonies and a labor shortage owing to years of overseas migration from a stagnating semifeudal agricultural sector gave Salazar more room to delay political changes, but Caetano moved toward liberalization after 1968, and by 1974 the domestic debate between "creeping Europeanization" or the "way to Africa" succumbed to the force of the African independence struggles, revealing the fragility of the former domestic peace. [13] In Greece, unrest because of rising unemployment at home and conflicts with the United States and Turkey over foreign policies aimed at supporting the shipping industry (over the Cuban blockade, diplomatic relations with East European countries, and Cyprus) led to a military coup in 1967, only to be reversed again in 1974 by renewed economic depression and the Cypriot crisis. [14] Between 1967 and 1971 in Yugoslavia, liberalizing reforms, increasing unemployment, and worker migration to northern Europe occasioned serious regional conflict that erupted into student demonstrations, in 1967 and 1968 over unemployment and in 1971 over the regional retention of foreign-exchange earnings. In Turkey, the military struck against the conservative government of Menderes in 1960 to push development policies, and it returned in 1971–72 and again in 1980 to repress the growing labor and student unrest over unemployment and increasing inequalities.

The strategy for growth had created social tensions, but further integration into the world of the developed countries for trade, investment, and credit required political stability. Trading partners and potential investors and creditors needed to be reassured that their investments would be safe and that labor pressures would not undercut their calculated gains. Certainly there were many other pressures for political change as well, but parliamentary democracy, with its method for regularizing class conflict in electoral contests among political parties and in social pacts between organized capital and organized labor, suited this need well. [15] The equal concern for political and economic peace can be seen in the near simultaneity of constitutional agreements and social pacts—for example, the Constituent Assembly and the Pacto de la Moncloa in Spain in 1976 or the 1974 Constitution and the 1976 Law on Associated Labor in Yugoslavia. That parliamentary democracy and concessions to labor of the rights to organize and to strike were the final step in a long process is also supported by what did not change in the mid-1970s. Everywhere the class composition of the ruling groups remained the same, the employees of state institutions (the civil service, the police, the public enterprises, the army [except for a selective purge in Greece]) were the same individuals, and the basic consensus on state policy in the economy continued as before. The early days of the Portuguese revolution did hold out a possibility that further alterations in the composition of ruling groups and in the fundamental policy direction in the economy might result from constitutional change, but external political and economic pressures, aided, according to some observers, by tactical errors of the Armed Forces Movement (MFA) leadership, assured continuity instead.

Politics in these countries continues to be shaped above all by this commitment to the strategy for economic growth chosen in the 1950s and by the consequences of that policy consensus. As before, government policy makers perceive their choices to be among those policies designed to increase the incentives to investors, both domestic and foreign, and those policies to further the successful integration of the economy into the world market—a search for better trade relations and for more foreign financing, continued attention to deflationary austerity measures designed to adjust balance-of-payments deficits and to stabilize the currency, and control over the cost of labor—to remain internationally competitive. If these policies have harmful consequences, such as unemployment, or lead to social unrest that might endanger the political stability necessary to both business confidence and healthy governance, only then do policy makers respond, accommodating as much as possible within the constraints they have already set. These constraints allow little room to maneuver, however. For example, as unemployment climbs but budgets must be cut, the government walks a tightrope by reworking the principles of unemployment compensation in social contracts pledging new jobs, as in Spain's tripartite social contract, the National Agreement on Employment of June 1981, where benefits are to be distributed according to length of employment, rewarding seniority with the smaller funds and encouraging job-search among the young. Or, if labor grows restless at its heavy burden, reactionary groups within the military also become more active (Sacred Union of Greek Officers [IDEA] in Greece in the 1960s, or the Spanish and Turkish officers in 1980-81), pressing the government from its right to stand firm. Any sign of political instability, from labor or from the military, can also provide foreign capital with an excuse not to invest. Policy makers once again are caught in a dilemma that they have created.

The economic consequences of these policies discussed earlier have their political consequences as well, particularly now that electoral results matter. The increasing freedom to private financial and industrial groups and the increasing income inequalities in their favor strengthen the political influence of those most in favor of current policies. Within these groups, those least in favor of continuing international integration, such as firms damaged by international competition or foreign investment, lose relative strength and are less able to protest expansion. Since the policies do not cure inflation or reduce the instabilities of economic activity vulnerable to international developments, business continues to withhold investment and pressures further for deflationary measures and lower costs, aimed both at labor and at government fiscal policy. If it succeeds in reducing the government's flexibility with fiscal instruments, it also increases the likelihood that conservative policies will continue.

While the resources of the right are being increased, the resources of those who might challenge these policies are being reduced by unemployment, smaller welfare benefits, declining real wages, and rising prices of necessities. Since the burden of these policies are borne by labor, that is, by the majority of the population, and these countries are now for the most part democracies, might one not expect rising pressures for radical change?

The contradictory pressures of these policies do result in highly unstable governments, but the revolutionary potential in southern Europe is actually very low. One reason for the low potential of revolutionary ferment is the legitimating force of democracy itself. There are few who did not welcome its return in Portugal, Greece, and Spain, who do not suffer under military rule in Turkey, or who do not show pride in the democratization of Yugoslav self-management. Furthermore, since the immediate alternative in all these countries except Yugoslavia has been dictatorship of the right, which aims its repression and often terror at the Left and at labor, any pressure for social change from the Left will be constrained for some time by the fear of return to dictatorial rule. Parliamentary democracy also means liberal freedoms, including the crucial basis for any power the Left does have—the right to organize. When this freedom is identified with open economic policies, when it seems to be part of a single package of liberalizing reforms in economy and society, any attack on the economic policies alone can easily appear to be an attack on democracy too. The caution of the Left under these circumstances is quite understandable, as were the development and appeal of their contribution to the Left debate on Eurocommunism. Competitors for power to their Right can make this position even more uncomfortable by repeating at every opportunity the "natural" relationship among the market, free trade, and democracy. Finally, to challenge so ingrained a consensus effectively requires coordination among many social groups, yet parliamentary democracy is based on their competition.

Social forces standing outside this consensus—the far Left on economic policy and the far Right on democracy—are relatively small, and of the two, only the far Right, especially the military, has the power to take effective action at the moment. In Downsian fashion, the main political parties are attracted toward the large popular mass at the center in order to maximize their electoral strength, competing with each other for the middle- and lower-middle-class groups who fall between their own class allies—those groups who are more equally affected by both inflation and employment than workers or business, who therefore could go either way with their support, and who are most likely to be averse to too much change, as mentioned earlier. The socialist parties are also pulled toward this central consensus by their own form of integration. Traveling northward to join conferences of the Socialist International, accepting financial assistance from socialist parties (especially the German SPD), with an eye to Europe generally after long years of isolation, they have abandoned their Marxism—it was long overdue, some said—for social democracy. The Party of Equilibrium, was the campaign slogan in 1974 of the Portuguese Socialists.

A third reason there is so little potential for serious protest is the weakness and internal division of the remaining Left forces. Unlike the Right, the Left has no traditional social institution, such as the church, to help recruit members. Its organization, in party and unions, had to be revived or created anew. Its potential base includes all those suffering from government policy, but that is a group that appears to have few other qualities in common. The Communist party has forged a class alliance of poorer industrial workers and agricultural laborers, joined occasionally by radical students. But class divides other radicals, such as women or ethnic nationalists, who split off to mobilize supporters into feminist movements or regionalist parties. To find a platform that unites most opponents of the current policy and that offers alternatives other than the protectionism of the inglorious past is difficult.

While broad consensus on the economic strategy, democratic institutions, redistribution of societal resources favoring the supporters of conservative policies, and the organizational and programmatic fragmentation of the Left make radical challenge in this region unlikely, they also increase the potential for governmental instability. The general agreement on policy and the greater organizational resources of the center-Right and Right have led to the creation of large coalitions, either as single parties such as the Union of the Democratic Center in Spain, or as groups of parties as in the Democratic Alliance of Portugal. These coalitions are subject, then, to the leadership quarrels and organizational breakdown of all coalitions that are larger than necessary to win, as has been the case of the Italian Christian Democratic party for so long and as was apparent in the defeat of Suarez in Spain in January 1981.[16] Parties on the center-Left are trapped in the contradictions of the economic policy. Accepting the general strategy of integration, they try to select policies of adjustment that will be least damaging to their constituency. For example, in the face of unemployment, they push to maintain levels of unemployment compensation as in Spain (unsuccessfully), or they counter IMF stabilization proposals with a program for democratic planning of the domestic economy as in Portugal (unsuccessfully). They try to reduce international vulnerability by juggling multiple international alliances, for example, as the Yugoslavs courting both the Council for Mutual Economic Assistance (COMECON) and the European Community (EC) (to declining advantage) or the Greeks balancing EC membership, reduced North Atlantic Treaty Organization (NATO) commitment, and trade with the Soviet bloc against the U.S. alliance (currently precarious with the revival of U.S. cold war policies, as the renewal of conflict with Turkey suggests). They continue labor migration to relieve unemployment and maintain the capital flow of remittances but try to ease the burden on workers by negotiating better work conditions in the receiving countries and cooperatives at home, where they can exchange foreign earnings for employment when they return, as in Yugoslavia or in Turkey under Ecervit (unsuccessful thus far). The policies of both center-Right and center-Left parties thus only reinforce the commitment to export-led growth and its structural contradictions—unemployment, inflation, further vulnerability, declining growth, austerity policies, and larger deficits. As output declines, the distribution problems increase: the policies of the center-Left parties are trapped in fiscal crises, and those of the Right in declining aggregate demand. Decision-making bodies confronted yearly with the same choices tend to find it more and more difficult to find resolutions quickly: if there is no deadlock, there is at least delay, and frequent cabinet crises and government turnover.

As voters try to make sense of these outcomes, electoral results also vacillate. Linz reports, on the basis of attitude surveys in Portugal, Greece, and Spain, that there is a "strong desire on the part of large segments of the population in these countries for stability, liberty within a democratic order, continuing growth, and some degree of social change."[17] According to the economic analysis above, these goals are increasingly incompatible. The result is frequent, small-scale shifts in electoral outcomes, and thus often narrow parliamentary majorities and the accompanying instability, as the people in the middle shift among center parties in search of a government that can satisfy them.

In an area where many regions have historical significance and retain a cultural identity, increasing regional inequalities further threaten stability. Wealthier regions with larger industrial bases and resources demand greater autonomy and poorer regions demand centrally financed redress (Spain and Yugoslavia are the most obvious cases, but autonomist movements in Turkey and the regional concentration of political alignments in Portugal and Greece are also troublesome). If the government grants more autonomy, it then loses control over regional development policy. As long as it continues to liberalize economic activity, the government must choose between the costs of increased repression and those of social disorder.

All these pressures add extra burden to these governments that cannot entice foreign investment or win additional credit, the twin hopes of their economic strategy, in a time of world recession, without political stability at home. The temptation on the right is to blame social unrest and democracy for the troubles and to engineer a military coup. That this is no real solution is particularly clear in Turkey today. Elsewhere, strong leaders with unusual political skill have been able to smooth over political crises—President Eanes in Portugal, King Juan Carlos (and his adviser Miranda) in Spain, President Tito in Yugoslavia, and, as he hopes clearly to do with the presidency, Karamanlis in Greece; but the passing of Tito is a reminder that such solutions are only temporary.

CONCLUSION

The current situation in southern Europe can be traced to a judgment by political leaders that internal economic development was essential. The strategy chosen between 1947 and 1960 has not led to self-sustaining, balanced growth, however, as the leaders claimed it would. Some industrialization took place and some infrastructural foundations—roads, schools, irrigation schemes—were laid. Significant social change occurred. But the internally generated, self-sufficient development on which political and economic sovereignty and domestic peace depend has not. In place of such development came increasing vulnerability to the economic conditions prevailing in and imposed by those developed countries and those economic actors who dominate international trade and its financing. As a result, the distortions in the domestic economy led to high rates of inflation and of unemployment, increasing inequalities in income and wealth, declining standards of living for the majority, and ever more remote domestic conditions for encouraging productive investment and growth. Social unrest on both left and right increased, and so did the pressure for more conservative policies. Facing these consequences, yet proposing no alternative, governments frequently fall or are deadlocked over budget debates and internal wrangling. Finally, the changes that might reverse these circumstances in world trade, in capital and labor flows, or in domestic ruling groups, are not likely. Nonetheless, proposals for remedies of the current difficulties suggest more of the same. Greece, Spain, and Portugal are entering the European Community. Portugal is rewriting its constitution to abolish the Council of the Revolution and reverse the postrevolution nationalizations. The Turkish military suggests it will hand over power to the center–Right Justice party along with restricted civil liberties, a strong executive, and even larger cuts in labor's share of productivity. Many talk about waiting until the world economy and domestic growth rates improve as if this will be soon and automatic. Others recommend a reassessment of their comparative advantage, switching from particular goods to trade itself. As age-old seafarers with commercial networks and cultural affiliations in many ports, each country could become an intermediary between the developed north and part of the less-developed south. In each proposal, however, these countries would continue to compete against each other while leaving their economic strategy untouched.

Few recommend domestic policies to redistribute income significantly in order to generate the effective demand necessary for local investment. Nor do many recommend that the government reassert some of its earlier control over the economy sufficient to guide investment for internal development. Yet, since these policies would address actual barriers to current growth, they might make real growth and development possible at least in the medium term. In the long run, however, vulnerability to the external sector must also be reduced by breaking out of the dependencies in the trade cycle. Since most of the countries around the Mediterranean do share the same current difficulties and their level of development and the extent of their natural resources together contain enormous potential, they could form a new trading community that they each enter on beneficial terms. Domestic development policy could be supplemented by shared expertise, needed resources, and useful economies of scale on mutually advantageous projects in a Mediterranean community for trade and development. This might even help loosen the bonds of the current world structure on other parts of the developing world and free them to do the same.

To create such a community and to abandon former policies requires political will. It requires first a willingness to see commonalities, not differences, in neighbors; to recognize that notions of interest, membership, salient traits, division, and consensus are politically defined and have their political uses. One could ask why outsiders tend to emphasize the differences in this region—Christians, Muslims, Jews; Europeans, Asians, Africans; parliamentary democracies, one-party states, monarchies; capitalists, socialists, mixed economies. They tend to define the area in relation to others—less developed than northern Europe, more developed than most of the Third World; outside the core but not in the periphery; returning to Europe; aligning with Arab states. They tend to offer solutions that are to be found elsewhere, too—in ties to Europe, or with the OPEC, or in military alliances with the two superpowers. The Mediterranean need not be so internally divided. It has an identity and a tradition of interchange, after all, that antedates Europe by thousands of years. A change in policy depends on a change in perspective; that requires an act of political will, and need not depend on conditions.

NOTES

1. Albania continued to follow a predominantly autarkic policy, although it did seek aid first from Yugoslavia, then from the Soviet Union, then from China. Since it now appears interested in trade with the European Community and Prime Minister Shehu's mysterious suicide in December 1981 came after losing a debate over further integration of the Albanian economy into the world division of labor—the timing could be coincidental—one can at least speculate that Albania may not remain an exception. Italy is not treated here as a Mediterranean country because of the political and economic dominance of its northern half; were its Mediterranean half a separate state, it would very likely belong to this group. Cyprus and Malta share many of the same problems in the 1980s, but are not directly considered here because of their later independence.

2. There was a serious attempt at land and tax reform in Turkey after the military ousted Menderes in 1960 and under the Republican People's party, 1961-64, but the Justice party effectively blocked its implementation.

3. See, among others, Ragnar Nurkse, "International Investment Today in the Light of Nineteenth-Century Experience," in Inter- national Finance: Selected Readings, ed. Richard N. Cooper (Har- mondsworth, England: Penguin, 1969), pp. 358-74.

4. Some useful studies of foreign investment in southern Europe and its consequences are Constantine Vaitsos, "A Note on Trans- national Corporations (TNCs) and the European Periphery," and Juan Muñoz, Santiago Roldán, and Angel Serrano, "The Growing Dependence of Spanish Industrialization on Foreign Investment," both in Under- developed Europe: Studies in Core-Periphery Relations, ed. Dudley Seers, Bernard Schaffer, and Marja-Lüsa (New York: Humanities Press, 1979), pp. 97-102 and 161-75; Volker Bornschier, "Dependent Industrialization in the World Economy: Some Comments and Results concerning a Recent Debate," Journal of Conflict Resolution 25 (Sep- tember 1981): 3; M. Serafetinidis, G. Serafetinidis, M. Lambrinides, and Z. Demathas, "The Development of Greek Shipping Capital and Its Implications for the Political Economy of Greece," Cambridge Journal of Economics 5 (1981): 289-310; and the chapter by Stefan Musto in this volume.

5. In some cases this process was actively assisted by governmental employment services, such as the service established by Turkey in 1961 to regulate the selection of workers permitted to migrate. A study by S. Lieberman and A. Gitmez revealed that the Turkish government chose more highly skilled workers with industrial experience over less-skilled workers underemployed in agriculture in order to relieve unemployment in relatively prosperous areas, even though this meant continuing low productivity in agriculture in the less-developed regions of the country and a loss of a skilled, experienced work force. See Samuel S. Lieberman and Ali S. Gitmez, "Turkey," in International Labor Migration in Europe, ed. Ronald E. Krane (New York: Praeger, 1979), pp. 201-20.

6. See Serafetinidis, Serafetinidis, Lambrinides, and Demathas, "The Development of Greek Shipping Capital," p. 301.

7. Useful introductions to balance-of-payments theory are B. J. Cohen, Balance-of-Payments Policy (Harmondsworth, England: Penguin, 1969); and H. G. Johnson, "Towards a General Theory of the Balance of Payments," and "Theoretical Problems of the International Monetary System," both in International Finance: Selected Readings, ed. Richard N. Cooper (Harmondsworth, England: Penguin, 1969), pp. 237-55 and 304-34.

8. On the Yugoslav case, see Cheryl Payer, The Debt Trap (New York: Monthly Review Press, 1974), pp. 117-42; and Susan L. Woodward, "Corporatist Authoritarianism vs. Socialist Authoritarianism in Yugoslavia: A Developmental Perspective" (unpublished manuscript); on the Portuguese case, see Norman Girvan, "Swallowing the IMF Medicine in the 'Seventies," Development Dialogue 2 (1980): 55-74.

9. Data are drawn from the Financial Times during 1981 and the Organization for Economic Cooperation and Development individual economic surveys of Portugal, Spain, Greece, Yugoslavia, and Turkey published almost annually.

10. See Suzanne Paine, "Replacement of the West European Migrant Labour System by Investment in the European Periphery," in Underdeveloped Europe: Studies in Core-Periphery Relations, ed. Dudley Seers, Bernard Schaffer, and Marja-Lüsa (New York: Humanities Press, 1979), pp. 65-95.

11. See the chapters by Musto and Pollis in this volume.

12. Salvador Giner and Eduardo Sevilla, "From Despotism to Parliamentarism: Class Domination and Political Order in the Spanish State," in The State in Western Europe, ed. Richard Scase (New York: St. Martin's Press, 1980), pp. 197-220.

13. See Nuno Portas and Serras Gago, "Some Preliminary Notes on the State in Contemporary Portugal," in The State in Western Europe, ed. Richard Scase (New York: St. Martin's Press, 1980), pp. 230-40.

14. The links between Greek foreign policy and the shipping industry are carefully laid out in Serafetinidis, Serafetinidis, Lambrinides, and Demathas, "The Development of Greek Shipping Capital."

15. Yugoslavia is, of course, an exception to this development: there is no electoral competition among political parties over who will form the government nor can one organize politically outside the League of Communists and its affiliated associations. Yet competitive elections do occur, there are genuinely democratic features of self-management, and the delegate system introduced in 1974-75 is an attempt to democratize government assemblies and to integrate political and economic decision makers in a way that serves the same function as parliamentary democracy and social pacts. Further, the pattern of governmental instability, policy conflicts, and political trends described here does apply to Yugoslavia. This instability is exacerbated by the confederal structure of power where regionally organized units of the League of Communists form coalitions and bargain over policy at the federal level.

16. See Lawrence Dodd's demonstration of the instability caused when governments are not formed by minimum winning coalitions in Coalitions in Parliamentary Government (Princeton, N.J.: Princeton University Press, 1976), pp. 73-96.

17. Juan Linz, "Europe's Southern Frontier: Evolving Trends toward What?" Daedalus, Winter 1979, p. 181.

11

THE SOUTHERN EUROPEAN SEMIPERIPHERY AND THE EUROPEAN COMMUNITY

ADAMANTIA POLLIS

INTRODUCTION

Societal organizations are inevitably premised on a particular set of doctrines that provides their rationale and indicates the broad parameters within which they operate. Particularly in instances such as that of the European Economic Community (EEC),* where organization was the result of a policy decision among sovereign states to create a particular social order embodied in a constitution or charter, a group's underlying precepts are determinant of the functioning of the system. In the case of the European Community, the goals of the Treaty of Rome, adopted in 1957, were first to create a common market, then to move to an economically integrated unified Europe, and last, to create a political union that would be triggered by the economically unified Western Europe. Attainment of these goals was predicated on a particular ideology and a particular theory regarding the processes of social change. Success in creating the united Europe that the ideologues among the instigators of the European Com-

---

*The EEC is one community in the larger complex that is officially called the European Communities. The term <u>European Community</u> is used with increasing frequency to refer either to the European Economic Community or the European Communities.

munity (EC) envisaged, is clearly dependent in large measure on the validity of the theoretical presuppositions underpinning its institutional framework and its mode of functioning.

The European Economic Community, whose initial membership consisted of the advanced industrialized states in the European continent, except for the United Kingdom, was premised on the doctrines of liberal economic thought and on the more contemporary theories of functionalism and neofunctionalism. By structuring the appropriate institutional framework enabling the free play of the market mechanism and the encouragement of competition, it was assumed that shared economic interests and interdependent linkages would lead to European economic integration. The presumably nonideological, technocratic managers of the European Economic Commission were empowered to adopt and implement appropriate policies within the guidelines set by the Council of Ministers. And as Europe gradually became economically integrated, an inevitable spillover effect would bring political integration. Underlying the ostensibly scientific, non-ideological foundation of the European Community is the unstated ideological presupposition that integration should and would be attained through the furtherance of capitalism. A united Europe, geared toward furtherance of the interests of the industrial class, would ipso facto bring about a better life for all classes and peoples.

Although the initial membership of the European club consisted of the industrialized states of Western Europe (except for southern Italy), the first enlargement in 1973 incorporated Ireland, which was hardly on a par with the other member states of the EC, and the United Kingdom, which by then was beginning to be seen as a post-industrial society in the throes of decline. More recently the EC has embarked on a second enlargement to consist of Greece, Spain, and Portugal. One of them, Greece, was admitted to full membership in January 1981, and negotiations are in progress for the accession of the other two. As a consequence of these enlargements, the cohesiveness of the European Community culturally, politically, and in terms of its level of economic development is being diluted. None of these three southern European Mediterranean countries are industrialized pluralist states conforming to the model of the Western European societies. Not only are they far less developed than the core members of the EC, but the form of their economic development is significantly at variance with that of the Western European states. Politically, they lack the historical legacy of pluralism and democratic rule, while culturally, the persistence of traditional values, attitudes, and behavior is far greater than in Western Europe.

The accession of these three Mediterranean countries poses fundamental questions vis-à-vis the complex of theoretical assumptions that govern the EC and its institutions. Despite vast differences

among Greece, Spain, and Portugal, all three share certain features—
in their recent political legacies, in the structure of their economies,
and in their social systems—that are of the utmost significance in
evaluating the potential impact on them of EC membership. Although
the long-run historical development of the three has been quite divergent, in modern times they find themselves similarly situated in relationship to the industrialized states. Greece became a modern
state in 1830 with the attainment of independence from Ottoman rule.
Its revolutionary political leadership espoused the goals of freedom,
laissez-faire, and democratic institutions, albeit the reality did not
conform to the rhetoric. Spain and Portugal, on the other hand, had
been empire builders with far-flung colonies during the mercantilist
era, but their power dwindled as a result of the rise of industrial
capitalism centered in Western Europe, particularly Great Britain,
which outstripped them to become the international center of military,
economic, and political power.

In the last two centuries none of the three, with the partial exception of Catalonia in Spain, became industrialized, none assimi-
lated the doctrines of liberalism, and none evolved into pluralist
democratic societies. In fact, in the post-World War II era, all
three have, until recently, been governed by one form or another of
authoritarian rule. [1] Their failure to industrialize or modernize in
the span of the two centuries since Western Europe's dominance became apparent casts serious doubts regarding the validity of development theories. Greece, Spain, and Portugal, in accordance with
these theories, should have become transformed into modern states
replicating the level of economic development and the politics of
pluralism of the Western states. Even a cursory view, however,
makes it apparent that despite sustained rates of economic growth in
recent years (which are currently declining), particularly in Spain,
none can be considered politically, culturally, or economically comparable to the United Kingdom, France, or West Germany. The consolidation of democratic politics with the collapse or overthrow of
authoritarian-military rule in the mid-1970s remains fragile. Clien-
telism persists as a dominant mode of political and social life. Despite growth, the industrial base remains narrow, with foreign ownership dominant in this sector. Under these conditions, membership
in the EC raises both pragmatic and theoretical concerns regarding
the impact of membership on the three. Will membership provide
the impetus and the necessary structural conditions and political options to bring about the hitherto elusive development?

The Treaty of Rome specifies that membership is open to all
European democratic states and the EC has satisfied itself that, in-
deed, Greece, Spain, and Portugal at present fulfill that requirement.
With respect to their economies, however, the EC is acutely aware of

disparities and, consequently, is concerned with the problems and
difficulties to which this may give rise. The reality of sharp differences in level of economic development between the three European
Mediterranean countries and the core members is attested to by the
European Economic Commission in its <u>Opinions on the Application
for Membership by Greece, Portugal and Spain</u>, the first issued in
1976 and the latter two in May and August 1978, respectively. In the
<u>Opinion on the Greek Application</u>, the commission expressed the ne-
cessity for "structural changes of a considerable magnitude,"[2] while
for Portugal it stated that "radical economic and social reforms to
bring about the necessary restructuring of the country" are necessary.[3] The commission also argued for restructuring in Spain, although it expressed greater optimism than was the case with regard
to the other two countries.[4] In other words, the commission recognized the wide disparity in the level of economic development between the three Mediterranean countries and the earlier members of
the EC and recommended extensive structural reforms to facilitate
rapid equalization and, hence, effective integration. With respect to
Greece, in fact, the commission had originally proposed a period of
time, even prior to the transition period leading to the assumption of
the obligations of full membership during which economic programs
would be undertaken, designed to accelerate essential reforms, a
recommendation that the Council of Ministers rejected.[5] Given this
action by the council, no such recommendation was made for Portugal
and Spain, although an extensive transition period after accession was
deemed as essential for Greece and, undoubtedly, for Spain and Portugal also.

ROAD TO DEVELOPMENT?

The entry of Greece (January 1981), Spain, and Portugal (both
forthcoming) as full members of the European Community clearly
raises critical theoretical and empirical issues regarding the impact
of membership both on the new members and on the earlier nine community members. The discussion that follows is limited to a consideration of the consequences of membership on the three. Stefan A.
Musto, in another chapter in this volume in which he discusses the
EC's Mediterranean policy (other than the three under consideration
here) states, "The question can be raised whether the Mediterranean
policy of the EC is to be considered as a contribution toward reinforcing the collective self-reliance capabilities of the structurally weaker
partners or rather as an instrument for strengthening the existing
links of dependence operated through the international system by the
dominant countries."[6] This question is of equal if not greater rele-

vance in view of the consequences of membership in the EC of Greece, Spain, and Portugal.

An evaluation of the impact of membership will be undertaken from two different theoretical perspectives: mainstream liberal and neofunctionalist theory and center-periphery or dependency theory. Looking first at expectations derived from mainstream theory, it is presumed that despite sectoral differences, through a series of domestic reforms, including structural changes, and through economic and technical aid from EC funds, Greece, Portugal, and Spain will be able to accelerate the domestic processes of modernization. Less productive, inefficient sectors would be negatively affected, but this is the inevitable price for progress. Furthermore, membership in the EC through the presumed economic integration that this entails will ensure the viability of the new-found democracy in these three states—an argument that clearly stems from the political spillover effect of neofunctionalist theory. Deleterious economic consequences would presumably result from sanctions imposed by the EC in the eventuality of the reimposition of authoritarian rule. In other words, despite interim difficulties the long-run effects on Greece, Spain, and Portugal will be attainment of those elusive goals of social, political, and economic modernization.

The premises of the European Community and, even more, those of its technocratic, decision-making organ, the European Economic Commission, are rooted in liberal economic thought, which provided in earlier centuries the philosophic rationale for emergent capitalism. [7] Through the free play of the marketplace within the arena of appropriate political institutions undergirding the capitalist economy, economic growth and democracy will prevail and, implicitly, the well-being of all will be served. Enlargement of the EC creates a larger market, rewards efficiency, increases productivity, and propels economic growth. With regard to developing countries or regions, the commission presumes that pursuit of appropriate domestic governmental policies geared toward reforming social and economic structures and buttressed by technical and financial aid from the EC will result in the modernization and development of Greece, Spain, and Portugal. The commission thus recommends that aid be given to the three from the EC's social fund and from the European Regional Development Fund (ERDF), a recommendation that ignores the intense competition that already exists for these limited monies among the developing regions within the EC. Perhaps of greater long-run importance, however, is the fact that the commission and the Council of Ministers have not taken into account the failure of bilateral and/or multilateral aid programs during the post-World War II period to reverse uneven development; instead, they recommend a continuation and an extension of those very failed policies, except that they are now to be implemented within the institutional framework of the EC. In the EC to date, its regional development policies, most notably in the Mezzogiorno in Italy and in Ireland, have not met with success. [8] Thus, while the EC is concerned with conflicting sectoral economic interests among member states, with the intensification of conflict accompanying enlargement, and with attempts to reconcile them, there is little skepticism or questioning regarding the fundamentals governing the appropriate policies for economic growth and development.

The European Economic Commission, it should be noted, is essentially a technocratic body whose analyses and recommendations focus on the effects of enlargement on the various sectors or interests, both in the EC and in the three Mediterranean countries. Loyal to the EC as currently constituted, commission members operate within a presumed value consensus on the superiority of a free market geared toward rational decision making that is aimed at increasing productivity and profits. As the Socialist group in the European Parliament stated, the purpose of the Treaty of Rome was to "increase European productivity through an international division of labor, the elimination of unnecessary production and the optimum employment of the labor force." [9] The goals expressed therein perforce further the interests of the European bourgeoisie, who acquired an expanded market, free capital, labor mobility, unencumbered transfer of funds, and so forth. The success and profitability of industry using advanced technology become the measures of well-being and prosperity.

Adoption and implementation of the requisite policies by the commission and the Council of Ministers to attain these ideal goals are constrained, however, by two considerations: first, the fact that the political leaders of member states represent domestic socioeconomic interests to whom they are accountable and on whom their electoral success is dependent and, second, the prevalence of notions of social justice that in the domestic arena have modified pure liberal economic thought. The commission therefore inevitably concerns itself with the impact of enlargement on different sectors and proposes measures to contain conflicting or competing interests among or between interests in the national economies, while the Council of Ministers negotiates, bargains, and compromises to protect or enhance state economic interests and entitlements for disadvantaged groups. [10] It is within this context that concern has been expressed and disputes have erupted, particularly in the agricultural sector, between southern France and Italy and between the older community members and Greece, Portugal, and Spain (particularly the latter).

Nevertheless, despite contradictions, tensions, and conflicts, the underlying ideology of economic integration along capitalist lines leading to eventual political union persists, and the conflicts are seen

as impediments to be overcome. The rapid rate of economic growth
experienced by the three Mediterranean countries during the 1960s
and early 1970s is frequently cited as evidence of the benefits to be
derived from EC membership. Greece is considered particularly
pertinent in view of its associate membership in the EC as of 1961.
The annual rate of economic growth for all three averaged from
6 to 8 percent. Recently, growth rates have dropped to between 2
and 3 percent, and there are indications that they may be dropping
further. [11] Concurrently, not only did Greek exports to the EC increase, but their composition shifted from agricultural to industrial
products. The industrialization of the Greek economy that this
seemed to reflect was largely an illusion; the growth was in basic
metals, particularly aluminum, petroleum products, and chemicals,
whereas for Spain it was in metal products and machinery. Although
the gross national product (GNP) increased, the contribution of manufacturing to this increase was negligible. Furthermore, basic metals,
petroleum products, and chemicals in Greece and manufacturing in
Spain absorbed little of the labor force, over 30 percent of which in
Greece continues to be engaged in agricultural activities. [12] In fact,
the industrial sector of the Greek economy was foreign owned and
export oriented. Greece's unemployment problem, meanwhile, was
exported, in the form of "guest workers," to the expanding factories
of Western Europe, as were Spain's and Portugal's. Imports into
Greece did not consist of capital goods for the growth of domestic industry but of foodstuffs and consumer goods. In fact, by the middle
of the 1970s, all three Mediterranean countries had abandoned the
policy of import substitution that had been adhered to with varying
degrees of commitment among the three. [13] By abandoning a policy
essential for self-sufficient industrial growth, Greece, Spain, and
Portugal placed themselves in a position of increased dependence on
the economies of Western Europe. Greece's associate membership,
in fact, seems to have solidified this dependence to a greater extent
than was the case with Spain and Portugal.

The economic prosperity of the three during the 1960s and early
1970s, as measured by such indexes as GNP and the growth of a more
affluent middle class, was at a time of rapid and impressive growth
in the industrialized countries. The current crisis in the international economic order and the recession and unemployment being experienced by the West European states challenge the validity of the
basic economic tenets of the European Community, while potentially
threatening disastrous consequences for the dependent states.

Aside from the possible consequences of the current world
economic crisis, the initial euphoria that economic integration would
have a spillover effect leading toward political integration has already
dissipated. The EC, while still seen as a framework for minimal

economic integration, is no longer perceived as moving inexorably toward political unity, despite the holding of direct elections of delegates to the European Parliament. Although political integration per se may not be in the offing, political considerations with significant economic implications did dominate the EC's decision to negotiate treaties of accession with the three, and to do so more rapidly than initially envisaged. In fact, the primacy of political concerns in the negotiations for enlargement are shared both by the governments of Greece (until the election of a socialist government in October 1981), Spain, and Portugal and by the member states of the EC.

The most forceful justification for incorporation of the three in the EC despite competing national economic interests and severe economic strains, a rationale articulated both by the Council of Ministers and by the governments of the three countries, is that membership in the EC will ensure the survival of parliamentary democracy, increase political stability, and strengthen the West's military defenses. All three had been under some form of dictatorial rule, Greece and Portugal until 1974 and Spain at least until Franco's death in 1975. The primacy of political considerations was clearly revealed in the EC's and the Spanish premier's and king's call for the speedy conclusion of negotiations for Spanish entry in reaction to the attempted right-wing military coup in Spain in March 1981. [14] Recently, the negotiating process has slowed in light of the threat Spanish agricultural products pose to French and Italian agriculture. Most political parties in Greece, Spain, and Portugal, the exceptions being the orthodox Communist parties in Portugal and Greece and the ambivalent stance of the Panhellenic Socialist Movement (PASOK), the Socialist party of Greece, have strongly advocated membership in the EC. Morbey Rodrigues, representing the Portuguese Industrial Association, in May 1981 listed the political factor related to the maintenance of democracy as the first of three reasons for accession to the EC, [15] while the Spanish foreign minister has been pushing for an acceleration of talks as a guarantee of Spanish democracy. [16]

The concern of both the Council of Ministers and the political leadership of Spain, Portugal, and until the elections in October 1981, Greece, with ensuring democracy through membership in the EC reflects underlying ideological and economic concerns. For both sets of actors, the Mediterranean countries and the European core, consolidation and stabilization of the postdictatorial conservative or center governments, which would minimize or eliminate the threat of left-wing or revolutionary upheaval as well as right-wing protectionist coups, has been a major consideration. It was these concerns that led to massive aid to Soares's party and his socialist government in Portugal after the overthrow of Caetano's dictatorship, while Greece's entry was accelerated in light of the possibility of a future electoral victory by the left-socialist Papandreou party. Nationalistic right-wing regimes that engage in protectionist policies threaten the economic interests of European industrialists whose access to markets is restricted, as is their ability to make investment decisions free of controls. Left-wing revolutionary regimes, on the other hand, would threaten the entire foundation of bourgeois states and the EC itself. From the Western European perspective, furthermore, enlargement would bring the three closer into the orbit of the North Atlantic Treaty Organization (NATO), thus diminishing the danger of right-wing or left-wing protectionist nationalist regimes, a development both economically and politically detrimental to the West. In fact, negotiations are in the final stages for the entry of Spain into NATO.[17] Whether membership in the European Community will ensure parliamentary regimes in Greece, Spain, and Portugal remains an untested hypothesis.

It is evident that both the EC and the governments of the three Mediterranean countries, at least prior to the socialist government in Greece, share the underlying economic and political premises regarding the desirability of enlargement as being beneficial to all in the long run. In 1977 the premier of Greece, Constantine Karamanlis, clearly articulated the expectations of all three Mediterranean countries when he stated, "Our integration with the united Europe is a grand policy which will change the destiny of our people. It will accelerate the economic and social development of our country; it will contribute to the safeguarding of our democratic institutions and, above all, it will strengthen the country's security against external dangers."[18]

THE ROAD TO UNDERDEVELOPMENT?

In the previous section, critical consideration was given to the anticipated benefits to be derived from EC membership by the Mediterranean countries. This long-run optimistic assessment, shared by the principal participants in the negotiations for entry, is founded on acceptance of a theoretical paradigm within which developing societies can accelerate the process of economic development and industrialization by adopting appropriate policies and by receiving aid from the more advanced industrial states. Alternative theoretical paradigms, however, lead to a far more pessimistic assessment of the impact of membership on the three. If the potential impact is analyzed from the perspective of world-system or dependency theories, incorporation of Greece, Spain, and Portugal into the institutional framework of the EC, adoption of the economic policies of the EC, and conformity to the regulations and decrees of the Economic Commission will result in further integration of the three into the world system in a peripheral position and reinforce their dependency. [19] Acceptance of free trade, adherence to the tenets of a competitive market system whereby resource allocation is determined by profitability and productivity or comparative advantage, and conformity to the product specifications and standards of the EC will thwart the Mediterranean countries' autonomous economic development and preclude the possibility of catching up with the West.

The economies of Greece, Spain, and Portugal, as indicated in the previous section, were already dependent on the West prior to negotiations for entry. Of the three, Spain's protectionist policies in the earlier Franco years had led to the rise of an indigenous industrial sector engaged in production for domestic consumption. The policy of support for a domestic industrial class protected against competition from the more advanced, technologically superior industry of Western Europe was abandoned, however, in the later Franco years with an opening to foreign capital. [20] Recently, domestic Spanish industrial investment has declined, while investment by foreign multinationals has mounted. Portuguese efforts, prior to and even during the colonial wars in Africa, to form a Portuguese economic union were a dismal failure; Portugal neither incorporated the colonies nor industrialized, although there was some expansion in the industrial sector. By the mid-1970s it was looking toward Western Europe, strengthening the already-existing dependency links to the United Kingdom and establishing links to other European countries. Greece's efforts at protectionism, reflected in high tariffs against imports of foreign manufactured and industrial goods, were belied by the enactment as early as 1953 of a law granting extraordinary special privileges and tax benefits to foreign investors. During the latter half of the 1960s and the early 1970s, coinciding with a high rate of economic growth and political rule by a chauvinistic military junta, foreign investment rapidly accelerated. [21]

By the time negotiations were seriously undertaken for entry of the three into the EC as full members, in addition to the existing structural integration, a formal relationship was in place: Greece had become an associate member in 1961, [22] Portugal, initially a member of the European Free Trade Association (EFTA), signed a free-trade area agreement in 1972, and Spain had signed a limited preferential agreement in 1970. The specific provisions differed in each case; for Greece it was clearly preparatory to eventual entry into the EC as a full member. The underlying principles, however, were similar: entry of Mediterranean exports to the EC without tariff restrictions (with some exceptions), while imports, in line with the infant-industry argument, would continue to be subject to tariff barriers. Greece's, Spain's, and Portugal's "economic miracle," as it has been labeled by one author,[23] during the years of their association with the EC, served to reinforce the arguments of procommunity forces, who attributed their apparent economic prosperity to EC association. Little attention was paid to the fact that in the industrialized capitalist world this was a period of economic expansion and prosperity; Western Europe faced a labor shortage and, hence, absorbed labor from these and other developing countries; a technological breakthrough revolutionized manufacturing and created new industries; and as markets were expanding, industries enjoyed high profits and states saw favorable balances of payments.

The apparent prosperity in Greece, Spain, and Portugal, the illusion of affluence, the availability and consumption of goods, and the increase in per capita income were all consequences of this overall worldwide prosperity emanating from the core. Prosperity was not the result of domestic developmental efforts and economic planning abetted by EC aid. At present, the three peripheral Mediterranean countries are particularly vulnerable to the effects of the recent disruption and stresses in the international economic order and the recession that has hit the core countries. The interdependence of the world economy is indisputable, but within this reality, the peripheral countries have few options compared with the industrialized ones. Western European countries and the United States presume that there are policy options for dealing with recession. They can debate and dispute among several alternatives: supply-side economics, Keynesianism, or monetarism. Greece, Spain, and Portugal, on the other hand, while profoundly affected by the economic crisis and by the policies adopted at the core, are severely constrained in their ability to respond or react in an autonomous fashion.

The structure of domestic economies in Greece, Spain, and Portugal, their trading patterns, and the policies they adopt and implement attest to their dependence on the industrialized states. The goal of autonomous economic development and modernization geared toward developing domestic productive capacities has not been materializing. Import substitution as a policy of economic development has been abandoned by all three, thus precluding further domestic industrial growth and subjecting whatever embryonic indigenous industry there is to competition from the industrial giants of the West. Furthermore, the industrialized sectors of these economies are heavily foreign owned, and in line with the EC's tenet of freedom of establishment, they can establish and disestablish plants at will, depending on profitability, with no legal or financial constraints.

The terms of trade are clearly unequal. Historically, a limited number of agricultural products have been exported by Greece, Spain, and Portugal to Western Europe at highly fluctuating world market prices in exchange for the higher-priced consumer and indus-

trial products of European factories. In the last decade, particularly
in Greece and Spain, nonagricultural products have been constituting
an increasing share of the export market. In Greece, aluminum,
petroleum products, and chemicals have become principal export
items, while in Spain, in sharp contrast to Greece, metal products
such as transport equipment and capital goods are principal exports.
Industrial exports from Greece jumped from 13.7 percent of total exports of goods in 1961 to over 65 percent in 1976, whereas in Spain
industrial exports were about 80 percent of the total. [24] These sectors of the Greek and, in recent years, the Spanish economies that
produce basic metals and manufactured or industrial goods (with the
partial exception of textiles) are largely foreign owned and export
oriented; they do not produce for the domestic market, and the earnings of the foreign owners are transferred abroad. Both the terms
of trade and the conditions for foreign investment result in exploitation of the Mediterranean countries and a transfer of surplus value
to the core. Nicos Mouzelis's description of the Greek economy, the
development of underdevelopment, typifies the situation in all three
countries. In the post-World War II era, Greece's underdevelopment,
he argues, "takes the form of a technologically advanced, highly dynamic manufacturing sector not organically linked with the rest of
the economy, so that the beneficial effects of its growth are not diffused over the small commodity agricultural and artisanal sectors
but are transferred abroad."[25]

Other features of Greece, Spain, and Portugal attest to the absence of self-sufficiency. A wide range of consumer goods, including appliances, that are not produced domestically are imported, as
are many basic staples. The paucity of domestic industry and manufacturing is also revealed in the composition of their labor forces:
approximately 30 percent of Greek labor is still engaged in agricultural pursuits. In Spain approximately 20 percent of the labor force
is in agriculture; in Portugal it is over 30 percent. In addition to the
fact that these are much higher percentages than in industrialized
states, where it is in the vicinity of 3 percent, the population that has
left the rural areas has not been absorbed in significant numbers by
industry in any of the three countries. [26] With the decline of economic prosperity in the metropolitan areas, guest workers have
stopped going to Western Europe and tourism has declined, as has
shipping in Greece. These services are critical for the Mediterranean economies both as components of the GNP and by providing
foreign-exchange earnings. Their current decline has been a major
contributing factor to Greece's, Spain's, and Portugal's deteriorating
balance-of-payments position.

The above discussion does not argue that the characteristics of
Greece, Spain, and Portugal are identical. Each has its own historic

legacy, its own variant social structure with a somewhat different class structure, and varying modes of production, while the specifics of their dependency relationship to the core differ from one to the other. Spain in particular was developing a national industrial bourgeoisie, a process that has been reversed in the last decade. An analysis of these differences, however, is beyond the scope of this inquiry. What is important for our purposes is that despite varying patterns, all three at present share the critical features of dependent economies.[27] Their economies are dual: one sector more traditional, agricultural or petty bourgeoisie; the other sector foreign owned and export oriented. The underlying economic doctrines of the European Community (a market economy, an international division of labor based on comparative advantage, and the goal of increasing productivity) result in exploitation of those regions that are economically less developed. In addition, the EC provides the institutional framework for the freedom of action of multinationals: mobility of capital, right of establishment, and absence of significant regulatory measures restraining their activities. Accession will symbolize abandonment of efforts at relative autonomy. For Spain, in particular, entry will constitute a solidification of its structural integration into the world economic system in a peripheral position. In fact, the European Economic Commission felt that Spain, of all the three, would be the most severely threatened because of its past protective policies against Western industrial goods. Specifically, it stated that "the dismantling of tariffs will have a greater effect on those firms which owed their development or their survival to the protection provided by the customs tariff or by non-tariff measures."[28]

As stated earlier, ironically, expansion of the EC is taking place just at the time when the world economy as currently constituted appears near collapse. The three Mediterranean countries are experiencing the problems all states are confronting in this period of recession: inflation, rising unemployment, decline in economic growth, balance-of-payments difficulties, and decline in tourism. Perhaps more significant as indicative of long-run trends, however, is that entry into the EC is taking place as Western European manufacturers are shifting production not to the Mediterranean periphery but to the Third World, where the cost of labor and the conditions of work are more favorable. The incipient trend toward establishing productive facilities in Greece, Spain, and Portugal is already shifting toward even less-developed countries despite integration. In fact, recent developments in Greece may well foreshadow the future. Not only has investment declined, but there are signs that disinvestment is taking place. All in all, the prospects of the impact of membership on the three are dim; they are not only being affected by the general economic crisis, but they do not possess enough self-sufficiency or control over their own resources to counter the deleterious consequences.

CONCLUSION

Mainstream theorists presume the autonomy of the nation-state and the capability of its political leaders to formulate policies geared toward the modernization and development of their economies and societies. The empirical realities of Greece, Spain, and Portugal bely the validity of this presumption. On the contrary, they are integrated into a world economic system within which their position is one of dependency in relation to the core members of the European Community. As a consequence, the development of their economies is determined by the needs and demands of the industrialized states, which inevitably impose structural constraints on the Mediterranean countries' policy options. Such a view of their relationship to the core, however, assumes a consensus among participants, particularly the political elites, regarding the inevitability and the long-run desirability of entry into the EC, whereby all the relevant actors behave accordingly.[29]

By and large, a consensus regarding membership in the EC has prevailed. Opposition in Greece, Spain, and Portugal, with one important exception, has either been muted, articulated in traditional ideological terms, or voiced in the context of pluralist interest-group politics. Thus, the orthodox communist parties in Greece and Portugal and the Catalan Communist party in Spain oppose membership on ideological grounds, contending that the EC is a further advance in the stage of monopoly capitalism. The leftist socialists express unease: some argue that membership should be opposed, others that the European Community should be transformed into a socialist community from within. Interest groups in all three countries, on the other hand, operate within a framework in which they accept the legitimacy of the basic tenets of the EC. Thus, they concede the long-run developmental potential of membership for their nations, but meanwhile, they operate as pressure groups on their respective governments to obtain terms in the agreement of incorporation beneficial to their particular interests. In fact, those segments of the Greek and Spanish bourgeoisie who are not tied to foreign capital express increasingly vociferous objections the more imminent EC membership appears to be.[30] Furthermore, they often argue for long transitional periods to protect themselves from the damaging effects of incorporation and to provide time for reforms and adjustments.

The only significant political leader in any of the three Mediterranean countries to reject mainstream theoretical constructs and to argue dependence as the grounds for opposition to membership in the European Community was the socialist opposition leader in Greece, Andreas Papandreou. After the collapse of the military junta in 1974, the return of parliamentary government, and Papandreou's return to

Greece from political exile, he organized the mass Left socialist
Greek political party. On the issue of the EC, he argued that Greece
was not equivalent to the industrialized states of the European Community and that membership would be detrimental to the interests of
the Greek people.[31] The EC, he argued, was an economic entity
designed for the benefit of European multinationals, and entry would
result in further exploitation of Greece. During the election campaign and after the overwhelming victory of his party in October 1981,
Papandreou, publicly at least, moderated his stance on the EC.
Faced with the constitutional reality that the president, Constantine
Karamanlis, the architect of Greece's entry, can veto any move for
withdrawal, Papandreou now argues for renegotiation of the accession agreement so that the terms can be altered in a manner more
favorable to Greece. If renegotiation is not successful, a likely
prospect in view of the sharply divergent perceptions regarding the
nature of the European Community and Greece's position and role
within it, Greece may feel compelled to withdraw. Papandreou's
government may well provide an empirical test of the feasibility of
a peripheral state detaching itself and pursuing autonomous policies.

In the latter part of 1981, two significant developments trans-
pired, both of which potentially pose severe challenges to the liberal
economic philosophy on which the EC is founded and to the future of
the EC as currently constituted. The election of François Mitterand
and the Socialist party in the last elections in France brought to power,
in one of the core countries of the EC, a leader who albeit committed
to European integration, seems equally committed to a socialist
ideology fundamentally at variance with the ideology of the EC. His
proposals for nationalizing additional sectors of the French economy
(adopted by the National Assembly in December 1981), his intent to
eliminate foreign control over French industries, and his domestic
economic policies to deal with recession, which differ sharply from
those of other EC countries, have elicited sharp opposition within the
EC. The consistency and compatability of the nationalization program
with the Treaty of Rome and with commission regulations have been
questioned both in the commission and in the European Parliament
and promise to remain a divisive issue.[32]

Papandreou, not only a socialist but a socialist who considers
dependency theory as explanatory of Greece's underdeveloped state,
potentially poses a more severe challenge to the EC by rejecting its
mainstream notions of economic development. His policy of social-
izing industries, geared primarily to those industrial sectors of the
Greek economy that are not competitive and would suffer most from
integration into the EC, and his intent to restrict foreign investment
to projects deemed beneficial to Greece challenge liberal economic
doctrines.[33] France, despite the interdependence of the world econ-

omy, possesses sufficient autonomy to formulate and implement policies protecting its home market from foreign competition, but the
structure of the Greek economy, its dependence, and the lack of domestic consumer industries may make a policy of autonomy difficult,
if not impossible, to successfully implement in Greece.

Future developments in Portugal and Spain cannot be foretold.
The political situation in both is highly fluid, and compared with
Greece, their parliamentary system is less institutionalized. Within
the current political spectrum there is no forceful political figure or
political party in Spain or Portugal arguing that their underdevelopment is a consequence of exploitation by the center. The conjunction,
however, of external economic crisis and internal social, economic,
and political turmoil, particularly in the eventuality that Greece's
policies may have a measure of success, makes Spain's and Portugal's
future relationship to the European Community unpredictable.

Enlargement of the EC was initiated at a time when continued
economic growth and prosperity were projected for the indefinite future. Even under the best of circumstances, the impact of membership on Greece, Spain, and Portugal would not have resulted in the
anticipated modernization and development. Under conditions of economic crisis, membership may be detrimental, and those very economic links that hitherto created an aura of relative prosperity may
be the economic undoing of Greece, Spain, and Portugal.

NOTES

1. Spain was ruled by the dictator Francisco Franco from 1935,
when the Royalists won the civil war, until 1975. Portugal was ruled
by Oliveira Salazar until 1968, when owing to internal strife, he was
replaced by Marcelo das Neves Caetano until the Portuguese Revolution of the Roses in 1974. Neither country has had an extended tradition of democratic norms and parliamentary rule. Greece has had
more experience with the formal institutional structures of parliamentary government. After a devastating civil war at the end of World
War II (1947-50), Greece was governed by authoritarian right-wing
governments until the military coup of 1967, which ruled until 1974.

2. Opinion on Greek Application for Membership, Bulletin of
the European Communities, supp. 2/76, p. 10.

3. Opinion on Portuguese Application for Membership, Bulletin
of the European Communities, supp. 5/78, p. 8.

4. Opinion on Spain's Application for Membership, Bulletin of
the European Communities, supp. 9/78, pp. 19, 21.

5. Opinion on Greek Application, p. 11.

6. See Chapter 9 above, pp. 151-52.

7. Stuart Holland, Uncommon Market: Capital, Class and Power in the European Community (London: St. Martin's Press, 1980). A thorough discussion of the capitalist ideological premises of the European Community and their implications and consequences.

8. For a discussion of the increasing regional imbalances within the EC as a result of integration, at least in the short run, and suggested strategies for regional policy, see Vincent E. McHale and Sandra Shaber, "Reflections on the Political Economy of Regional Development in Western Europe," Journal of Common Market Studies 15, no. 3 (March 1977): 180-98. In 1978 socialists in the European Parliament in their monthly review, Euso, no. 2 (March 1978), lament the failure of the EC's regional aid policy in not reducing disparities. Specifically cited are the south of Italy and the west of Ireland. The situation has not improved since then; in November 1981 a question posed to the European Commission attacked it for the failure of the ERDF to improve conditions in Calabria, Italy, which has the lowest per capita domestic product in the EC; see New York, Commission of the European Communities, Spokesman's Group, 1981, file no. 4416.11.

9. New York, Commission, Spokesman's Group.

10. The Common Agricultural Policy (CAP) of the European Economic Community is the prime example of the interlacing of national interests responsive to sectoral demands within states that rebounds to the disadvantage of some states, such as the United Kingdom.

11. The exact figures are cited in Loukas Tsoukalis, The European Community and Its Mediterranean Enlargement (London: George Allen & Unwin, 1981), p. 26.

12. See Europe, January-February 1981, pp. 15-16.

13. For a detailed historical discussion of the shifts in policy in all three countries, see Tsoukalis, The European Community, pt. 1 in particular.

14. See "Spain: Business Resists Drive to Join the EEC," International Business, Business Week, May 18, 1981, p. 65; "Political Day," Europe, March 14, 1981, discusses Parliament's request for speeding up entry negotiations as a way of strengthening democracy in Spain.

15. New York, Commission of the European Communities, Delegation to the United Nations Library, "EEC/Portugal: Prospects for Portuguese Industry within the EEC," Business Brief, May 21, 1981, file no. 44102(62).

16. The then Spanish Prime Minister Leopoldo Calvo-Sotelo Bustelo clearly looked to European Community membership as a

guarantee of Spanish democracy. See World Business Weekly, June 15, 1981, p. 12.

17. Since this writing, Spain has become a member of NATO.

18. Speech by Constantine Karamanlis, cited in International Herald Tribune, September 1977, p. 3.

19. Although there is no thorough theoretical analysis using dependency theory, political critiques do exist; see Holland, Uncommon Market; see also a publication of the Left socialists in the Community, Agenor (Brussels), no. 76 (May 1979), p. 17, which stated, "The admission of new members for enlargement to Greece, Spain and Portugal will render more acute the problems of regional imbalance which exist in the present Community of nine, whilst worsening problems of inequality within the new countries. Only if the Community faced up to the need for a major effort of regional development and resources transfer could it be otherwise"; see also "Joining the Sinking Ship," Agenor, no. 65 (October 1977), pp. 3-24.

20. "After 1959 there was a dramatic change in government policy towards foreign investment," in Tsoukalis, The European Community, p. 82.

21. A. K. Giannitsis, "Problems of Greek Development" (in Greek), Oikonomia Kai Koinonia [Economy and society], May 1979; see also Tsoukalis, The European Community, pp. 45-49.

22. For an analysis of Greece's relationship to the EEC as an associate member during the military junta, see Van Coufoudakis, "The European Economic Community and the 'Freezing' of the Greek Association, 1967-1974," Journal of Common Market Studies 16, no. 2 (December 1977): 114-31.

23. Tsoukalis, The European Community, pp. 17-27. This is the term used by the author to describe the expansion during the 1960s and early 1970s.

24. See United Nations, Department of Economic and Social Affairs, Statistical Office, Yearbook of International Trade Statistics (1975, 1978).

25. Nicos Mouzelis, Modern Greece: Facets of Under-development (London: Macmillan Press, 1978), p. 29.

26. See Organization for Economic Cooperation and Development, Labour Force Statistics Report (Paris: OECD, 1980), p. 190.

27. Papademetriou, in a study that is critical of a simplistic use of dependency theory as applied to migration, nevertheless states, "Greece and Spain appear to be in the threshold of becoming incorporated into the politico-economic structure of the west." The author discusses the problems and suggests strategies for dealing with return migration, a phenomenon of recent years. Demetrios G. Papademetriou, Emigration and Return in the Mediterranean Littoral: Conceptual Research and Policy Agendas (New York: Center for Migration Studies, 1981), p. 21.

28. Opinion on Spain's Application for Membership, p. 15.

29. It is important to point out that the relationship of the periphery to the core is not as deterministic as capitalist world systems theorists assume. Domestic social, economic, and political dynamics must be accounted for in order to develop a more meaningful theory.

30. In a meeting with members of the European Commission in 1981, Carlos Ferrer, president of the Spanish Employee Association, while stating his organization's acceptance in principle of Spanish entry, asked for a long transition period of more than ten years. New York, Commission of the European Communities, Delegation to the United Nations Library, EEC/Spain, Institution and Policy Coordination, July 28, 1981, file no. 441.2(47), p. 4.

31. Panhellenic Socialist Movement, "Metropolis, Periphery Independent Development and Socialist Change," Socialist Transformation (International Relations Committee Series B, no. 2), September 1977, pp. 23-31. See also Nicholas Papandreou, "Greece," In These Times (Chicago: Mid-America, 1979), pp. 10-11.

32. See Jane P. Sweeney, "Mitterand's Economic Program: The Constraints of European Community Membership" (Paper presented at Conference of Europeanists, Washington, D. C., April 30–May 1, 1981), for a discussion of efforts made by financial and industrial circles to have the French nationalization program deemed in violation of the EC's precepts of private enterprise.

33. Legislative program presented to Greek Vouli by Prime Minister Papandreou on November 22, 1981.

12

MONETARY INTEGRATION IN A 12-MEMBER EUROPEAN COMMUNITY: THE IMPLICATIONS FOR GREECE, PORTUGAL, AND SPAIN

ROBERT A. BAADE
JONATHAN F. GALLOWAY

With Greece already in the fold and the likely accession of two other Mediterranean nations, Spain and Portugal, the European Community (EC) will grow to 12 members, probably by the mid-1980s. In anticipation of the Twelve, some effort has been expended analyzing the political economy of the expanded EC. One dimension of EC political economy is the monetary arrangement. The current Nine have struggled for monetary union with some measure of success, though much still needs to be done. Some architects of EC integration have promoted the idea that monetary union will spur consolidation in other aspects of the European Community's political economy. Such beliefs undoubtedly will persist through expansion. Should they persist? There has been debate on whether the Nine can be thought of as an optimum currency area, and the introduction of three developing (by EC standards) Mediterranean economies promises to heighten the controversy about just what the optimal currency arrangement for the EC ought to be. Two issues will be examined in this regard. First, can the Twelve be thought of as an optimum currency area or even a viable currency area from the point of view of the three Mediterranean economies? Second, if EC welfare is increased through monetary integration, will the three Mediterranean nations share equitably in the benefits? Or will there be master, or top, currencies in the arrangement that will skew economic benefits and bestow political power on a few core economies?

The theoretical literature on the optimum currency question[1] and on economic dependency[2] vis-à-vis currency dependency is quite extensive. A currency area is composed of several regimes or countries with a common currency unit or permanently fixed exchange rates among the members. This area can be thought of as optimum in membership when it includes the number of regimes or countries that will minimize the total adjustment cost via exchange-rate changes and other adjustment policies. While the minimizing of adjustment costs defines the optimum currency area, there may be negative political and economic consequences from creating such an arrangement. Basically, interdependence may become a mask for various forms of asymmetrical dependencies, both short-term policy costs and long-term structural costs. These costs and benefits must be weighed. To this effect, the most pertinent literature is reviewed and the organizational scheme for this analysis is developed in the first section of this chapter, precipitating an analysis of the issues from which a number of conclusions will be drawn.

THEORY

Theoretically, the optimum-currency-area issue has at least a 20-year history, and the seminal work in the field was produced by Robert Mundell (1961) and Ronald McKinnon (1963). This issue grew from the larger debate on whether fixed or free-floating exchange rates contribute more to world welfare. If free-floating exchange rates provide fiscal and monetary authorities with the freedom to pursue stabilization policies, would not stabilization in every small region in the world be promoted through currency fluctuations? Where unemployment is pronounced, fiscal or monetary stimulus is deemed appropriate, and the depreciation in the country's currency would contribute to the full-employment effort. To carry the argument to its logical extreme, Why not shrink currency areas to the size of villages or even individual firms? Such an argument ignores costs that must be included in the social calculus in determining viable or optimum currency areas.[3] The quintessence of the issue is whether monetary integration contributes more to the individual country's welfare through promoting a more stable, coordinated economy than it diminishes welfare by eliminating economic autonomy in targeting unemployment and levels of economic growth. Such an assessment for an expanded European Community of twelve is the subject of this chapter.

In organizing the analysis, consider a currency welfare function for each member of the currency area of the following form:

$$W_n = W (Y, \dot{P}, \dot{Y})^2$$

where

W_n = welfare for the nth country
Y = real income for the nth country
$\dot{P}$ = price changes for the nth country
$\dot{Y}$ = real income changes for the nth country

From a neoclassical economic perspective, the welfare of any country is an increasing function of its real income, its price stability, and its income stability. This is to say, mathematically,

$$W'(Y) > 0, \ W'(\dot{P}) > 0, \ \text{and} \ W'(\dot{Y}) > 0$$

The formation of a currency area consisting of the Twelve will affect the welfare of Greece, Portugal, and Spain through its effects on the three variables discussed above.

ANALYSIS

Our first task is to assess the impact that monetary integration will have on the real incomes of Greece, Portugal, and Spain. It is useful in this regard to distinguish between the short-term and long-term impacts. Though the stock of labor, capital, and land in the area will not be altered in the short run, neoclassical economists would argue that the efficiency with which these resources are used throughout the currency area will improve. If monetary integration is one aspect of more complete integration, labor and capital resources will gravitate toward locations that promise the greatest return. In addition, the permanent pegging of exchange rates among the members of a currency area eliminates the instability and risk in interstate commerce that result from free-floating currency prices. In a free-floating system, forward foreign-exchange markets alleviate the risk; but every insurance service has a cost. Eliminating this cost theoretically implies increased resource mobility. Thus, specialization in production will be more complete and consistent with comparative advantage, and area welfare should increase.

In the longer run, increased income will induce additional saving, investment, profit, research and development, and technological progress. Further, a more competitive financial environment may result in a reduction of interest rates, which in turn, will generate greater capital formation. For the currency area, then, dynamic efficiency will improve, real income will increase, and welfare will increase.

This neoclassical vision of efficiency-inspired real-income gains is not universally endorsed for a variety of reasons. Foremost

among them is the accuracy of the neoclassical description of resource movements and the subsequent distribution of real-income
gains. Neoclassical economists visualize resources moving from
areas where they are abundant to areas where they are scarce. In
the jargon of developmental literature, unskilled labor moves from
the periphery to the core or center with capital moving in the opposite
direction. Such movements would ultimately lead to equal development in the periphery and core and an equitable sharing of the real-
income gains. The neoclassical core-periphery relationship assumes,
among other things, full employment and competitive labor markets
in the center. If such is not the case, it is conceivable that the core
will select a migration pattern that exacerbates the degree of unequal
development through the depletion of human capital available to the
periphery. Proponents of the view that unequal development is
chronic include Galtung, Myrdal, and Shiller.[4] In Table 12.1, emigration statistics and net capital flows are recorded for Greece,
Portugal, and Spain.

Information presented in Table 12.1 detracts from the neoclassical argument. Emigration from Greece, Portugal, and Spain declined significantly during and after the 1973-75 recession, and it
still shows no real signs of recovery. Since 7.4 percent, 17.4 percent, and 3.4 percent of the work forces of Greece, Portugal, and
Spain worked in EC member states in 1976, the economies of the
three must have been hard pressed to find employment for the significant number of workers no longer employable in the EC.[5] Capital
flows, on the other hand, were not affected adversely by the recession.
In all three countries, net capital flows increased from 1975 to 1979.
Greece even showed an increase from 1973 to 1975. The figures suggest that unemployment in the core decreases the flow of labor from
the periphery without affecting the flow of capital in the reverse direction. Further, a review of labor migration policies in some core
countries suggests that periphery labor is welcome only if a job has
been secured by the migrant worker.[6] Migration occurs more as a
result of demand from industrial sectors in the core than oversupply
in the periphery. Since skilled professions are generally more re-
cession-proof and more needed by the core, it is probably true that
the proportion of skilled to unskilled labor migrating to the EC has
increased since the mid-1970s.

To ascertain more directly the impact monetary integration
would have on resource flows, Table 12.2 was developed. Since monetary integration means permanently pegging exchange rates, migration and net-capital-flow statistics were regressed on exchange rates.
The years used in the analysis were 1967 through 1976 for all three
countries for emigration. Quarterly data for the years 1975 through
1979 were used for all three countries for net-capital-flow analysis.
The results are recorded in Table 12.2.

TABLE 12.1

Net Capital Flows and Emigration for Greece, Portugal, and Spain for Selected Years

Year	Greece		Portugal		Spain	
	Net K – Flow (in millions of U.S. dollars)	Emigration	Net K – Flow (in billions of escudos)	Emigration	Net K – Flow (in billions of pesetas)	Emigration
1967		42,730		92,502		45,169
1968		50,866		80,452		86,104
1969		91,552		70,165		120,885
1970		92,681		66,360		97,657
1971		61,475		50,400		128,139
1972		43,397		54,084		110,143
1973	1,057	27,525		79,517		101,144
1974	1,039	24,448		43,397		55,281
1975	1,082	20,330	-6.2	24,811	143.7	24,477
1976	1,089	20,394	8.8	19,469	136.8	15,496
1977	1,549		24.3	19,543	263.9	
1978	1,640		33.4		142.3	
1979	1,499		68.4		169.3	

Note: Net capital flow (Net K – Flow) is the difference between capital flows into (+) and out of (–) the country; emigration figures are for persons gone longer than one year; Portuguese emigration figures are for metropolitan Portugal only.

Sources: Organization for Economic Cooperation and Development, Main Economic Indicators (Paris: OECD, 1977–81; December 1977, p. 90; December 1978, p. 106; December 1979, pp. 136–37; and July 1981, pp. 116, 136–39; The Europa Yearbook (London: Europa, 1981), 1:1027; United Nations, The Demographic Yearbook (New York: United Nations, 1977), no. 29, pp. 494–95.

TABLE 12.2

Correlation Statistics for Labor and Capital Migration and Exchange-Rate Variation for Greece, Portugal, and Spain

Country	a_1^n	b_1^n	R_2	a_2^n	b_2^n	R_2
Greece	191,043	-4,190.4	.58	-130.09	14.1*	.08
		(1,257.6)			(13.3)	
Portugal	216,731	-5,415.5	.48	-52.5	1.47*	.14
		(2,515.5)			(1.06)	
Spain	490,464	-5,835.2	.21	-78.1	1.8*	.10
		(3,946.2)			(1.59)	

*Specifying a lag structure does not affect the results significantly.

Note: Equations are of the form:

$L_e^n = a_1^n + b_1^n$ (exchange rate) and

$K_m^n = a_2^n + b_2^n$ (exchange rate), where

L_e^n = labor emigration from the nth country,

K_m^n = net capital flows for the nth country, and

R_2 = squared correlation coefficient

If neoclassical analysis is correct, and if resource migration is significantly affected by exchange-rate fluctuations, a pegging of exchange rates should encourage additional resource flows. Though parameters relating labor emigration and exchange-rate variation for both Greece and Portugal are significant, the correlation coefficients are uniformly small, and all other parameters computed are insignificant. It is conceivable that other competitive market forces overwhelm the impact of exchange rates or that noncompetitive market forces tend to dominate. Given the policies adopted in the EC toward migration since 1974, the noncompetitive influences are undoubtedly significant, if not preeminent.

Public policy of various types can counteract the tendencies toward inequality that result from the play of noncompetitive markets. The MacDougall report estimated that in the very long run, if the EC is to become a fully integrated economy, a commitment of 5 to 7 percent of gross domestic product (GDP) is necessary (estimates for the Nine).[7] To fully integrate the economies of Greece, Portugal, and

Spain into the EC would necessitate a higher percentage. The trend of the general budget of the EC, as indicated in Table 12.3, offers limited hope for policy-inspired egalitarianism. Though the budget percentage is increasing, it is nowhere near the 5-to-7-percent range, and the budget reduction for the recession years 1973-75 does not portend well for the future of EC stabilization policy.

Considering the longer term, monetary integration may result in a reduction in interest rates through an elimination of costs that result from exchange-rate variation and through increased competition. Such a development will increase capital formation substantially in Greece, Spain, and Portugal only if there is a significant relationship between gross-fixed-capital formation and interest rates. In Table 12.4, the results of regressing gross-fixed-capital-formation figures on the discount rates for Greece, Portugal, and Spain for 1967-76 are recorded.

From a neoclassical perspective, the results are anomalous. While there is a significant relationship between capital formation and interest rates in Greece and Spain, it is positive; conventional neoclassical wisdom, rather, would describe a negative relationship between the two. The results may suggest that capital formation is financed by foreign capital sources, and that the higher interest rates reflect robust economic activity, which attracts foreign capital.

TABLE 12.3

General Budget of the EC as a Percentage of EC Gross Domestic Product, 1973-79

Year	General Budget as a Percentage of GDP
1973	.53
1974	.51
1975	.35
1976	.62
1977	.63
1978	.81
1979	.81

Source: European Economic Community, Commission, "Global Appraisals of the Budgetary Problems of the Community," mimeographed (Brussels: EEC, March 12, 1979), no. 85, pp. 1-27.

TABLE 12.4

Results from Regressing Gross-Fixed-Capital Formation on
Discount Rates for Greece, Portugal, and Spain, 1967-76

Country	a^n	b^n	R^2
Greece	-53.5	21.15	.95
		(2.5)	
Portugal	34.2	1.93*	.10
		(1.64)	
Spain	-1,481.7	395.5	.70
		(92.1)	

*Results are sensitive to the lag structure specified, though the signs remain the same.

Note: Equations are of the form $KF^n = a^n + b^n$ (interest rate), where $\overline{KF^n}$ = gross-fixed-capital formation in the nth country.

Price stability was the second item included in the welfare equation. The more stable are prices, the more efficiently money performs. In the absence of money, resources are utilized in bartering that could otherwise be used in productive activity. Monetary integration increases price stability and thus reduces the need to resources to facilitate exchange for at least two reasons. First, in a larger area, random disturbances that induce price changes are more likely to be moderated than would be true in a smaller area. The extent to which this is valid generally is determined by the geographic and economic diversity of the area. Second, a permanent pegging of exchange rates among countries eliminates fluctuations in price indexes that result from changes in currency prices. Domestic prices of tradables are determined by the price prevailing in the rest of the world times the exchange rate. Thus, the more trade-dependent countries forming a currency area are, the more likely price stability will be promoted by monetary integration.

In Table 12.5, Greek, Portuguese, and Spanish imports from the European Community as a fraction of their GDPs for the 1970s are recorded in column 1. These fractions were then multiplied by the change in their currency prices in Special Drawing Rights (SDRs) (column 2). The information in column 1 is then multiplied by the data compiled in column 2 to give a measure of import-induced inflation that results from currency fluctuations. The information in Table 12.5 on EC imports to the three countries as a fraction of world imports allows an es-

Relevant Price Stability Statistics for Greece, Portugal, and Spain

	Greece				Portugal			
Year	(1) Imports/GDP	(2) Annual Change in D/SDR	(3) (1)+(2) Percent	(4) Imports and Exports/GDP	(1) Imports/GDP	(2) Annual Change in E/SDR	(3) (1)+(2) Percent	(4) Imports and Exports/GDP
1970	0.184	0.0	0.0	0.284	0.303	0.10	3.00	0.539
1971	0.184	8.60	1.58	0.288	0.315	0.04	1.28	0.557
1972	0.20	0.0	0.0	0.318	0.314	-0.0007	-0.02	0.577
1973	0.25	0.10	2.50	0.395	0.331	0.064	2.10	0.589
1974	0.256	0.025	0.64	0.412	0.414	-0.034	-0.014	0.674
1975	0.269	0.136	3.70	0.432	0.322	0.068	2.19	0.519
1976	0.258	0.031	0.80	0.427	0.303	0.14	4.24	0.471
1977	0.252	0.0024	0.06	0.413	0.33	0.321	10.58	0.508
1978	0.247	0.088	2.16	0.414	0.315	0.238	7.50	0.515
1979	0.255	0.075	1.92	0.421				

	Spain			
Year	(1) Imports/GDP	(2) Annual Change in P/SDR	(3) (1)+(2) Percent	(4) Imports and Exports/GDP
1970	0.144	-0.005	-0.07	0.28
1971	0.135	0.028	0.38	0.277
1972	0.145	-0.039	-0.54	0.29
1973	0.155	-0.005	-0.07	0.30
1974	0.192	0.0	0.0	0.336
1975	0.172	0.019	0.32	0.305
1976	0.181	0.134	2.42	0.321
1977	0.167	0.239	3.99	0.312
1978	0.143	-0.071	-1.01	0.293
1979	0.145	-0.046	-0.67	0.295

	Greece, EEC/World$_m$	Portugal, EEC/World$_m$	Spain, EEC/World$_m$
1970	0.50	0.48	0.40
1971	0.51	0.48	0.42
1972	0.55	0.46	0.43
1973	0.50	0.45	0.43
1974	0.43	0.44	0.36
1975	0.42	0.40	0.34
1976	0.40	0.41	0.33
1977	0.42	0.44	0.34
1978	0.44	0.46	0.35
1979	0.44	0.42	0.36

SDR = Special Drawing Rights GDP = gross domestic product E = escudo P = peseta D = drachma

EEC/World$_m$ = imports from the European Economic Community members as a percentage of world imports to Greece, Portugal, and Spain

Sources: International Monetary Fund, International Financial Statistics, Yearbook (Washington, D.C.: IMF, 1980), 33:195–97, 349–51, 377–79; United Nations, Monthly Bulletin of Statistics 27, no. 12 (December 1973): 193–96; 29, no. 12 (December 1975): 193–96; 32, no. 12 (December 1978): 195–98; 35, no. 12 (December 1981): lix–lxii.

timate to be made on the reduction in inflation that will occur directly from monetary integration. It could mean as much as a 1.9 percent, 5 percent, or 1.9 percent reduction in inflation in Greece, Portugal, and Spain, respectively. The information in column 4 in Table 12.5 and the statistics on EC imports as a percentage of world imports to the three imply that the three would contribute significantly to the diversity of the EC, and that a monetary union consisting of the Nine and the three new countries would promote price stability through increasing the area's capacity for diluting regional, random shocks.

Welfare generally increases with an increase in economic security or a reduction in risk. When income varies, it means unemployment or the fear of unemployment. How will monetary integration affect income variability in Greece, Portugal, and Spain?

While it is obvious that monetary unification cannot do anything about preventing real random shocks to employment equilibrium, it does have an impact on the ability of a member of a currency area to adopt effective countercyclical action. On the positive side, if exchange rates floated, a downturn in economic activity and employment in one region would foster an appreciation in the region's currency price, and the subsequent deterioration in the balance-of-payments position would exacerbate the downturn. If there was a system of fixed exchange rates and area reserves were sufficient, the deflation would induce an improvement in the balance-of-payments position, and the disturbance would be exported in part to a nation's trading partners. Thus, monetary integration leads to a sharing of the burden produced by any random disturbance, while a system of free-floating exchange rates bottles the disturbance at the point where it developed. Given the extent to which Greece, Portugal, and Spain depend on trade, and particularly trade with the European Community, for their livelihood, monetary integration would serve to promote less income variability in the three countries.

The situation described above implies that the Nine would benignly accept a burden imposed on them by the three Mediterranean nations. This seems doubtful. Whenever a conflict of interest does exist, the rules of the game will be modified in a manner that reflects political power. It may be, however, that no conflict of interest would result between the Nine and the new three from a permanent pegging of exchange rates. If we characterized the Nine as collectively experiencing domestic unemployment and a surplus in trade with Greece, Portugal, and Spain, while Greece, Portugal, and Spain are experiencing inflation domestically, a flow of reserves from the three countries to the Nine (adjustment in a fixed exchange-rate system) would serve the interests of both. This nondilemma situation, while it roughly characterizes the collective present, ignores the special problems that the individual countries face. For example, it would be more accurate to characterize Spain and Portugal as experiencing stagflation internally and a deficit externally. Should unemployment increase, the pareto optimum situation could be transformed quickly into a dilemma. In Table 12.6, unemployment statistics and inflation statistics are recorded for the three countries.

Full monetary integration means a sacrifice of monetary autonomy for Greece, Spain, and Portugal. Monetary policy will be determined jointly, and the collective policy will be more than likely less well suited to their needs. If monetary policy is conducted de facto by the area's reserve currency country, as was true in the case of the Bretton Woods system, what would the implications be? If Germany possesses the master currency, monetary policy for the three countries will be tighter than it has been. In Table 12.7, monetary growth rates for Greece, Portugal, and Spain are recorded, along with Germany's rate of monetary expansion.

Whether or not the loss of monetary autonomy poses a problem for Greece, Spain, and Portugal depends on two things. First, it depends on the compatability of the goals of the monetary authority with the individual country in question. Second, it is a function of the effectiveness of monetary policy in facilitating change. Operating under the supposition that the deutsche mark will function as the reserve currency, the three Mediterranean countries may be forced to accept higher rates of unemployment, if a trade-off does indeed exist between inflation and unemployment. This could be vital to the interests of these struggling democracies, particularly in Portugal and Spain, in view of their unemployment problems. Monetary integration could conceivably defeat a primary purpose of the three in joining the Nine— a strengthening of their democracies. Given the importance of this issue, it demands further consideration.

Brendan Brown has noted that if countries follow similar monetary policies, exchange-rate stability in periods over a month is likely.[8] The greater the exchange-rate stability, the less disruptive and problematic the formation of a monetary union is. In Table 12.8, correlation coefficients between drachma-dollar, escudo-dollar, peseta-dollar, and drachma-deutsche mark, escudo-deutsche mark, and peseta-deutsche mark currency prices and the deutsche mark-dollar currency price are recorded.

The results indicate that monetary policies and inflation paths for Germany and the three Mediterranean countries differ, and in the case of Spain, differ dramatically. It is interesting to note that the monetary policies of the United States and the three countries are far more consistent. In any event, the imposition of Germany's monetary policy on Spain, Greece, and Portugal is likely to produce some significant economic changes, particularly in Spain.

The likelihood that changes in the monetary policy imposed on the three Mediterranean countries by the area's monetary authority

TABLE 12.6

Inflation and Unemployment Statistics for Greece, Portugal, and Spain for Selected Years

Year	Greece Unemployment (in thousands)	Greece CPI Change	Portugal Unemployment (in percent)	Portugal CPI Change	Spain Unemployment (in thousands)	Spain CPI Change
1965	—	3.1	—	3.4	—	13.0
1966	—	4.8	—	4.9	—	6.3
1967	83.5	1.7	—	5.5	146.0	6.6
1968	73.7	0.4	—	5.9	182.0	4.8
1969	66.5	2.4	—	8.9	159.0	2.1
1970	48.7	2.9	—	6.4	146.0	5.8
1971	30.3	3.0	—	11.9	190.0	8.1
1972	23.8	4.2	—	10.6	191.0	8.3
1973	21.4	15.6	—	12.9	149.6	11.4
1974	27.1	26.9	1.8	25.3	150.3	15.7
1975	35.0	13.4	4.5	15.2	256.6	16.8
1976	28.5	13.3	6.4	21.1	376.4	15.1
1977	27.7	12.2	7.5	24.4	539.6	24.5
1978	30.9	12.5	8.1	22.5	818.5	19.8
1979	18.6	19.0	8.2	23.8	1,037.2	15.6

CPI = consumer price index

Sources: International Monetary Fund, International Financial Statistics, Yearbook (Washington, D. C. : IMF, 1980), 33:59, 61; United Nations, Monthly Bulletin of Statistics 26, no. 12 (December 1972): 18; 29, no. 12 (December 1975): 18; 32, no. 12 (December 1978): 18; 33, no. 12 (December 1979): 12–13; 34, no. 12 (December 1980): 12–13.

TABLE 12.7

Monetary Growth Rates for Greece, Portugal, Spain, and Germany for 1965–79
(in percent)

Year	Greece	Portugal	Spain	Germany
1965	12.5	10.8	16.4	9.5
1966	14.3	4.7	11.1	4.4
1967	14.1	7.6	6.9	3.3
1968	11.8	4.9	17.2	7.9
1969	5.8	9.6	13.9	10.0
1970	8.0	9.0	8.5	6.4
1971	13.3	4.9	14.8	12.4
1972	16.4	10.6	24.2	13.7
1973	25.4	30.1	26.1	5.3
1974	19.7	16.9	18.8	5.9
1975	18.5	20.7	16.6	14.1
1976	17.5	13.1	20.9	10.2
1977	19.6	14.9	20.8	8.3
1978	19.4	10.2	18.9	13.8
1979	19.8	18.4	13.8	7.5
Average	15.74	12.43	16.59	8.85

Source: International Monetary Fund, International Financial Statistics, Yearbook (Washington, D.C.: IMF, 1980), 33:55, 57.

233

TABLE 12.8

Currency Correlation Results, Quarterly Observations, 1970–80

Other Currency Values	Deutsche Mark-Dollar (R^2)
Drachma–dollar	.55
Drachma–deutsche mark	.85
Escudo–dollar	.49
Escudo–deutsche mark	.704
Peseta–dollar	.025
Peseta–deutsche mark	.77

R^2 = correlation coefficient

TABLE 12.9

Relationships between Unemployment and Monetary Growth Rates in Spain, Greece, and Portugal

Country	a^n	b^n	R^2
Greece	82.2	-2.55 (1.07)	.41
Portugal	99.71	-.09 (.05)	.38
Spain	136.14	3.487 (3.96)	.09

a^n = intercept for the unemployment/monetary growth rate equation
b^n = slope for the unemployment/monetary growth rate equation

Note: The relationship between employment and monetary growth rates for Portugal is calculated; parentheses indicate the standard deviation.

will influence economic activity depends on the sensitivity of their economies to monetary policy. In Table 12.9, the results of regressing unemployment on monetary growth rates are recorded.

Our computations imply that there is a significant relationship between unemployment and monetary growth rates only in Greece. The paucity of data on employment in Portugal and Spain may account for the results in those two countries, however. If the results are valid, the implication is that a tighter monetary policy imposed by Germany on the three Mediterranean countries would have a positive impact on inflation without a substantial loss of jobs.

CONCLUSION

In examining a monetary union consisting of the Nine and Greece, Portugal, and Spain, some light has been shed on the viability and desirability of such an arrangement from the point of view of the three Mediterranean nations. The evidence does not unambiguously favor accession. The greatest benefit to the three countries appears to be on the inflationary front, as a tighter monetary policy imposed by the area monetary authority may moderate inflation without inducing significantly greater unemployment. However, to the extent that unequal development between the core and the periphery is chronic, more complete economic integration may serve to frustrate the longer-term economic aspirations of Greece, Portugal, and Spain. Their ability to pursue independent monetary policies would be weakened, and they would be forced to follow deflationary policies imposed on them by the master currency country, Germany. Further, the cumulative effects of monetary integration on these three countries would change their tendency to pursue monetary policies consistent with those of the United States. Whether this will be optimum development or dependent underdevelopment for Greece, Spain, and Portugal awaits future analysis.

NOTES

1. See, for example, Arthur B. Laffer, "Two Arguments for Fixed Rates," and Robert A. Mundell, "Uncommon Arguments for Common Currencies," both in The Economics of Common Currencies, ed. Harry G. Johnson and Alexander K. Swoboda (Cambridge, Mass.: Harvard University Press, 1973), pp. 25-34 and 114-32, respectively.
2. See, for example, Susan Strange, "The Politics of International Currencies," World Politics 23, no. 2 (January 1971): 215-31.

3. For a more extensive discussion of the theory of currency area information, see the following: H. G. Grubel, "The Theory of Optimum Currency Areas," Canadian Journal of Economics 3, no. 2 (May 1970); and idem, "The Theory of Optimal Regional Associations," in The Economics of Common Currencies, ed. H. G. Johnson and Alexander K. Swoboda (Oxford: Alden Press, 1973), pp. 99–118.

4. J. Galtung, "A Structural Theory of Imperialism," Journal of Peace Research 8, no. 2 (1971): 81–117; Gunnar Myrdal, Economic Theory and Underdeveloped Regions (London: C. Duckworth, 1957); and G. Schiller, "Channeling Migration: A Review of Policy," International Labor Review 3, no. 4 (1975): 335–55.

5. Horst Reichenbach, "A Politico-Economic Overview," in Integration and Unequal Development: The Experience of the EEC, ed. Dudley Seers and Constantine Vaitsos (New York: St. Martin's Press, 1980), p. 93.

6. Michael J. Piore, Birds of Passage: Migrant Labor and Industrial Societies (New York: Cambridge University Press, 1979).

7. European Economic Community, Commission, General Report, Economic and Financial Series, A13 (Brussels: EEC, 1977), 1:27 (McDougall report).

8. Brendan Brown, The Dollar-Mark Axis (New York: St. Martin's Press, 1979), p. 5.

13

A RETROSPECTIVE LOOK AT MEDITERRANEAN LABOR MIGRATION TO EUROPE

DEMETRIOS G. PAPADEMETRIOU

INTRODUCTION

It has been over 20 years since workers from many Mediterranean countries began to arrive in substantial numbers in western and northern European advanced industrial societies. This manpower movement was initially hailed as a new and glorious chapter in European cooperation. Unutilized and underutilized labor from southern Europe and, gradually, the Maghreb countries was transferred to democratic Europe's industrial heartland to fuel its economic engine that was experiencing substantial labor shortages. The sojourn was to be explicitly temporary. Upon return to the countries of origin, workers were to provide the human capital essential for the economic transformation of their countries.

It is by now a truism that most expectations about the worker migration phenomenon have failed the test of time. Although most observers would agree with Kindleberger's assessment that Europe's growth during the 1960s was premised on the availability of an infinitely flexible supply of migrant labor,[1] the end of that decade saw the beginnings of a more sober and critical mood in scholarly treatises on the subject.[2]

According to these studies, advanced industrial Europe's decision to respond to its labor bottlenecks by instituting bilateral agreements with countries in the European periphery for a controlled importation of labor was increasingly shortsighted. In fact, this process

237

was leading into a condition in which the labor importers became increasingly dependent on a constant supply of foreign labor while, concomitantly and imperceptibly, the labor supply was becoming more and more independent of the actual labor needs of the host economies. In other words, what had always been assumed to be the biggest asset of the foreign-labor-recruitment course, that is, its ability to act as a flexible cushion that could be called upon as a temporary expedient with which to overcome unusual labor-demand pressure (what the Germans call the Konjunkturpuffer function), became increasingly less reflective of the actual situation. As the migration flows became more mature and labor demand pressure persisted, the temporary and presumably revocable nature of the arrangement began to fade away and, in the resulting policy void, was replaced by the de facto expansion of opportunities for longer-term stays, family reunions, some modest occupational mobility, and all but the formal establishment of an immigration flow. These events, in turn, brought about a significant qualitative change in the character of the labor migration phenomenon as receiving societies found themselves confronting a heretofore virtually unknown—and totally unexpected—ethnic-minorities problem.

It is now abundantly clear that industrial Europe's recourse to foreign labor has not only failed to solve the structural problems of the labor-scarce industrial societies but has actually contributed toward maintaining and aggravating them. In fact, it would be reasonable to argue that except for the obvious economic benefit accruing to the workers themselves and their significant contribution to the short-term profitability of certain classes of private capital[3] (to which they have become indispensable), the importation of labor has otherwise given rise to additional serious economic, political, and social problems.[4] The latter two have been only inadequately and slowly appreciated by most labor-receiving countries.[5]

As a result, the often ad hoc measures by policy makers to address the foreign-worker issue have been repeatedly frustrated. The problems, of course, stem from the fact that many of the premises that have undergirded the recruitment of foreign labor in Europe have been of questionable validity and have unmistakably fostered significant untoward socioeconomic and political repercussions. Labor importers had been reluctant to recognize this, although most have officially discontinued the importation of labor since 1973-74. This posture has been adopted even though the relevant decision makers are fully aware that such restrictions may not only ultimately distort the optimal economic allocation of labor and the international distribution of income but may also polarize center-periphery relations in Europe.

As Table 13.1 indicates, the official discontinuation of immigration mentioned above has had an almost imperceptible impact on the

TABLE 13.1

Migrants in Continental Europe by Major Countries of Origin and Destination, December 1975 and December 1980
(in thousands)

Country of Origin	Austria 1975	Austria 1980	Belgium 1975	Belgium 1980	France 1975	France 1980	Germany 1975	Germany 1980	Luxembourg 1975	Luxembourg 1980	Netherlands 1975	Netherlands 1980	Sweden 1975	Sweden 1980	Switzerland 1975	Switzerland 1980	Total 1975	Total 1980
Algeria			3.0	3.2	420.0	382.1	2.0	2.7					0.2	0.0			425.7	388.0
Austria							78.0	87.2							21.0	19.5	99.0	106.7
Finland							–	3.6					103.0	108.0			103.0	111.6
Greece	2.0	2.1	8.0	10.7	5.0	–	212.0	138.4			2.0	1.3	8.0	7.5	–	4.8	235.0	162.7
Italy			85.0	90.5	210.0	157.6	318.0	324.3	10.7	11.2	10.0	10.0	2.5	–	281.0	233.8	919.2	829.5
Morocco			60.0	37.3	165.0	171.9	18.0	16.6			28.0	34.2	0.5	–			271.5	260.0
Portugal			3.0	6.2	430.0	434.6	70.0	59.9	12.5	13.7	5.0	4.3	1.0	–	4.0	7.5	525.5	526.2
Spain	–	0.2	30.0	32.0	250.0	128.9	132.0	89.3	1.9	2.3	18.0	10.6	2.0	–	72.0	62.1	505.9	325.4
Tunisia			–	4.7	90.0	73.2	15.0	–			1.0	1.1	0.2	–			106.2	79.0
Turkey	26.2	30.1	10.0	23.0	35.0	–	582.0	623.9			38.0	53.8	4.0	–	16.0	20.7	711.2	751.5
Yugoslavia	136.0	120.9	3.0	3.1	60.0	–	436.0	367.0	0.6	0.6	10.0	6.8	23.0	24.0	24.0	30.7	692.6	553.1
Other	21.0	30.8	76.0	212.0	235.0	243.6	328.0	464.8	21.1	24.1	104.0	88.9	60.0	126.1	135.0	122.1	980.1	1,312.4
Total	185.0	184.1	278.0	332.7	1,900.0	1,591.9	2,171.0	2,168.8	46.8	51.9	216.0	211.0	204.0	234.1	553.0	501.2	5,553.8	5,276.3
Total foreign population	244.0	275.0	835.4	903.7	4,100.0	4,148.0	4,100.0	4,454.0	83.5	90.0	280.0	537.8	401.0	421.7	1,013.0	892.8	11,056.9	11,723.0

Sources: Organization for Economic Cooperation and Development, Manpower and Social Affairs Directorate, SOPEMI (Continuous Reporting System on Migration) (Paris: OECD, 1973–81), selected years, 1975–80; Papademetriou, 1976, 1978.

stock of foreign migrants in Europe. Furthermore, in spite of this policy, new work permits have been issued rather liberally, and in certain years, return flows have been smaller than new inflows, particularly because of liberalized family-reunion legislation.

Labor importers clearly recognize that, along with the internationalization of capital, labor is also becoming, by mutual necessity, increasingly internationalized. It is largely because of this development that the Europeans have been unable to control effectively either the size or the composition of their foreign-labor forces (see Table 13. 1), in spite of their inflow control measures and efforts to maintain the temporariness of recruited foreign workers. These efforts have been directed at both the psychological and the administrative levels. With regard to the former, most receivers have proclaimed with regularity that theirs is not an immigration country and have implemented measures that continuously reinforce the tenuous position of their foreign residents. (A notable exception to this rule has been Sweden.) With regard to the administration of often not particularly illiberal legislation, the arbitrary and intemperate behavior of the police authorities charged with responsibility over foreign residents further reinforces these feelings of insecurity and uncertainty.

On balance, then, advanced Europe's migration policies can be said to have been (1) marred by persisting oscillation, superficiality, and ad hoc responses to changing conditions; (2) based on an inadequate appreciation of the almost-inevitable "self-feeding" character of all migratory flows;[6] (3) characterized by overly optimistic expectations of the economic advantages of the recruitment of foreign labor and a corresponding remarkable naïveté with regard to the longer-term social, political, and economic costs of the process; and (4) pervaded by an unwarranted confidence in the importers' ability to manipulate the worker inflow valve at will. For most European countries, such optimism has been unsupported by the evidence. In fact, the suspension of all recruitment since 1974 notwithstanding, the total foreign population in these countries has not diminished (in the Federal Republic of Germany [FRG], for instance, it has increased by about 15 percent, the size of the Turkish community in particular having exploded as a result of liberal family-reunion legislation)[7] despite attrition, often aggressive nonrenewal of residence permits, substantial unemployment, a persistent economic downturn, and some cash bonus programs for those foreign workers who agree to quit their jobs and return to their own countries.

These realities have now led to wide recognition that migration is going to continue to be an independent international force to be reckoned with. Regardless of Europe's future posture vis-à-vis this flow, the migrants already there will continue to present severe systemic challenges for which Europe seems as unprepared as it has ever been during the past decade. Questions still in need of scholarly and, subsequently, policy attention include the impacts of foreign labor on the receivers' labor markets; the effects of large inflows of relatively cheap labor on capital investment; the social intrastructure costs of foreign workers and their families on host societies; the implications of protracted political disenfranchisement of large numbers of people in the midst of democratic societies; the second-generation-migrant problems in such matters as political participation, social advancement, language, and education; the protection of the economic, social, and human rights of foreign migrants; and, finally, the interface between voluntary and involuntary migration and international economic and political relations.

THE MEDITERRANEAN LITTORAL

While the appraisal of labor importation by the European advanced industrial societies has provided a new perspective on migration (see Figure 13.1),[8] the developing countries along the Mediterranean littoral have been much slower in becoming committed to a similar fundamental reevaluation of the process. These countries became involved in the migration process amid expectations of the dawning of a new era. Enveloped in the convulsions attendant to the struggle for socioeconomic development, they saw the industrial societies' offer for a regulated emigration of their unemployed and underemployed workers as an unqualified blessing. What the countries of worker origin did not anticipate, however, was the evolution of the process into an uncontrolled depletion of their already meager supplies of skilled manpower, the most healthy, dynamic, and productive members of their populations; the negative demographic and, consequently, socioeconomic effects of emigration; and, finally, the very marginal socioeconomic gains from the skills and remittances of emigrants. Thus, the initially buoyant coincidence of needs between labor-short and labor-surplus societies has gradually been replaced at both ends of the flow by the view that emigration may have actually contributed to the maintenance and aggravation of the structural conditions to which it had been expected to respond.

Many of the critics of the international politicoeconomic system view emigration principally as a human response to international uneven socioeconomic development and as a tangible expression of dependence.[9] In the words of a recent Organization for Economic Cooperation and Development (OECD) report, what for the receiving countries "answers a need arising from the prevailing international economic system, which it freely accepts, results for . . . [the sending coun-

FIGURE 13.1

A Detailed Causal Model of Migration

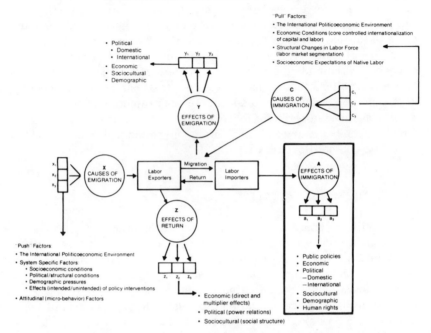

Note: This model offers a detailed and comprehensive view of the causes and consequences of international migration for both sending and receiving countries. It follows the partial least squares (PLS) technique developed by Wold and his associates. The PLS is a latent variable causal modeling approach that occupies a midpoint between data-oriented, narrowly inductive analytical strategies and more sophisticated hard modeling. The softness and paucity of the migration data make PLS an appropriate research tool for studying international migration. The arrow scheme involves manifest (directly observed) variables, which are depicted as squares, and latent (indirectly observed) variables, shown as circles. Analytical complexity is reduced by treating blocks of observables as the structural units of the model. Each block is assumed to have a block structure according to which the manifest variables are treated as linear indicators of a latent variable; the latter is estimated as a weighted aggregate of the indicators. The arrows of the scheme characterize the model's structural relations.

Source: D. G. Papademetriou and G. Hopple, "Causal Modelling in International Migration Research: A Methodological Prolegomenon," Quality and Quantity, in press, fig. 5.

FIGURE 13.2

Determinants of Outmigrations: A Preliminary Latent Variable
Path Model

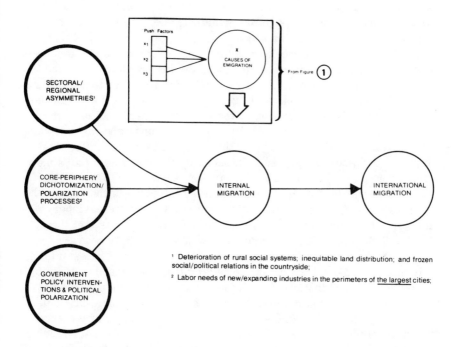

Source: D. G. Papademetriou and G. Hopple, "Causal Modelling
in International Migration Research: A Methodological Prolegomenon,"
Quality and Quantity, in press, fig. 9.

tries] from necessities that are imposed on it by the same system."[10]
Significantly, the developmental asymmetry between the two universes
often widens, in spite (or, perhaps, because) of migration. In effect,
this asymmetry appears to assign developing societies as satellites
to the industrial metropoles and adds migration as a structural component of a world economy (see Figure 13.1) which, especially since
the end of World War II, has undergone an unremitting internationalization of capital. It was within this European framework of uneven
development that a reserve labor army evolved, "raised, paid for,
and maintained by the periphery and placed at the disposal of international capital at a critical juncture in the development of the center."[11]
    The uneven development paradigm's usefulness in the study of
migration becomes even more striking when one looks at sectoral

and regional asymmetries and the consequent dichotomization and polarization between core and periphery within the countries of worker origin (see Figure 13.2). Such intrasystemic conditions release powerful internal migration forces that, in turn, contribute to the social, cultural, and economic isolation and impoverishment of the countryside and, with a time lag, bring about acute socioeconomic deprivations and overload the services in the principal poles of attraction for internal migrants, the perimeters of the largest, and particularly the capital, cities. Such uncontrolled internal migration and the consequent runaway urbanization exacerbate the incongruence between the center's social, economic, and cultural attraction and that of the rest of the country, while constantly reinforcing the latter's dependence on the center and contributing to the persistent penetration of the periphery's social values and cultural institutions, and the undercutting of its economic viability, by the metropolis. [12] Finally, this asymmetry becomes further reinforced by the consistent focusing of most technical and economic innovations, as well as of most significant industrial initiatives, in the capital region (as is the case with most labor surplus societies), thus contributing to the periphery's further stagnation and isolation and denying it the fruits that the effective diffusion of economic development would likely bear. [13]

The severity of this asymmetry can be assessed by examining internal region-by-region data regarding living standards, availability of services, distribution of manufacturing establishments, employment structures, and the structure of the gross domestic product (GDP). These asymmetries place uneven development at the very heart of internal and, together with its operation in the international economy, international migration (see Figure 13.2). However, while the flight from the countryside appears to be a condition pandemic to societies at a certain level of development, other conditions also contribute to this process.

Limited amounts and inequitable and fragmented distribution of arable land
Low rates of return on labor in agriculture
Low levels of agricultural mechanization and subsequent gains in productivity
The socioeconomic and cultural isolation and deterioration of the countryside and small towns

These structural conditions are often further compounded with

High levels of unemployment and underemployment
Low industrialization
High birthrates (in most labor exporters)

An entrenched social and frequently repressive political structure
An overgrown and distorted tertiary sector

While the intensity of these conditions varies with the different labor-exporting countries, they all contribute to and facilitate international emigration.

To recapitulate, the axiom that development and labor migration are intimately related has been an ubiquitous but uncritical fundamental premise of the literature.[14] A necessary corollary to this premise is that as individual labor-exporting countries approach a theoretical threshold in their socioeconomic development, the worker outflow valve will gradually be shut. In fact, such socioeconomic gains are expected to facilitate the evolution of an environment that would provide both social-psychological-cultural disincentives for external migration and socioeconomic (and often political) incentives to return. Equally crucial, attaining this dynamic developmental goal will likely allow a system to gradually depart from the universe of the less-developed labor senders in the periphery and become integrated into the center—a transition that will likely be witnessed by the end of the next decade in both Greece and Spain. The exact place of labor migration in this process, however, has yet to be empirically ascertained.

The task of the analyst, then, is to outline the conditions under which emigration can become a vehicle for economic development. Accordingly, if that relationship is to be confirmed, countries of worker origin must

1. Make certain the emigration draws exclusively on either the unemployed and underemployed or those whose positions can easily be filled by unemployed workers;
2. Prevent the emigration of workers whose skills are in short supply and/or are projected to become in short supply and/or are deemed essential to the sending country's future development, even though at the present they may be abundant;
3. In collaboration with receiving countries, encourage workers to acquire new relevant skills abroad and, through aggressive incentive programs, attract such workers to return to their home countries and induce them to place their skills in the service of their countries' development efforts;
4. Encourage migrant workers to save and remit heavily through such incentives as favorable exchange rates, preferential loan terms, and intensive and free managerial training and to invest in new or existing producer-goods industries at home, rather than spending such money on consumer goods (thus fueling inflation) or on imported commodities (thus depressing the psychological appeal of similar com-

modities produced at home and contributing to further disequilibrium in their countries' balance of payments); and

5. Finally, governments of worker origin must develop an overall development strategy complete with an employment policy that takes into account the potential contributions migrants can make and a socioeconomic development policy that, through various incentives, focuses on the revitalization of the countryside and on industrial decentralization.

Although this list is by no means exhaustive, it is broad enough and flexible enough to allow for a thorough investigation of the migration process for the labor exporters and to facilitate the formulation of concrete policy options.

THE EFFECTS OF EMIGRATION ON SENDING SOCIETIES

It is now clear that the institution of labor migration has been founded and allowed to flourish on assumptions about which significant questions can be raised. Yet, very few efforts have been undertaken that either systematically identify and then offer imaginative ways to improve the structural socioeconomic conditions that fuel emigration or place labor migration at the nucleus of a developing country's development plans. What is needed is the clearing of the underbrush of ad hoc and marginal solutions to emigration in favor of public policies that emanate from a clear understanding of the central position that emigration can occupy in the sending country's total development picture.

On balance, the demographic and socioeconomic effects of emigration have so far been, at best, mixed. The reference is not simply to changes in the composition of the population of entire regions and the depletion of the population base of many villages. One must also take into account the altering of regional demographic profiles by a substantial skewing in both sex and age cohorts as a result of the emigration-induced, high dependency ratios. More important, however, such changes have significant, though not necessarily always negative, implications for the social organization and power relationships in the life of villages and small towns. In view of the fact that, in spite of some persistent assumptions to the contrary that appear in the literature on return migration, many European migrants opt to return to the rural or small-town communities from where they came (regardless of whether they emigrated directly from there or with an intermediate stop in a larger city),[15] one must study the linkages between remittances and actual returns, on the one hand, and changes in the social and political power structure of these communities, on the other.

While the impact of individual returnees on such relationships has been correctly discounted in the literature,[16] a significant and regular return flow will likely strengthen the cumulative potential for change by creating an atmosphere initially of confrontation and eventually of change as modernity meshes with tradition. Although there are very few studies to support this admittedly rather optimistic view, the reason may be the dearth of anthropological studies focusing on the effects of permanent return migration (rather than emigration) on communities. This void can be attributed to the rather recent character of substantial return flows (see Table 13.2). In any event, the preconditions for a return movement to become an agent of change include that the return should be voluntary, to the community of origin, and part of a regular return flow. Furthermore, returnees should have maintained an active interest in the affairs of the community reinforced both by regular visits and through the proxy of an activist family. Finally, once a tradition of return is established (where only one of emigration existed before), the architects of the transition in social and power relationships must show both financial success abroad and financial responsibility upon return.

Another major component of the international migration literature addresses the acquisition of skills by emigrants. The question of skill acquisition or skill enhancement is clearly part of the mythology of migration. While it cannot be denied that emigrant workers receive exposure to the discipline and rhythm of industrial life, such qualifications are only a passive byproduct of industrial life, and, in the majority of cases, Mediterranean emigrants have received such exposure prior to emigration. This is not to deny that this gain might not be more relevant for emigrants originating in countries considerably less developed than most of those in the Mediterranean littoral and where, presumably, industrialization and the resulting proletarianization of the work force are only marginal.

With regard to the enhancement of skills and/or acquisition of new technical skills, the myth confronts an unpleasant reality at various levels. The labor markets of the receiving countries are not usually in need of skilled workers. Instead, they need workers able to perform repetitive tasks at the lower end of a highly articulated industrial-jobs hierarchy. Thus, it is not in the interest of employers to offer foreign workers skill-enhancement opportunities and especially formal technical training. Furthermore, even if such training was more liberally available, many foreign workers would eschew it unless it was offered at full pay during the work day. The operative variable here is the goals of migrants. They have not emigrated in order to advance their skills—only their economic position back home.

TABLE 13.2

Departures from Selected Immigration Countries, 1975-80
(in thousands)

Country	1975	1976	1977	1978	1979	1980
Belgium	40.7	42.1	39.7	42.4	41.6	41.3
Germany	600.1	515.4	452.2	405.8	366.0	385.8
Netherlands	22.1	25.7	24.7	28.1	24.4	23.7
Sweden	21.4	18.7	14.9	15.6	16.3	20.8
Switzerland	121.1	110.3	84.3	63.8	55.8	63.7

Source: Organization for Economic Cooperation and Development, Manpower and Social Affairs Directorate, SOPEMI (Continuous Reporting System on Migration) (Paris: OECD, 1973-81), selected years 1975-80.

Their gaze is usually fixed on their home community. Deferring gratification (by investing in their future in the receiving society through participation in training courses) is unacceptable as prolonging the attainment of their goals. So, if they will work in the evenings, they will be likely to opt for overtime work. In addition, investing in their own human capital through training presupposes that they aspire to work in industrial jobs upon return to their countries. This is also not borne out by the evidence.[17] Finally, even in the few cases where technical skills are acquired, one of three things is likely to occur. First, the worker will often be reluctant to return because of his or her superior economic integration into the host society, especially since, ironically, such workers are more than likely to be subject to concerted integrationist attempts by that society. Second, if they return, they will be reluctant to become engaged in industrial work. One must note that if a worker has acquired substantial new skills and holds a responsible position abroad, he or she is likely to have been a long-term migrant and to have made a clear economic improvement of his or her lot. If this is the case, one is unwilling to work in the home country under remuneration and work conditions that are significantly inferior to what he or she has become accustomed. And last, even if both the first and second circumstances were inoperative, the skills that a returnee is likely to bring back home are often irrelevant to the country of origin, which is not likely to have the type of technologically advanced industry that could use the technical skills of returnees. To this scenario one must add

TABLE 13.3

Foreign-Worker Remittances for Selected Countries and Years
(in millions of U.S. dollars)

Country	1972	1974	1976	1978	1980
Greece	575	645	803	984	—
Italy	844	753	1,370	2,440	—
Portugal	882	1,059	1,014	1,695	2,926
Spain	865	1,071	853	946	1,155
Turkey	740	1,425	983	983	2,071
Yugoslavia	963	1,621	1,878	2,899	4,791

Source: Organization for Economic Cooperation and Development, Manpower and Social Affairs Directorate, SOPEMI (Continuous Reporting System on Migration) (Paris: OECD, 1973-81), selected years 1972-80.

the frequent reluctance of industrial management in the home country to offer responsible positions to returnees for fear that they have been contaminated with Western Europe's syndicalist ideas. [18]

The last major reputed advantage of emigration—the question of utilization of remittances—is equally surrounded by controversy. The dispute is not about the role of remittances in the balance-of-payments situation of the countries of worker origin (see Table 13.3); this role is indeed significant. [19] The difficulty begins when one examines whether or not such funds are placed into productive use and the broad socioeconomic impact of such uses on the sending societies. Research in actual remittance usage shows investment in housing and land purchases accounting for between two-thirds and three-quarters of all remittances. The remaining tends to go toward purchasing consumer goods, retirement of debts, and other family-centered activities. Only a small fraction of the total goes toward investment in productive activities—mostly toward the purchase of agricultural equipment and the financing of service-sector activities (opening of small stores, garages, or purchasing buses and taxis). The problem here is that most of these activities are prototypical examples of clashing individual and societal goals. Migrants utilize their earnings in a manner essentially consistent with their goals for emigration. Yet, such spending behavior has unintended and often serious negative consequences, including

1. That investments in housing distort the real estate market and are responsible for serious inflationary pressure in the building sector. Although such activity does generate significant employment opportunities for sending countries and does have helpful multiplier effects, its intermittent character has few lasting economic advantages;

2. That the propensity to purchase consumer goods with relative abandon in a limited market is responsible for broad demand-pull and, gradually, cost-push inflation along the entire economic structure of the country;

3. That the increasing pressure for luxury imports to satisfy the substantial consumption appetites of emigrant families for advanced services and products often leads to similar behavior by non-migrant households, the latter having had their consumption aspirations raised via the demonstration effect of remittances;

4. That, finally, a corollary to these changing consumption patterns is the inflation of the economic and psychological value of foreign products (and the concomitant depression of that of domestic goods) and the increasing allocation of foreign currency reserves to import such products.[20]

CONCLUSION

The need for developing and implementing systematic and comprehensive programs that would reflect a commitment to making emigration into a sound development instrument must by now be clear. Such programs must focus on occupying the unemployed and helping to raise the standard of living for everyone through productively invested remitted and transferred savings while, in the longer run, not causing any irreparable damage to the society. Such programs should be based on the conviction that emigration can be a temporary phenomenon and can help bring about the transformation of the skill and occupational levels of returning emigrants. Both assumptions may appear unduly optimistic, particularly in view of the record of benevolent neglect so characteristic of countries of worker origin, which have consistently failed either to initiate and implement any comprehensive programs designed to allow themselves to benefit from the savings or the experiences of their returning citizens or to provide for the integration of returnees into their social, economic, and political systems.

It is with an eye toward assisting all countries of worker origin, then, that one must assess empirically the costs and benefits of labor migration and sketch the outlines of a return-migration policy that would begin to prepare these societies for their nationals' return even

before they leave for their trip abroad. Key elements to this policy
would involve the social-psychological-occupational preparation of
the emigrant prior to departure, the aggressive protection of his
rights and continuous contact with him while abroad through the spon-
soring of educational-cultural activities, and the development of long-
term policies that will allow sending countries not only to attract
emigrants back and absorb them upon their return but also to benefit
most from their experiences, thus effectively linking the migration
process closely with national economic and social development strategies.[21]

Although the problem of return migration is not yet particularly
acute or urgent, it is important to anticipate and prepare for its consequences. In developing contingency plans for return migration,
the developing countries of worker origin must recognize it as a crucial public-policy and public-private management problem that, although analytically distinct from emigration, is intimately related,
since many of the conditions that would moderate emigration would
also tend to initiate and maintain a return flow. In today's uncertain
international economic environment, none of the labor-exporting
countries are sufficiently socioeconomically advanced or confident
enough about their short- and medium-term prospects for advancement to embark upon aggressive programs of return migration.
Furthermore, many of the migrants best prepared to make a strong
social, economic, and occupational impact on their mother countries
have been successful enough abroad that their return is highly unlikely. What countries of origin must focus on, then, is building sufficient incentive programs, which might attract some of the more
dynamic and qualified among their emigrants to return.

In closing, one is compelled to observe that 20 years of intense
scholarly attention to intra-European labor-migration movements
have failed to reach a consensus on the effects of these movements
on either the receiving or the sending societies. The reasons are
many. International migration is an analytically complex phenomenon
that has systematically defied long-term cost-benefit analyses and
predictions based on such analyses. The confusion is not only over
the distributional effects of the process and the frequent incompati-
bility between economic and noneconomic societal goals and expectations but also about the economic and noneconomic behavior of migrants. As a result, policy intervention at both ends of the flow has
been haphazard and inept, and both sending and receiving countries
have typically surrendered all potentially useful initiatives to employers and the migrants.

NOTES

1. C. P. Kindleberger, Europe's Postwar Growth: The Role of Labor Supply (Cambridge, Mass.: Harvard University Press, 1967).

2. I have reviewed that literature elsewhere. See D. G. Papademetriou, "The Social and Political Implications of Labor Migration in Europe: A Reappraisal and Some Policy Recommendations" (Ph.D. diss., University of Maryland, 1976), chaps. 2-4.

3. This capital is invested in peripheral industries (which have a thin profit margin) or in previously robust industries that are devolving under the pressure of precipitous demand declines. Such demand declines are usually the result of high labor costs (which make the product uncompetitive in the world market) or of failure to invest in new technology.

4. The economic problems include the failure of industries with unlimited access to foreign workers to capitalize sufficiently so as to remain competitive in the world market and the infrastructure costs associated with large numbers of foreign stocks. The social and political problems are not only a function of the persistent economic downturn but include such matters as cultural heterogeneity, political demands, and perceptions of job and general resource competition.

5. See D. G. Papademetriou, "Rethinking International Migration: A Review and Critique," Comparative Political Studies, Winter 1983, for a discussion of these themes within the context of the segmented-labor-market theory. On the political dimension of foreign labor, see M. J. Miller, "The Political Impact of Foreign Labor: A Reevaluation of the Western European Experience," International Migration Review 16, no. 1 (Spring 1982): 27-60; and G. Minet, "Spectators or Participants? Immigrants and Industrial Relations in Western Europe," International Labour Review 117, no. 1 (1978): 21-36.

6. By "self-feeding" I mean principally two things. First, the gradual formation of ethnic communities requires its own service infrastructure that serves almost exclusively the needs of the foreign residents, rather than those of the labor-receiving societies. Second, industries become dependent on foreign labor. This dependence takes several forms. Peripheral industries depend on low-wage labor for their survival. Natives will often not work there because of low wage levels, poor working conditions, and, gradually, the additional status problem of working with or in jobs that are "foreign-worker jobs." The same holds true with many of the public-sector service jobs that are shunned by natives (such as streetcar drivers, sanitation workers, and so forth). A more serious form of dependence,

however, occurs when major industries postpone the cost of needed technological innovation in favor of employing more foreign migrants by breaking complex jobs down to their simplest components so that they can be done by semiskilled personnel.

7. Week in Review 13, no. 18 (May 14, 1982): 3.

8. Papademetriou, "Rethinking Migration."

9. See especially A. Portes and J. Walton, Labor, Class and the International System (New York: Academic Press, 1981), chap. 2; S. Castles and G. Kosack, Immigrant Workers and Class Structures in Western Europe (London: Oxford University Press, 1973); and S. Amin, Unequal Development: An Essay on the Social Formations of Peripheral Capitalism (New York: Monthly Review, 1976).

10. Organization for Economic Cooperation and Development, The OECD and International Migration (Paris: OECD, 1975), p. 15.

11. D. G. Papademetriou, "European Labor Migration: Consequences for the Countries of Worker Origin," International Studies Quarterly 22, no. 3 (1978): 381; see also on this broad theme the various chapters in D. Seers, B. Schaffer, and M. Kiljunen, eds. Underdeveloped Europe: Studies in Core-Periphery Relations (Atlantic Highlands, N.J.: Humanities Press, 1979); J. Valenzuela and A. Valenzuela, "Modernization and Dependency: Alternative Perspectives in the Study of Latin American Underdevelopment," Comparative Politics 19, no. 3 (1978): 535-57; F. Cardoso, Dependency and Development in Latin America, trans. M. M. Urquidi (Berkeley: University of California Press, 1979); J. Caporaso, "Dependence, Dependency and Power in the Global System: A Structural and Behavioral Analysis," International Organization 32, no. 1 (1978): 13-43; T. Smith, "The Underdevelopment of Development Literature: The Case of Dependency Theory," World Politics 31, no. 2 (1979): 247-88; and R. Duvall, "Dependence and Dependencia Theory: Notes toward Precision of Concept and Argument," International Organization 32, no. 1 (1978): 51-78.

12. For a full development of these themes, see D. G. Papademetriou, "Emigration and Return in the Mediterranean Littoral: Conceptual, Research, and Policy Agendas" (Paper presented at the First European Conference on International Return Migration, Rome, November 14, 1981).

13. See H. Entzinger, Return Migration from West European to Mediterranean Countries," World Employment Program, Migration for Employment Project, Working Paper WEP-2-26/WP 23 (Geneva: International Labour Organisation, 1978); M. Negreponte-Delivani, Analysis of the Greek Economy: Problems and Prospects (in Greek) (Athens: Papazisis, 1981); Z. Baletic, "International Migration in Modern Economic Development: With Special Reference to Yugoslavia," mimeographed (Zagreb: Economic Institute of the

University of Zagreb, 1982); K. Ebiri, "The Impact of Labor Migration on the Turkish Economy" (Paper presented at the World Peace Foundation Conference on Temporary Labor Migration in Europe: Lessons for the American Policy Debate, Elkridge, Md., June 1980); D. Gregory, La odisea andalusa: Una emigración intereuropea (Madrid: Editorial Tecnos, 1978); K. Hoepfner and M. Huber, Regulating International Migration in the Interest of the Developing Countries: With Particular Reference to Mediterranean Countries, World Employment Program, Migration for Employment Project, Working Paper WEP-2-26/WP 21 (Geneva: International Labour Organisation, 1978); and G. Kottis, Industrial Decentralization and Regional Development (in Greek) (Athens: Institute of Economic and Industrial Studies, 1980).

14. For an often critical evaluation of this axiom, consult W. R. Boehning, Return Migrants' Contribution to the Development Process: The Issues Involved, World Employment Program, Migration for Employment Project, Working Paper WEP-2-26/WP 2 (Geneva: International Labour Organisation, 1976); J. Birks and C. Sinclair, "The International Migration Project: An Enquiry into the Middle East Labor Market," International Migration Review 13, no. 1 (1979): 122-35; Papademetriou, Social and Political Implications; R. Lucas, "International Migration and Economic Development: An Overview," in Internal Migration: A Comparative Perspective, ed. A. Brown and E. Neuberger (New York: Academic Press, 1977), pp. 37-60; J. Connell, B. Dasgupta, R. Laishley, and M. Lipton, Migration from Rural Areas: The Evidence from Village Studies (Delhi: Oxford University Press, 1976); S. Diaz-Briquets, "International Migration within Latin America and the Caribbean: A Review of Available Evidence," mimeographed (Washington, D.C.: Population Reference Bureau, 1980); C. Kindleberger, "Migration, Growth, and Development," OECD Observer 93 (July, 1978): 19-27; J. Mendez and O. Moro, "The Relation between Migration Policy and Economic Development," International Migration 14, no. 1-2 (1976): 134-61; and A. Simmons, S. Diaz-Briquets, and A. Laquian, Social Change and Internal Migration: A Review of Research Findings from Africa, Asia and Latin America (Ottawa: International Development Center, 1977).

15. The few recent surveys of return migrants attest to this. See F. Cerase, "Expectations and Reality: A Case Study of Return Migration from the U.S. to Southern Italy," International Migration Review 10, no. 1 (1974): 13-27; T. Collaros and L. Moussourou, The Return Home: Socioeconomic Aspects of Reintegration of Greek Migrant Workers Returning from Germany (Athens, Greece: Reintegration Center for Migration Workers, 1978); H. Geck, Die griechische Arbeitsmigration (Koenigstein: Hanstein, 1979); E. Mendonsa, "Search for Security: Migration, Modernization, and Stratification

in Nazare, Portugal," International Migration Review 16, no. 3 (1982); D. Seferagic, "Scientific Work in Yugoslavia on Migrant Returnees and Their Impact on the Mother Country," International Migration Review 11, no. 3 (1977): 363-74; L. Velzen, Peripheral Production in Kayseri, Turkey (Ankara: Ajansturk Press, 1977); and K. Unger, "Determinants of the Occupational Composition of Returning Migrants in Urban Greece" (Paper presented at the First European Conference on International Return Migration, Rome, November 14, 1981).

16. See especially Cerase, "Expectations and Reality"; Boehning, Return Migrants' Contribution; and Mendez and Moro, "The Relation between Migration Policy."

17. See the citations in note 13; also, see Papademetriou, "Rethinking Migration"; J. Widgren, The Migratory Chain (Paris: Organization for Economic Cooperation and Development, 1976); M. Trébous, Migration and Development: The Case of Algeria (Paris: Organization for Economic Cooperation and Development, 1970); S. Paine, Exporting Workers (Cambridge: At the University Press, 1974); and M. Werth and N. Yalcintas, Migration and Reintegration: Transferability of the Turkish Model of Return Migration and Self-Help Organizations to Other Labor-Exporting Countries, World Employment Program, Migration for Employment Project, Working Paper WEP-2-26/WP 29 (Geneva: International Labour Organisation, 1978).

18. See Papademetriou, "Emigration and Return," pp. 32-33.

19. See Organization for Economic Cooperation and Development, Manpower and Social Affairs Directorate, SOPEMI (Continuous Reporting System on Migration) (Paris: OECD, 1973-81), various years.

20. On these points, see the following works: for Greece, R. Fakiolas, "Problems and Opportunities of the Greek Migrants Returning from Western Europe," mimeographed (Athens: Center for Planning and Economic Research, 1980); D. G. Papademetriou, "Greece," in International Labor Migration in Europe, ed. R. Krane (New York: Praeger, 1979); X. Zolotas, International Labor Migration and Economic Development (Athens: Bank of Greece, 1966); and Collaros and Moussourou, The Return Home. For Turkey, see S. Adler, A Turkish Conundrum: Emigration, Politics and Development, 1961-80, World Employment Program, Migration for Employment Project, Working Paper WEP-2-26/WP 52 (Geneva: International Labour Organisation, 1981); Velzen, Peripheral Production; Paine, Exporting Workers; and R. Penninx, "Migration and Development: A Critical Review of Theory and Practice: The Case of Turkey," mimeographed (The Hague: Ministry of Culture, Recreation, and Social Affairs, 1982). For Yugoslavia, see I. Baucic, "Some Economic Consequences

of Yugoslav External Migrations" (Paper presented at the Coloque sur les Travailleurs Immigres en Europe Occidentale, Paris, June 1974); Baletic, "International Migration"; and O. Haberl, Die Abwanderung von Arbeitskraeftern aus Jugoslawien (Munich: R. Oldenburg, 1978). For the Iberian Peninsula, see Mendez and Moro, "The Relation between Migration Policy"; D. Gregory and J. Perez, "Intra-. European Migration and Regional Development: Spain and Portugal" (Paper presented at the World Peace Foundation Conference on Temporary Labor Migration in Europe: Lessons for the American Policy Debate, Elkridge, Md., June 1980); Gregory, La odisea andalusa; and Mendonsa, "Search for Security." For Algeria, see S. Adler, Swallows' Children: Emigration and Development in Algeria, World Employment Program, Migration for Employment Project, Working Paper WEP-2-26/WP 46 (Geneva: International Labour Organisation, 1980); and Trébous, Migration and Development. For the Middle East area, see Birks and Sinclair, "The International Migration Project"; G. Pennisi, Development, Manpower and Migration in the Red Sea Region (Hamburg: Deutsches Orient Institut, 1981); F. Kirwan, "The Impact of Labor Migration on the Jordanian Economy," International Migration Review 15, no. 4 (1981): 671-95; and J. Swanson, Emigration and Economic Development: The Case of the Yemen Arab Republic (Boulder, Colo.: Westview Press, 1979).

21. See especially Kindleberger, "Migration"; Boehning, "Returning Migrants"; Paine, Exporting Workers; Adler, "A Turkish Conundrum"; Adler, Swallows' Children; Hoepfner and Huber, Regulating International Migration; Penninx, "Migration and Development"; and Papademetriou, "Emigration and Return."

14

CAMP DAVID AND AFTER:
FOREIGN POLICY IN AN
INTERDEPENDENT ENVIRONMENT

SALUA NOUR
CARL F. PINKELE

Hardly a week goes by when the brutal, bitter conflict between Israel and one or another of the Arab states or peoples is not brought again to center stage. To hold that we have the answers to resolving this multifaceted, multidimensional dispute would be both wrong and presumptuous in the extreme. The intent in this chapter is to bring into sharp focus the forces that brought about the one major past or present effort at a peaceful resolution of the conflict.

The Camp David accords, process, and personalities continue to hold the fascination—positive as well as negative—of commentators and political actors alike. At the time of this writing, the Sinai aspects of the agreements are being finalized, and, even in the face of intense opposition on the part of ultraorthodox and opposition Jewish settlers, Egypt has recovered territory from Israel.

The interconnected points of departure in this discussion are (1) that the Camp David agenda and product appeared upon the Egyptian, Israeli, and United States individual agendas for quite different reasons, and (2) that Camp David was unique because of the fact that historical conditions or circumstances provided fertile terrain from which to reap a modicum of conflict resolution.

Perhaps the most difficult single item to handle in any historical-political analysis such as this has to do with assessing the role of the individual personalities involved. The debate over "person" as set against "process" has been a long and necessary one, although often tedious and needlessly mystical. It will surely not end with either the writing or the reading of these pages. The pitfalls of engaging in a brief discussion of the question must be laid aside and the debate joined, at least briefly.

Almost invariably, discussions of the larger Arab-Israeli conflict call attention to its historical, evolutionary dimensions. Whether one starts with Zionist images of a homeland, the Balflour Declaration, the post-World War I role of Great Britain, or, abbreviating history more, World War II and after, most analyses at a minimum establish a historical context for any aspect of the controversy with which they deal. However, having done a bit of historical analysis, when one comes finally to the designated topic, the historical dimension is dispensed with and the focus switches (often without prior warning) to the individual actors involved. No better example of this flip-flop mind set exists than the abundant and misdirected literature concerned with the whys and wherefores of Camp David.

Sadat proclaimed at the time that Camp David was Jimmy Carter's show. Since the Camp David meetings (the accords were signed on September 17, 1978), it appears that what has been remembered most is the three principal actors—Carter, Sadat, and Begin—clasping hands; thus, Camp David is the product less of historical forces than the much more dramatic instance of these three people coming together and "making history." Camp David became and continues to be a cheery symbol of the triumph of the will over the otherwise dark forces of the past and present.

Without continuing (or, perhaps, prolonging) this discussion too much further, several additional items should be added toward completion of this ahistorical, quite liberal picture of the present Egyptian-Israeli reconciliation. It is hard to overstate the hyperbole and loud applause given to Sadat for his November 19, 1977, Jerusalem gambit. "A man making history" was heard repeatedly. Then, on October 6, 1981, Anwar Sadat was assassinated, and one of the history makers was gone from the play. From his death forward, speculations abound concerning his legacy, or the lack of such. Another piece in the puzzle is the Israeli leader Menachem Begin. Begin's political and psychological personalities are much-discussed items; his dedication to points of view and his seemingly solidly entrenched combativeness are ever present in many serious (as well as almost all) news accounts of Israeli affairs.

The suggestion here is that to focus upon individuals by reducing one's analysis of macro political events to an investigation of the personality of the participants misses the more significant point. That is, individuals—and particularly individuals who make decisions for nation-states—are creatures, if not captives, of historical situations and, more important, historical forces. It is the contention of the authors that processes prefigure events and, thus, the behavior of individuals. This does not preclude the fact that individuals cannot make, do not make, or have not made a difference by framing a particular policy at a particular time. Without capitulating to the argument that departs from an individual-will or great-person perspective, it can be identified that individual actors often account for the timing of events. But this is a far cry from the argument that, in a longer-term view, individuals make history.[1] Paraphrasing Edward Friedman, there are prefiguring limits placed upon the part by the whole.[2] In other words, in the longer course of things, individual options are penetrated by and, indeed, are largely the product of systemic process variables and not micro personality variables. Translating this conceptual-theoretical discussion into a discussion of the Egyptian-Israeli reconciliation, it cannot be argued that Camp David was not to a degree the product of Sadat, Carter, and Begin; it was. However, it is far more significant that Egyptian-Israeli reconciliation was emerging irrespective of the three individuals.

THE CAMP DAVID AGREEMENT

There are three significant aspects of the Camp David Agreement: (1) why it came about; (2) what it contained; and (3) what it signals in terms of its own future and in the broader context—not only of Egyptian-Israeli relations but in terms of Egyptian-Arab relations and in the context of U.S. involvement in other arenas of Mediterranean conflict. Because the heart of this analysis is concerned particularly with the first of these aspects (and to a lesser extent the third) the focus initially and quite briefly will be on the second aspect—what was agreed to at Camp David.

Camp David involved a twofold process. On the one hand, there was the now-completed agreement for Israeli withdrawal from the Sinai and a peace treaty with Egypt. This was the easy part, with the more difficult part being the evacuation of Jewish settlers from such Sinai towns as Yemit. There was to be a United Nations (UN) peacekeeping force in the Sinai (now in place), and restrictions were placed upon the degree of Egyptian military use of the Sinai. Also, of great symbolic importance, Egypt established normal state-to-state diplomatic relations with Israel. As difficult as these points in the agreement were to arrive at and to accomplish, they have proved to be almost simple to execute in comparison with the parts relating to West Bank-Gaza accommodations and the Palestinians.

President Sadat felt compelled to go beyond matters directly affecting Egypt into, specifically, the Sinai. However, the Sinai part of the agreement was of paramount geographical concern for Sadat,

as it would have been for any Egyptian leader in the late 1970s.[3] This
is not to hold that other matters were or are unimportant (for different
reasons) in Egyptian eyes—only that they are less consequential, particularly on a day-to-day basis, than the Sinai. Basically the West
Bank-Gaza and Palestine sections of the Camp David Agreement called
for Palestinian self-government, a reduction and then withdrawal of
Israeli troops, and Israeli-Jordanian political interaction and cooperation and adherence to UN Security Council Resolution 242. Furthermore, the Camp David sessions led to additional U.S. military commitments to Israel, as well as placing U.S. prestige squarely on the
line with respect to a particular reconciliation process and product.[4]

Camp David has, to date, clearly resulted in an impressive
partial success. There has been a meaningful reduction of tension
between Egypt and Israel—one is even tempted to say peace. On the
other hand, there has been everything but success for the sections
dealing with the West Bank and Gaza or the Palestinian question.[5]
There are several reasons why the Camp David accords are only a
partial success and, more interestingly perhaps, why one part was
completed and the other was not.

THE PROCESS

Prior to Camp David, there were two events of almost seismic
proportions to Egyptian policies. The first of these was Egypt's defeat in the 1967 Six Day War; the second was Egypt's virtual defeat in
the 1973 Yom Kippur War, which was for solid domestic reasons con-
verted to a symbolic victory.

The defeat of 1967 had definitely shown that Egypt could no
longer hope to achieve its ambitious anti-status-quo aims, even with
the aid of the Soviet Union. The Soviet Union did, it is true, rebuild
the Egyptian military machine in the months following the defeat, but,
at the same time, the Nasser regime began to understand the extent
to which the Soviet Union intended to keep it on a string.[6] From 1967
on, the Soviet Union adopted the standpoint that no further war could
be permitted in the Middle East. For the Soviets, the risks and costs
of such a war had become unacceptable. Soviet military aid to Egypt
was confined to building up Egypt's defensive capacity. In the event
of an Israeli attack, therefore, Egypt would not be so easily defeated
by Israel and consequently slip into the U.S. sphere of influence. The
Soviet contribution to building up Egypt's defensive forces was, therefore, dependent on the attitude and position of the United States. When
the relations between the United States and the Soviet Union improved,
arms supplies to Egypt were braked; when relations worsened, arms
transfer was increased.[7]

It was clear to the Nasser regime that it could no longer solve the Middle East conflict on its terms by military means. Both the Soviet Union and the United States were agreed that the conflict must be solved by negotiation. The dream of the Nasser regime of Egypt's ascent in the regional and international hierarchy by means of a victory over Israel had come to nothing. Israel's supremacy had become an accomplished fact. If the Nasser regime were to avoid collapse, then it had to try at least to abolish the direct results of the defeat of 1967—the loss of Egyptian territory to Israel. This meant it was necessary to create conditions under which negotiations with Israel on this subject would become possible and to devise a strategy of withdrawal that would allow it to keep face despite the capitulation. However, by the same token, the regime could not enter into immediate and direct open negotiations with Israel.

In any such negotiations from a position of obvious Egyptian weakness, Israel would have dictated its conditions for establishing peace between the two countries, and the Nasser regime would have had to accept these conditions without discussion. Rapprochement with the United States offered a way out of this dilemma, because only the United States was in a position to exert pressure on Israel and to force Israel to make concessions to Egypt. Furthermore, rapprochement with the United States and the West had meanwhile become a necessity for the Nasser regime because it had begun to look for an opportunity to escape from dependence upon the Soviet Union that was increasingly experienced as being oppressive.

By means of its trade and aid relations with Egypt, the Soviet Union had obtained prerogatives, particularly in the fields of foreign policy, military strategy, and economic policy, which it utilized to limit the regime's freedom of action. [8] As long as the regime had been confident that the relationship with the Soviet Union would help it to achieve its aims, Nasser had tolerated these limitations. After 1967, however, there was no prospect of such help because the Soviet Union could not give more economic aid and would not give more military aid. Therefore, Nasser saw little reason to tolerate further the claims of Soviet agenda penetration. In the situation as it emerged after 1967, emancipation from Soviet penetration wrongly suggested that Egypt would regain the degree of independence it had formerly achieved and would be able to pursue a seesaw policy between the great powers; the domestic and international conditions favoring such a policy no longer existed. A turning away from the Soviet Union, therefore, necessarily implied rapprochement with the West, amounting to an exchange of dependence upon the Soviet Union for dependence upon the United States and the Western industrial countries. Under the conditions imposed by dependence upon the West, Egypt could no longer attempt to rise in the regional and international hierarchy by

means of the struggle against Israel. The Nasser regime had recognized this in November 1967 when it accepted UN Security Council
Resolution 242, which both ensured Israel the right to exist and pre-
scribed the ending of the state of belligerency in the Middle East and
the return of Arab territory occupied by Israel in the course of the
June War. Yet, what amounted to an informal capitulation on Egypt's
part did not move Israel to return its territorial booty; neither did
the United States exercise any pressures on Israel to make it comply
with regard to the territorial issue. Israel, on the contrary, proceeded now to convert the ceasefire borders into de facto frontiers by
means of progressive settlement of the occupied areas. In view of
these facts, the Nasser regime was forced to launch what became
the War of Attrition against Israel on the Suez front. This war was
designed not only to bring movement into the conflict, which had stagnated as a result of the 1967 experience, but also to conjure up the
danger of escalation, which would force the United States to abandon
its passive attitude and to bring pressure to bear upon Israel, thus
making a political solution possible.

The Nasser regime additionally attempted thereby to interdict
both the wave of radicalization that had accompanied the formation
and the activities of the Palestine Liberation Organization (PLO) in
Egypt and the threat of an internal rebellion against the regime. This
new offensive demonstrated to the masses that the regime was fully
engaged in the struggle for the national cause. Any opposition and
any form of protest could be declared by the regime to be treason to
that cause.

Without the aid of the Soviet Union, however, the War of Attrition, launched in 1969, would have ended once again with a tragedy
for Egypt. The Egyptian forces were not able to prevent Israeli air
raids on the cities of the Delta and the industrial centers in Upper
Egypt or to make any reprisals against Israel. The Soviet Union,
through its intervention at this point (the installation of SAM 3 antiair-
craft missiles and the dispatch of 1,500 Soviet troops to Egypt in
March 1970), displayed its intention to reinforce Egyptian defense
capacities but emphasized that it was not ready to restore the total
fighting capacity of Egypt to an extent that it could wage a new general
conflict with Israel. [9] The Soviet reinforcement of Egyptian defense
capacity was to be understood instead as a contribution to the attempt
to achieve a political solution to the Middle East conflict. Such a
solution appeared to be hindered mainly by the lack of Israeli readiness to grant concessions and the lack of U.S. readiness to exert
pressure upon Israel to force such concessions.

A new destabilization element in the situation emerged with the
growing strength of the PLO and the radicalization of the Arab masses.
In view of the inability of the Arab regimes to avenge the defeat of

1967, the United States, which wished to maintain a balance in favor of the moderate regimes in the area (to which the Nasser regime officially belonged after 1967), issued the Rogers Plan in June 1970. This plan restated UN Security Council Resolution 242 and included additional provisions for a ceasefire on the Canal front. It was intended, in the main, to influence Israel to abandon its claim for a formal capitulation on the part of Egypt and to make those concessions that would enable the Nasser regime to impose a political solution on the masses without provoking mass uprisings resulting from wounded feelings of national pride. Egypt, ironically, under pressure from the Soviet Union, accepted the Rogers Plan. After some hesitation, it was later accepted by Israel. The conflict between the two countries, though not yet formally ended, had been transferred through these maneuvers to a new plane. Egypt's recognition of Israel's right to exist transformed the conflict into a form of frontier dispute between two sovereign states, and for Egypt the Palestinian problem was thereby temporarily settled.

After 1967 Egypt gave up its policy of challenging the status quo regimes in the Middle East. In so doing, Egypt became reconciled with Saudi Arabia, which together with Kuwait and the Gulf Emirates, was from that point to help carry Egypt's war debts. Because the expansionist thrusts of the Nasser regime had been stopped once and for all, there was no more reason for animosity between Egypt and the status-quo regimes. The moderate oil-rich Arab states paid their share of Egypt's war debts willingly because they thus displayed their readiness to make sacrifices for the "national struggle against Zionism" without having to expose themselves to the dangers of such a struggle. [10]

With Egypt's 1967 defeat, the offensive part of the struggle was already over. Only the PLO was interested in continuing. This is the reason why the PLO was exposed, in the following period, to massive repression by the Arab regimes, extending, in the case of Jordan in 1970, to attempts at physical liquidation. The Nasser regime, nevertheless, retained its revolutionary, pan-Arab rhetoric and continued to show solidarity with the progressive regimes in the Middle East by supporting the coups in Libya and Sudan in 1969 and in the Third World. But (because Egypt had not stood up to the crucial test) its radical posture was no longer convincing. Nasser's regime was no longer the symbol of hope for emancipation from backwardness and alien dominance. In fact, it had degenerated into a symbol of the inability of the late-comers ever to surmount the restraints imposed upon them by their unfavorable dependency situation. The defeat of the Nasser regime was, in the main, positively received in the West. It was seen as a just and exemplary punishment meted out for a regime having the arrogance to attempt to put the regional status quo, and thus potentially the international status quo, into question. One Western commentator remarked that this defeat had the advantage of thoroughly destroying the myth of "Arab Prussia." The Six Day War had "demonstrated that Egypt was not Prussia, and Nasser's death had removed the prospective Arab Bismarck as well." As a result of the shattering of the pan-Arab Nasserist myth, the Arab governments would adopt a "more modest, restrained, sober view of their role and will survey the record and draw the appropriate conclusion—that where Nasser failed they would be advised not to try."[11] Nothing needs to be added to this commentary other than this rhetorical question: What can and will these countries derive from this "modest, restrained, sober view," apart from the perpetuation of their disunity, their increasing development problems, and their dependence upon the industrial countries?

Nasser's death in 1970 brought Anwar al-Sadat to power. As is indicated in the following chapter in this volume, Sadat not only adopted but embraced and intensified the movement toward being a status-quo power. He also fell into step with the shift toward the United States and the moderate Arab states (particularly Saudi Arabia) and away from the Soviet Union. The Soviet attitude toward the Sadat regime was, from the outset, one of distrust. Although the Soviets continued to supply Egypt with arms, they insisted on merely raising Egypt's defensive capacity and not the system's overall fighting power. The Soviet posture had already led to considerable tensions between the Soviet Union and the Nasser regime. In the case of the Sadat regime, which was seeking an opportunity to turn away from the Soviet Union anyway, this tension increased to serious conflict proportions.

In May 1971 Sadat removed the pro-Soviet faction of his regime from power, and in the summer of 1972, although it had conceded to a friendship pact with the Soviet Union in the wake of this purge, the Egyptian government threw out all the Soviet advisers and declared the agreement null and void on the grounds that Egypt had been forced into it.[12] In spite of these acts, Sadat's dramatic efforts to win U.S. goodwill were unsuccessful until mid-1973. The United States made no attempt to force Israel to adopt a more flexible posture toward Egypt, and thus a no-peace-no-war situation persisted in the Middle East. Egypt's debt to the Soviet Union thus continued, and relations with the West did not show any significant improvement. Meanwhile, domestic problems and pressures increased to such an extent that the Sadat government found itself forced to act to put an end to the deteriorating conditions. The 1973 October war was launched in large measure with the intention of overcoming the disabling and, for Sadat, delegitimizing stalemate.

In this 1973 war, the same mechanisms were at work as in the War of Attrition that had continued since 1969. Most of Egypt's war materials came from the Soviet Union, without the help of which Egypt still could not have hoped to achieve the slightest degree of success in its military engagement with Israel. Moreover, the 1973 war, like the War of Attrition, was in the final analysis made feasible because of the support of the Arab oil countries. They had assumed a large part of Egypt's military expenditures since 1967, when it had become certain that Egypt would no longer be in a position to call the status quo in the Middle East into question militarily and that Egypt's interest in a political solution to the Middle East conflict corresponded to theirs. Because any solution could only be achieved by forcing Israel to make some concessions to the Arab countries, it seemed to be necessary to increase Egypt's capacity to exert a certain amount of pressure on Israel; that is, enough leverage to make Egypt once again able to face Israel as a fighting factor, but still not enough for it to become a threat to the general status quo in the Middle East. Thus, although it was started by Egypt, the October war was not part of an offensive strategy comparable to that of the Nasser regime. It was a reflection of, and justification for, Egypt's capitulation to the regional and international status-quo forces. [13]

With regard to avoiding the radicalization of the masses, the 1973 October war, like the 1969 War of Attrition preceding it, was carried out solely by the military apparatus, without the help of a mass mobilization. The status-quo interests that prompted this war were further revealed by the fact that Egypt, to prepare for it, chose to cooperate with moderate countries like Saudi Arabia and the Gulf Emirates rather than with radical Libya or the PLO. [14]

The October war was, of course, portrayed by the Sadat regime as its own great triumph and was used to justify, from that point, Sadat's domestic and foreign policies as well as subsequent measures aiming at the creation of conditions under which the ruling class would be more able to realize its privileges. Therefore, the regime held from the public two facts that could have destroyed the image of Egypt's (or even Sadat's personal) heroic victory over Israel: first, Syria's large contribution to the partial victory that Egypt did achieve, which cost Syria 50 percent of its industrial potential and the bombing of all its major cities (a fate that Israel thought better to spare Egypt);[15] and second, the fact that Egypt would no doubt have definitively lost this war as well had not the United States intervened in time to force Israel to accept a ceasefire. To be sure, Egyptian troops had crossed the Suez Canal and occupied portions of the East Bank; but two days later Israelis had also crossed the Canal in the opposite direction and occupied parts of its West Bank. Whether the Egyptian army would have been able to carry out the war to any decisive conclusion remains unclear and speculative. What matters in this context is the fact that the United States, in October 1973, forced a ceasefire between Egypt and Israel at a time when Egypt's position had significantly deteriorated.

The reason for U.S. intervention was, apparently, that it was not interested in any settlement of issues in this conflict as a result of which the equilibrium between Egypt and Israel, inherited from the Six Day War, would be disrupted. Disequilibrium would have meant a further protracting of the general Middle East conflict. The United States favored a solution preserving the Sadat regime, which had declared itself a status quo regime and had put Egypt under the protection of the United States. Thus, any solution had to exert leverage on Israel to make some concessions toward Egypt as a first step on the way to establishing a new equilibrium in the region.

In downplaying Syria's role in securing partial success in the 1973 October war and in partly concealing the U.S. role in limiting this war, the Sadat regime presented to the Egyptian people and to the world a picture that in fact was not complete. Sadat portrayed it as a liberation struggle and a proof of the system's ability to challenge Israel's superiority and his own ability to restore Egypt's flagging national dignity. The October war became the fulfillment of the conditions for abandoning the long-standing, challenging attitude toward the regional and international status quo forces and for the regime to concentrate "all its efforts on building up the economy." In fact, Egypt in this war gained neither an increase in its capacity to stand up to Israel with regard to the situation that the latter had created in the area nor a growth in the degree of its independence that would entitle it to be treated equally by the regional and international status quo forces. Initiating the war was rather a desperate act on the part of the Sadat regime. In view of the deteriorating economic situation, it had no other choice but to take a considerable risk in order to force the issue, in the hope of resolving the whole Middle East quagmire without having to lose too much face. Had it not been for Syrian assistance and U.S. intervention, 1973 would have witnessed a repeat of the Egyptian failure in 1967. Given the fact that this 1973 war did not have any unifying effects on the Arab countries and that Egypt's perspectives for overcoming its dependence on Western industrial countries by intensifying its economic cooperation with Arab countries did not improve by this war either, it cannot be maintained that through it Egypt had achieved any additional power vis-à-vis Israel or the status-quo forces. To the contrary, the Sadat regime only made itself more dependent on these forces. [16]

The Sadat regime was basically willing to put through any plausible measure that it seemed might bring it nearer to achieving its goals. It adopted the regional strategy of the United States toward the Middle East, offered the United States restricted use of its military bases, compacted with Israel with regard to pro-Americanism, and altered domestic policies, at least formally, in accordance with Western norms—democratization of political life and liberalization of the economy.

On the diplomatic level, the West did display a positive reaction to these efforts. However, the Western aid that Sadat had hoped for in the form of credits and weapons did not automatically or abundantly materialize. Western private investors were hesitant to move into Egypt because the investment climate there was considered to be too insecure; plus it was no longer necessary to win Egypt over to the Western camp. Egypt's relations with the Soviet Union had deteriorated beyond repair by this time, so that even in the event that Western aid would prove to be insufficient to meet Egypt's needs, it was impossible for Egypt now to leave the Western camp. Sadat's 1973 gambit had basically failed. Israel was not forced to make concessions, and the desired political solution did not prevail. In order to prompt the Western industrialized countries to supply Egypt with more foreign aid and investors, Sadat had to locate another means to soften up Israel's inflexibility. This was a tall order, given Israel's considerable position of military muscle and wherewithall that was buttressed even more by U.S. commitments.

To achieve his regime's stability in the face of the 1967 and 1973 failures, Sadat had to face up to what was basically the only means still available to any Egyptian leader—capitulation to Israeli and U.S. designs. This capitulation involved the official abandonment of the Arab cause and submission to the conditions that Israel, under U.S. protection, set for peace. It was a reflection of the extensive degree to which both Israel and Egypt, but especially the latter, had had their own agenda-setting calculi penetrated by the interests and agenda of the United States. Sadat's decision to travel to Jerusalem in 1977 in order to demonstrate his willingness for peace proceeded from these considerations and realities.

Israel's claims could not be broken by force under the Nasser regime, and it was not possible to overcome Israel's resistance to granting concessions to a defeated Egypt. The Sadat regime had to justify the capitulation in the eyes of the people, either through a limited war or with the help of goodwill. The regime, confronted with growing economic problems and social unrest (most notably the January 1977 uprising), could only try to achieve peace with Israel on the terms the latter would choose to dictate. 17

Starting with Sadat's visit to Jerusalem in October 1977, Egypt's external relations showed an increasingly clear pattern: Egypt tended to become more and more friendly toward the West and Israel, while relations deteriorated precipitously with the Soviet Union and the Arab countries. The Sadat regime also abandoned one by one the goals for which the Nasser regime had struggled and the population had mobilized during two decades, without obtaining anything from

Israel in return. The regime assiduously distanced itself from the cause of the Palestinians, from Arab unity, and from challenging Israel's claim to domination over occupied Arab territories. Now the Palestinians were portrayed in Egypt as "greedy ingrates willing to fight to the last drop of Egyptian blood." Sadat's initiative for the "unification of the three monotheistic religions in the Middle East to face the atheistic threat [of communism]" was really designed to overcome the basic contradiction between Zionism and Arab nationalism and represented a religious variant of the U.S. concept of the "strategic consensus" between Israel and the moderate Arabs.[18]

The population was exposed to a massive propaganda campaign designed to drum into them the idea that Egypt's only hope for a better future lay in maintaining the status quo. Approval by the population —numbed by increasing poverty—was secured through the use of tried and proven techniques: demonstrations orchestrated and financed by the government, repression and imprisonment of right- and left-wing opposition forces who protested against the government's "peace course," and suppression of opposition in the army and the removal of higher officers who rejected peace with Israel through their transfer to civilian positions.[19] The government received genuine support only from the traditional bourgeoisie, who were represented by the Wafd party, and from those factions of the new bourgeoisie for whom this course signaled potential chances for securing still more privileges than those which they already enjoyed.

Although Sadat's dramatic visit to Jerusalem had a surprise effect on Arab and world opinion, it did not affect in the least Egypt's disadvantageous situation with regard to Israel. It did not alter the basic structures that had produced and perpetuated the Middle East conflict—ultimately the superiority of Israel and the inferiority of Arab countries in respect to their capacity for adopting Western industrial civilization.

Egypt had lost every war against Israel, and it had finally recognized the necessity of giving up the struggle against Israel on principle. Sooner or later, therefore, and whether in the form of an officially declared capitulation or in some other form, Egypt would have had to submit to Israel's conditions for peace between them, to put itself in the hands of the United States, and to bury the dream of Arab unity. This meant that the separate peace treaty concluded with Israel at Camp David would have come about with or without the Jerusalem overture, since it was dictated by Egypt's situation.

Sadat's Jerusalem visit was not of the slightest consequence to Israel's inflexible attitude with regard to the peace conditions that it was ready to accept. Given its superior position, that is, its capacity to set its own conditions for peace with any Arab country on the basis of the take-it-or-leave-it principle, Israel had no reason whatsoever, Jerusalem visit or not, to make any concessions to Egypt. Israel's inflexibility, as a matter of fact, was not just an attitude that could potentially be changed by persuasion and goodwill, but rather the necessary expression of its superiority. Begin's position indicated the existence of specific structures and mechanisms of power that could not be changed by Sadat's goodwill and histrionics. Also, the attitude of the Sadat regime did not result from a developed will for conciliation but was simply the necessary reaction on its part to the disadvantageous situation in which Egypt found itself.

Western aid did not compensate Egypt for the loss of Soviet aid, and in attempting singlehandedly to force a solution to the conflict with Israel, Sadat had forfeited further support of Arab countries. In distancing itself from these Arab countries, Egypt destroyed the very minimal basis of unity that had allowed it to achieve partial success in the October war and forfeited the only means by which it later could put pressure upon the United States and Israel—the petroleum card. In view of all this, Egypt had no other choice but to come to terms with Israel under the conditions the latter desired—by concluding the Camp David accords, as it turned out.

The Camp David accords reflected the inequality between the parties who signed them. They contained long-term disadvantages for Egypt, while the long- and short-term advantages went almost exclusively to Israel and the United States. Egypt's short-term interests were harmed by the dissociation of the Arab countries from it in the wake of Camp David: transfer of the seat of the Arab League from Cairo to Tunis, suspension of Egypt's membership in all Arab regional organizations, and loss of Arab credits and aid to Egypt (only Oman and the Sudan, the regimes that depended on Egyptian military aid for stabilizing their rule, continue to support it). The disadvantages will be, however, of a short duration, because the majority of Arab regimes, according to the dictates of their own international situations—the Arab countries are disunited and incapable of uniting under the existing conditions; they are for the most part underdeveloped and totally dependent on Western industrialized countries; and their social and class structures do not allow the existence of any but status-quo-oriented regimes (which are threatened by internal changes or potential changes)—cannot radically change the situation, and they too will have to submit to the status-quo pressures and will sooner or later accept Israel's peace dictate and seek conciliation with Egypt.

Further damage incurred by Egypt through the Camp David accords involved the loss of considerable sovereignty over the Sinai. Egypt did regain control over the oil fields located there, at the cost of losing development aid from the Arab oil countries that was higher than the income that could be obtained through the export or utiliza-

tion of the Sinai oil.[20] It is of critical significance to note that Egypt
was forced to relinquish military control over the area to a multilateral armed corps charged with overseeing compliance with those
terms of the treaty that forbade warlike activities between Israel and
Egypt.

In addition, two other advantages that the Sadat regime hoped
to obtain for Egypt through the Camp David accords have not significantly materialized. First, the United States was expected, in return
for Egypt's extreme readiness for cooperation and for respecting
peace, to reconsider its relations with Israel and to dramatically improve its relations with Egypt. Second, Egypt's military expenditures
were to be drastically reduced upon termination of the state of war
by means of the accords.

The relations between the United States and Israel proceed in
accordance with principles that, because they are determined by the
vital interests of both countries, could not be repealed by any amount
of Sadat's cunning or goodwill. Neither Sadat's conciliatory attitude
nor Begin's uncompromising behavior had much effect on these relations. Israel remained the most dependable partner for the United
States in the Middle East owing to the fact that, given its political
culture and level of technological development, it constitutes a part
of the Western industrialized world, while Egypt remains an unstable,
beleagured, and dependent system that is plagued by economic and
social problems and contradictions. Ultimately, the United States
can realize its vital interests in the Middle East as much without as
with Egypt's help, while Israeli cooperation remains indispensable
for U.S. strategic purposes.[21]

The further assumption on Sadat's part that Egypt's economic
situation would improve as a result of the reduction of military expenditures following the conclusion of the peace treaty with Israel
also turned out to be wrong. In the first place, military expenditures
were not reduced, because Egypt still had to maintain its defense
capacity. In the second place, signing the treaty with Israel isolated
Egypt in the Arab world and even exposed it to the hostility of the
more radical countries (especially Libya), which is a reason why
Egypt is forced to keep its military apparatus fully equipped and even
to develop and expand it. In addition, after becoming a status-quo
power, Egypt was charged with police and security tasks in the Persian Gulf and Africa. In order to fulfill these new obligations, it has
been necessary to continue building up the armed forces. Besides,
although Egypt concluded a peace treaty with Israel, it has still not
overcome the structural differences that exist between them because
of unequal levels of development, which had earlier contributed to
the conflict between them and which possibly could produce another
such conflict in the future. In view of this potential conflict, Egypt

(notwithstanding the peace treaty) must continue to arm itself and to be ready to face any military eventuality. As a matter of fact, though military expenditures had constituted an excessive burden for the weak Egyptian economy prior to the treaty, afterward they tended to become an even greater burden. Prior to Camp David, Egypt received considerable military aid from the Arab countries (7 billion dollars within six years), and such aid stopped after the conclusion of the treaty, while the expenditures continued to grow and Western aid was not raised commensurately. [22]

The country that obviously profited most from Camp David was Israel. Not so obvious, but just as real, was the success of the United States, which, as was indicated, found in Israel its primary ally and in the achievement of a status quo its principal Mediterranean and Middle Eastern goals. The collapse of the Arab alliance against it led automatically to the strengthening of the Israeli position in the Middle East. From that point on, Israel could pursue an unrestrained policy of "what is annexed is annexed, as long as we can hold on militarily," in other words, the establishment of settler colonies in the occupied areas, annexation of the Golan Heights, and transformation of Jerusalem into the eternal capital of Israel. This attitude was manifested especially in Israel's categorical refusal to meaningfully discuss the problem of the Palestinians, which is the central issue of the Middle East conflict. At Camp David, Israel would only consider the formula concerning the Palestinians that it had already put forth in 1967 (autonomy and self-rule), according to which the Palestinians were to be integrated into the Israeli state as second-class citizens. [23]

Indeed, before the dust had but settled after Israeli troops and settlers had withdrawn from the Sinai, the Begin administration announced that this was the last return of territory anywhere and that Israel's position on further talks was one of resoluteness and no compromises. Also, for his part, Mubarak talked tough, and while the Egyptians tried to show the face of a victor, the desperateness of their position continued to be in evidence.

As a direct result of the Camp David Agreement, Egypt forfeited its position of leadership in the Arab world. It was replaced in the main and for the time being by Saudi Arabia, which has attempted, with little success to date, to make all the moderate regimes adopt a common basis of action with regard to the Middle East conflict. The moderate regimes are those Arab regimes—with the exception of Syria, Algeria, Libya, and the radical faction of the PLO that had come together in the Rejection Front—that constituted the Front of Silence, which united the majority of the Arab regimes against Sadat's peace policy. [24]

In November 1981 Saudi Arabia took the political initiative with the Fahd Plan. This plan was received enthusiastically in the West

because it displayed new signs of a readiness in the Middle East for
negotiations with regard to the conditions under which the Arab regimes would formally accept the status quo in their relations with Israel. The Saudi leaders had long informally practiced their international relations from the standpoint of preserving the status quo but
the Fahd initiative was thought to signal a willingness to justify it to
the Arab masses. In fact, however, the Fahd Plan contained nothing
that went beyond the plans put forward by the Nasser and the Sadat regimes since 1967. Those plans essentially demanded that Israel should
make concessions to its Arab neighbors—withdrawal from the territories occupied in 1967, liquidation of the newly built settlements in
these territories, and creation of a Palestinian state of which Jerusalem should become the capital—so that the Arab regimes would be
in a position officially to recognize its existence, justifying before the
Arab peoples the ending of the national Arab struggle against Zionism.
To date efforts to sell the Fahd Plan among Arab leaders have not
been successful, for example, the failure of the Rabat summit.

Israel has already rejected earlier proposals of the sort included
in the Fahd Plan when they were made by Nasser and Sadat. And, at
Camp David, Israel had shown that it was capable of imposing upon
Egypt, the most powerful Arab country, a dictated peace that accorded
almost exclusively with Israel's interests and by which Egypt had
nearly nothing long term to gain. After Egypt had left the Arab Front
and this front had crumbled into radical and moderate sections, there
remained in the Arab world no force that could stand up to Israel to
the extent to which Egypt had done that during the Nasser regime. So,
given the fact that Israel had refused to make any concessions to the
Arab countries when it was confronted with Egypt, which was an op-
ponent to be (comparatively) reckoned with, now that there was no
force on the Arab side that could bring pressure to bear, Israel actually had absolutely no reason to make such concessions. Thus the
Fahd Plan was doomed to failure from the very start. It was categorically rejected by Israel, which could be neither induced nor
forced to accept it. For Israel, it is only a question of time until the
Arab countries recognize its existence under the conditions it will
choose to set without the slightest concessions being made on its part.
Therefore Israel has decided to just wait until this time arrives,
being in a superior power position that imposes no reason to hurry.
This Israeli assessment is probably based upon a correct reading of
the political trends across the region. However, it should be qualified
in one respect, and that involves the case of the Arab oil-producing
countries that have potentially significant roles, particularly in U.S.
and Western core nations' eyes.

The political-economic equilibrium in key countries does not
at present depend upon a solution to the general Middle East conflict.

A new and lengthy period of uneasy status quo in this region lies ahead. The termination of the status quo neither depends upon the goodwill of anybody nor could be brought about by means of any number of plans and initiatives. Changes will result directly from the future development of interdependent conditions within all the countries in this region. The more Arab countries will be forced to open up and are penetrated by Western industrial civilization and subjected to its contradictory effects—particularly the increasing polarization of wealth and need and the disintegration of traditional structures that formerly stabilized political and social life—and the more they will be confronted with the industrial advancement of the more developed countries and their own failure in this respect, the greater will be the impulses and aspirations to change the status quo. However, it remains to be seen whether or not the status quo can really be changed by means of these impulses and aspirations. This is especially the case given the very poor chance most of these countries have to enjoy economic improvement, as so many of the chapters in this volume correctly point out.

NOTES

1. For a most perceptive discussion of this position see Laird Addis, The Logic of Society: A Philosophic Analysis (Minneapolis: University of Minnesota Press, 1975), pp. 159-77.
2. Edward Friedman, "Maoist Conceptualizations of the Capitalist World-System," in Process of the World System, ed. Terence Hopkins and Immanuel Wallerstein (Beverly Hills: Sage, 1980), p. 181.
3. See the discussion in P. J. Vatikiotis, A History of Egypt: From Muhammad Ali to Sadat (Baltimore: Johns Hopkins University Press, 1980), pp. 412-13.
4. See The Middle East, 5th ed. (Washington, D.C.: Congressional Quarterly, 1981), pp. 24-29.
5. See Russell Stone's chapter in this volume to get a sense of just how far Israeli public opinion is away from accepting the Palestinians, let alone a Palestinian state.
6. M. H. Heikal, The Road to Ramadan (London: Collins, 1975), pp. 78-83; idem, Sphinx and the Commissar: The Rise and Fall of Soviet Influence in the Middle East (London: Collins, 1978), pp. 172-89.
7. Heikal, Sphinx, p. 193.
8. M. Hussein, La lutte des classes en Egypte, 1945-1970 [Class struggle in Egypt, 1945-1970], 2d ed. (Paris: François Maspero, 1971), pp. 322, 335; M. Heikal, Ramadan, pp. 79-93, 162;

idem, Sphinx, pp. 212, 221-28, 242-55; and G. Golan, Yomkippur and After: The Soviet Union and the Middle East Crisis (Cambridge: At the University Press, 1977), pp. 14-15, 21-28, 34-42, 63-73.

9. Hiekal, Ramadan, pp. 78-86; Hussein, La Lutte, pp. 317-26; and P. Johnson, "Egypt under Nasser," MERIP Reports, no. 10 (July 1972), p. 13.

10. Hussein, La Lutte, p. 318; P. J. Vatikiotis, "Inter-Arab Relations," and Shimon Shamir, "The Arab-Israeli Conflict," both in The Middle East: Oil, Conflict and Hope, ed. A. L. Udovitch (Lexington, Mass.: Lexington, 1976), pp. 158-59 and 208.

11. M. Kerr, "The Arabs and Israelis," in The Middle East, ed. W. A. Beling, 1973, p. 55.

12. Sadat's speech to the party Congress in Neue Züricher Zeitung, July 26, 1972; see also Soviet comments on Sadat's memoirs in Neue Züricher Zeitung, February 22, 1977.

13. This point is more fully developed in Nour and Pinkele, The Costs of Interdependence: Egyptian Politics from Nasser to Mubarak (in preparation).

14. R. Heacock, B. Poncel, and A. Kielmansegg, "Les revirements politiques de l'Egypte depuis la guerre d'Octobre" [Political changes in Egypt since the October war], Le monde diplomatique, June 1974, p. 15.

15. Ibid., p. 15.

16. See Nour and Pinkele, Costs.

17. Ibid.

18. J. Tucker, "While Sadat Shuffles: Economic Decay and Political Ferment in Egypt," MERIP Reports, no. 65 (March 1978), p. 8.

19. "Class Roots of the Sadat Regime: Reflections of an Egyptian Leftist," MERIP Reports, no. 56 (April 1977), p. 5; J. Stork, "Sadat's Desperate Mission," MERIP Reports, no. 64 (February 1978), p. 12; M. C. Aulas, "Egypt Confronts Peace," MERIP Reports, no. 72 (November 1978), p. 9; and Tucker, "While Sadat Shuffles," p. 69.

20. Arab development aid to Egypt amounted to $7 billion in the six years from 1973 to 1978. Long-term credits and balance-of-payments assistance amounted to an additional average of $1.5 billion a year during the same period. After the return of the Sinai oil fields to Egypt, its oil production amounted, by the end of the 1970s, to 35 million tons a year and generated an income of $2.6 billion per year. Frankfurter Allgemeine Zeitung, June 16, 1979; "Egypt Out in the Cold," The Middle East, May 1979, p. 12; A. Shohayeb, Muhakamat al initah al iqtisadi fi misr [Open door policy on trial] (Beirut, 1979), p. 232.

21. S. Turquie, "Une nouvelle corte: Le relais saoudien," [A new card: the Saudi connection], Le monde diplomatique, No-

vember 1981, p. 3. The whole business of the Saudi connection or the Saudi card in U.S. policy toward Egypt is examined in Nour and Pinkele, Costs.

22. Frankfurter Allgemeine Zeitung, June 16, 1979.

23. Stork, "Sadat's Desperate Mission," pp. 12-14; see also Chapter 17 in this volume.

24. Turquie, "Une nouvelle corte," p. 3; and T. Sommer, "Werben für den Frieden," Die Zeit, November 20, 1981, p. 7.

15

EGYPT UNDER SADAT:
THE CONTOURS OF
ASYMMETRICAL INTERDEPENDENCE

SALUA NOUR
CARL F. PINKELE

The contemporary circumstances obtaining for Egyptian domestic politics and the position of Egypt in the broader context of Mediterranean affairs are critically significant pieces of the general Mediterranean puzzle. In Egyptian politics, indeed, one finds almost a show-
case example of both the interaction between domestic and international
politics and how, because of that linkage, the options available for
policy makers with respect to either dimension of decision making are
severely limited. Furthermore, the Egyptian case demonstrates in
sharp relief the shape and extent of asymmetrical interdependence.

Comparing Egypt's situation in 1982 with its situation in 1952,
it must be said that there has been little change for the better. On
the level of rhetoric, there have been a number of radical changes
in its domestic, economic, and foreign policy in these 30 years. In
practice, however, the continuity has resulted in slow unguided
change, to Egypt's great damage and growing disadvantage.

THE CONTEXT OF INTERDEPENDENCE

The 1967 war, more precisely the defeat in that war, precipi-
tated a watershed in Egyptian policy making. As a result of his crush-
ing defeat, Nasser was virtually forced to turn from a policy that
challenged the status quo forces in the Middle East and the Western

powers toward policies of capitulation and accommodation to those forces.

Upon ascendance, the Sadat regime was confronted with similar problems to those facing the Nasser regime after 1967. Its room for maneuver was limited by the same factors that had determined the room for maneuver of the post-1967 Nasser regime. Basically, Sadat could only continue what Nasser had started; but Sadat had to proceed under steadily worsening economic conditions. The smaller the regime's margin for maneuver became, the more significant were the determining effects of foreign policy on the internal political and economic processes. While the resources the Nasser regime had at its disposal in the early 1960s had allowed it to keep some of its freedom of action on the domestic level, in spite of the existing external constraints, or even occasionally, to act upon and modify these constraints, after 1967 the resources were exhausted. Thus, the regime's margin for maneuver became so limited that it could, from now on, only assume a reactive attitude toward the internal and external problems with which the system continued to be confronted. In a more dependent, less-developed polity, such a loss of initiative necessarily leads to a rapid growth of external constraints. Little by little, as the system became even more dependent, the interests of economic and domestic policy were increasingly subordinated to the concerns of foreign policy and economic relations. A point was reached wherein an external orientation became the most important determinant of the domestic process.

The concept or goal of socialism was gradually set aside as the Nasser and then the Sadat regimes increasingly concentrated their efforts on finding quick solutions to the two problems that were considered then as the most pressing—the rising external debt and the Israeli occupation of Egyptian territory—even at the cost of giving up planning for long-term development. These problems not only forced abandonment of all considerations regarding the requirements of development on the domestic level but also the revision of its foreign-policy strategy. Between 1967 and 1970, the Nasser regime was no longer concerned with realizing Egypt's ascent in the international hierarchy, but only tried to restore the balance of the system after its devastating defeat in the Six Day War.[1]

The Sadat regime, following its predecessor, energetically pursued a course toward accommodation with the West and particularly Saudi Arabia. It was even more pressed than the Nasser regime to come up with quick solutions for the economic bomb that was ticking away. Like Nasser, Sadat was not in a position to correct the negative production-consumption ratio or to reduce nongrowth expenditures, especially for the expanding bureaucracy. Owing largely to the resistance of its supporting class, which profited from these expenditures, Sadat had to try to maintain the system's equilibrium by other means, such as military expenditures and increasing the supply of foreign capital to Egypt. In order to accomplish this, however, Egypt had to be transformed into a reliable status quo system. The majority of the internal or external policies adopted or abandoned by the regime during the 11 years of Sadat's rule were designed in the main to serve this purpose. The regime's attempt to stabilize Egypt in the eyes of the West, that is, to make it credit-worthy, included domestic political and economic measures, the effects of which will be analyzed later because compared with the foreign policy measures, they were of secondary importance to the development process. On the foreign-policy level, this attempt implied that Egypt would press for improving its relations with the United States, which seemed to be alone in holding the key to the solution of the conflict with Israel. In Sadat's opinion, the United States was the only country capable of forcing the latter to make concessions in favor of Egypt.

In order to positively orient the United States toward Egypt, the Sadat regime did not shy at expending almost any amount of effort. Sadat reversed the friendly relations that Egypt had been pursuing with the Soviet Union. In one limited sense, the costs of this move were minimal because Soviet aid had not helped Egypt gain victory over Israel, nor had it helped in realizing development or ascent in the international hierarchy. Soviet help could also not be expected to bring about an acceptable political solution to the Middle East conflict because the Soviet Union was not in a position to put any pressure on Israel. Moreover, relations with the Soviet Union could only destroy the chances for improving relations with the United States, whose help had by then become urgently needed.[2]

THE DOMESTIC LINK

Parallel to the attempt to solve Egypt's external problems by abandoning the role of the challenger of the regional and international status quo, the Sadat regime initiated (on the domestic front) an attempt to revive the economy by dropping the ambitious development plans of the Nasser regime and by creating conditions that favored the activities of private capital. Prior experiences had shown private capital was unable to generate spontaneous economic movement and tended to produce moments of inertia within the system that blocked development. But, in what Sadat perceived to be the existing conditions and objectives, it had grown into the only force in Egypt that seemed to be capable of desirable results and was therefore to be promoted. In addition, the forces associated with private capital had become so powerful that they could not be restrained any more by forces within the ruling class, which still focused upon planned development of the system as a whole.

After Sadat's purge of the Left faction of the regime in May 1971, the switches were set for an unlimited course of liberalization, engagement in which meant that there could be no return for Egypt to Nasser's (or any similar) economic development policies. Liberalization was the reaction of the dominant sections of the ruling class to the fact that the available resources were no longer sufficient to develop the system as a whole and at the same time to implement their own particular interests. Consequently, the system-wide perspective and planning goals were dropped, and top priority was henceforth directed toward the pursuit of self-interest. In the course of this alteration in the policy-direction process, that section of the ruling class that had stood for the acceleration of general development was gradually removed from power (that is, the Nasserist forces that had already been decimated by Nasser's death and by the conversion of many of his supporters inside the ruling class to the particularist standpoint at the beginning of the 1970s). This power switch was not because Nasser's ideas had proved to be ambiguous or ill-suited for transformation of the system, but essentially because the resource basis of the Nasser model had been exhausted once and for all, and because those sections of the ruling class that had now won the upper hand rejected his model only after having achieved a position of power. [3]

The shift away from Nasser's rather more system-wide developmental scheme to a privatist perspective was not a sudden reversal. Indeed, quite to the contrary, the forces favoring a movement toward a private, self-interest view of the future were in place and growing throughout the Nasser regime.

In the first stage of their installation in power after 1952, all sections of the ruling class had had to put aside their particularist interests in order to strengthen their common material basis of power, that is, to develop the industrial apparatus that they were going to control and that was supposed to constitute this base. It is true that during this first phase they also profited from any opportunity to make easy money; however, they refrained in general from activities that would have sabotaged the position of the whole class. In a second phase, the various sections of the ruling class had established themselves in power (in the leadership of the public sector, in the army, in the administration), and each of them sought to grab the largest possible share of the national surplus product. During the third phase, which began in the mid-1960s, after having accumulated great private wealth through their activities in the state and production apparatus, they sought increasingly for a way to use this wealth in private accumulation and thus had to press for the liberalization of the economy. This launched an irreversible trend for the release of the forces of private capital in Egypt; irreversible because there were no forces left in the system to oppose it, a fact that was reflected by Sadat's Opening Policy. [4]

For the sections of the dominant class that were represented by the Sadat regime, the problems facing Egypt that had not been solved by the Nasser regime were: its economic growth had stagnated, its industrial potential was not used to capacity, its negative balance of payments and foreign indebtedness had grown to an alarming degree, its investment and consumption funds were shrinking, unemployment was increasing badly, the living standards of the population were falling constantly, and the potential for social conflict increased in the same measure in which these problems were aggravated and intensified. By pointing out that it had not proved possible to deal with these problems in the years 1961-71, the new government under Sadat justified its rejection of Nasser's political and economic concepts, and launched the so-called Opening Policy. The legitimizing phraseology of "Arab socialism" was retained, but its earlier content was adapted broadly to what the Sadat government now understood to be the pressures for finding a solution to the problems facing the country. The central idea of Arab socialism had been that national capital should be concentrated in the hands of the government, which would itself develop the basis for an independent national process of accumulation because private capital was not capable of constructing such a basis. According to Sadat's criticism, the operationalization of Arab socialism had not produced the desired answers to Egypt's aggravating problems.

As a representative of the Egyptian bourgeoisie, it was not his task to analyze in detail the reasons for the failure of the concept of Arab socialism, but only to draw the simple conclusion from this failure that it was necessary to abolish the principle of the concentration of capital in the hands of the government. That is to say, Sadat felt that the economy should be reprivatized and made healthy by the reintroduction of the principle of competition and by the limitation of the power of the bureaucracy that had retarded development. The private sector was now to be expanded at the cost of the public sector, the inefficiency of which the new government orientation explained by the fact that the activities within its sphere were not based upon economic rationale, but were largely determined by noneconomic considerations. Under Sadat the planning of the economic process was to borrow the form of the indicative planning used in the Western industrialized countries and to avoid hindering the forces of market economy that were to be established in Egypt. In general, the control of government over the economy was to be reduced and the managers of the public enterprises were to have full discretion regarding prices, size of the working force, inputs, and investments. In addition, the state export monopoly was to be abolished, so that there were to be no limitations on the movements of capital and goods across the frontiers. The Opening Policy, introduced in 1974, aimed, in short, at mobilizing national and foreign private capital in the sense of encouraging it to contribute to the development of Egypt's productive capacity. Foreign investments were to create new jobs and thus more income and to help stabilize the balance of payments. The increase in production capacity brought about by these investments was to result in increased exports, and thus an increase in import capacity. This in turn would make a further increase in production capacity possible.[5]

The Opening Policy had really begun in 1971 with the creation of a General Authority for Arab and Foreign Investments and Free Zones (Law 65, 1971), intended to prepare and ease the activities of foreign capital in Egypt. However, the Sadat regime (to justify it to the population) had to wait for a decision in the conflict with Israel before it could fully adopt this policy, through which the Nasserist development model was to be finally dismantled.

Immediately after the 1973 war, the regime hastened to announce the new policy by Sadat's so-called October Paper and by the promulgation of a series of new laws that created attractive investment conditions for private domestic and foreign capital in Egypt.[6] The time had now come to reorganize the public sector, in fact, to dissolve it as far as was necessary with a view to the interests of private capital, and to open up new fields of activity for the latter. The duties of the public sector and those of the private sector were newly defined in such a way that the expansion opportunities of the private sector increased massively. Sadat continued to speak of socialism; however, his socialism was to be a socialism of proprietors and not a socialism of the dispossessed. Competition and state control were to be brought into a balanced relationship, which meant that state control would be confined to those fields of production and distribution that did not interest private capital.[7] In Sadat's socialism of prosperity, the freedom of the individual was declared to be inviolable, but, at the same time, strikes and demonstrations were banned and the state promised the employers that it would ensure law and order among the working population.[8]

It is most significant to note that in the framework of the Opening Policy, foreign capital also was to enjoy the privileges of the private sector, now favored and fostered by the state. For instance, it was to be freed from the laws that guaranteed the employees in the public sector certain social benefits (job security, profit sharing) and enterprise codetermination (two worker representatives on the board) at the cost of the employer, even if it participated in enterprises of the public sector. In addition, stipulations on the transfer of profits were to be liberalized, and the expropriation of foreign capital was declared illegal. [9]

In connection with the Opening Policy, there was also talk of reform within and reduction of the bureaucracy, because the regime had more or less abandoned planning activities. The Ten-Year Plan (1973-82) never materialized, the Transition Plan (1974-76) and the annual plans indicated only a better utilization of underutilized capac- ity, and the reconstruction of ruined facilities and the rebuilding of the Suez Canal Zone consisted of generalized development targets that could hardly be implemented in view of the apparent lack of means. [10] Actually, however, expenditures for the bureaucracy continued to grow under the Sadat regime at a pace surpassing that at which they had grown previously. [11]

In most Western literature, the Opening Policy has generally been treated in favorable if sometimes cautious light. It had sup- posedly led to an acceleration of growth in Egypt (the indicator taken for this being the increase in imports, which includes the increase in imports of production goods, and deficits in the balance of payments being an evidence of a booming economy); to the revival of the private sector, which had shown itself to be "vital and capable of development and which created the impetus for further expansion and liberaliza- tion"; and finally, to an extension of government social welfare poli- cies that "emphasize the distributive equity in the system."[12] How- ever, such a favorable assessment is misleading. Growth has been judged on the basis of inflated figures. The deficit in the balance of payments and the increase in imports should not be taken to display an increase in the production capacity of the system, since the basi- cally negative consumption-production ratio worsened massively after 1974, as is illustrated by the indebtedness increasing to an exorbitant extent. It is true that investments increased, but they went mainly into parasitic fields and tended thus to heighten the propensity to con- sume without increasing the production capacity. As to the "extension of government social welfare policies," the observers must have been victims of crude misinformation, since social benefits were drasti- cally reduced as a result of the Opening Policy, [13] quite apart from the fact that in the years following 1974 the living standards of the majority of the population fell by an alarming degree as a result of inflation, rising prices, and the freeze in wage increases. [14]

The state withdrew from the economic process, accepting re- sponsibility only for expansion of the infrastructure and the service sector. Because there was no longer any authority that coordinated and supervised economic activities with the long-term interests of the country in view, the negative effects of the Opening Policy soon became evident. No additional production capacity came into being as a result of foreign investments. In the free-production zones, the investments served to increase the storage and transport capacity but not the production capacity, and therefore had no significant impact on employment there.[15] The goods produced in the free-production zones were not exported as originally agreed upon, which would have improved the Egyptian balance of payments, but were in fact sold on the domestic market, which caused the ruin of entire branches of the Egyptian consumer-goods industry (particularly the textile, cigarettes, tube, and pharmaceutical industries) that could not stand up to the competition with the more concentrated and highly developed foreign capital.[16]

The liberalization and increase of imports were not compensated by an increase in exports and did not serve to expand production capacity. The state had withdrawn from the economic process, which led to investments in long-term industrialization projects showing a relative decline. There was also a reduction in the import of investment goods and, as a result of liberalization, an upsurge in the imports of consumer goods and luxuries. Large deficits in the balance of payments came into being as the growth of the imports of such goods exceeded the increase in production capacity and, thus, export capacity. The liberalization of imports also led to a flooding of the Egyptian market by foreign products, which drove out locally produced articles. Unemployment became threatening, and the redistribution of income to the disadvantage of the poorer social strata went very fast (500 new millionaires in the period between 1973 and 1977). Increases in wages and salaries lagged well behind prices, and the real income of the major part of the population declined rapidly after 1974.[17]

The government's high hopes for a positive reaction by foreign capital to the Opening Policy were actually extremely exaggerated. (It had predicted a capital inflow of $2 billion and the creation of 120,000 new jobs in the period 1974-77.) The actual capital inflow amounted to $160 million, and only 20,000 new jobs were created during this period. Four-fifths of total foreign investments flowed into the tourism sector and the building and oil industries. Arab capital mainly entered the real estate sector, and other foreign capital was mainly invested in profitable operations outside the agrarian and industrial sectors. Foreign capital showed no interest at all in the capital-goods industry. Despite the Opening Policy, foreign capital found an unfavorable investment climate in Egypt—high inflation rate; mass poverty; political oppression; inefficient bureaucracy; lack of infrastructure; low building capacity; corruption; obstructive customs practices; and anachronistic monetary, commercial, and social legislation—and therefore refrained from becoming too deeply involved there.

The projects financed by direct foreign investment did not meet
the urgent needs of the population (insufficient supplies, deficient
means of transport, rising living costs) and did not increase the production capacity of the system. There was an increasing lack of those
less-profitable investments that served long-term development toward
self-sufficiency. As a result, Egypt's trade and balance-of-payments
deficit, together with its indebtedness, increased to an alarming extent. In 1964 the balance-of-payments deficit amounted to $241 million;
by 1977 it had risen to $7.1 billion. In that same year, 1977, Egypt's
indebtedness reached $12.2 billion and had more than doubled in ten
years.[18]

Up to 1978, credits from the Arab oil countries had prevented
the collapse of the Egyptian economy, but they were no help in preventing the rapid degeneration of economic conditions. By the end
of the decade, Egypt was devoting 77 percent of the capital aid received to servicing its debts. At the beginning of the decade of the
1980s, it needed $2.5 billion yearly in loans and aid simply to repay
old debts.[19] At the end of 1976, Egypt was unable to obtain the necessary credits from the World Bank and the International Monetary Fund
(IMF) without granting these institutions rights of control over the
economy, which were hardly less stringent than those that Great
Britain and France had exercised over the Egyptian economy in the
nineteenth century.[20] Before it would grant Egypt new credits, the
IMF, for example, demanded that the government should effectuate a
monetary reform and abolish the enormous subsidies to stabilize the
prices for basic foodstuffs. These subsidies had been badly misused
by the privileged and by the higher-income strata. However, the
survival of the great majority of the (especially urban) population depended ultimately on them. By announcing measures corresponding
to these demands (abolition of the subsidies), the Sadat regime provoked a mass upheaval in Egyptian cities (in January 1977), which
was to become one of the bloodiest and most dangerous threats to the
stability of the ruling order that Egypt had seen since 1952.

Egypt's economic problems steadily worsened after the liberalization policy was initiated. The economy became increasingly dependent on foreign investments and credits; the investment structures
became more distorted and showed more anomalies. Planning was
gradually abandoned because this new policy implied no development
strategy. The solution (or nonsolution) of such urgent problems as
the provision of food, housing, and employment for the abnormally
increasing population, quite aside from the problems of the system's
overall development, spoke loudly of its mistakes.

The problems created by the Opening Policy (explosion of consumption growth, relative fall in production capacity, increase of
corruption, and illegal personal enrichment) were so enormous that

they could not be solved by means of the income from petroleum export ($2.65 billion in 1980), or the currency transfers sent home by the one million Egyptian workers and skilled technicians employed in the Arab petroleum countries ($2.2 billion), or the income derived from tourism and fees for use of the Suez Canal ($1.4 billion).[21] These sources of income supply only a small portion of what Egypt needs in order to service its debts, and therefore cannot be devoted to increasing the system's production capacity. In addition, unless they are kept under control through planned intervention by a central authority into the economic processes (especially with regard to the rational allocation of scarce resources), the problems in question result from the release of those social and economic mechanisms that have been proved to necessarily produce imbalances in an undeveloped system. Therefore, these types of problems could not be solved by simply pumping more money into the system. What was and is needed to overcome them is, rather, a type of reform that, as already shown, was impossible in Egypt but that does not become less necessary in view of the regime's inability to put it through. Even the U.S. General Accounting Office (GAO) seemed to recognize in 1977 the necessity of such reform when it stressed that "economic and organizational difficulties severely limited Egypt's ability to absorb U.S. aid."[22] Owing to reduced planning and the resulting growing chaos in Egypt's economy, however, this reform seems to become less possible each day.[23]

With regard to social and political policies, the system's structures and functions remained, on the whole, unchanged, in spite of various formal and institutional alterations that took place during this period. While social unrest increased under Sadat and religious movements won mounting support, these phenomena did not indicate any essential change in existing power relations. The relations between rulers and ruled, which had been established during the Nasser regime (and which, although in another form, had also existed earlier), were not subjected to the slightest modification, even after Sadat's democratizing measures had been initiated. The system remained a one-man dictatorship, supported by a ruling class in which the right- and left-wing opposition were equally suppressed—the Left maybe a bit more suppressed. What is regarded by some Western observers as proof of the change of Egypt's political culture with the help of Sadat's introduction of democratization policies does not stand up to any analysis that concentrates on the processes occurring below the formal surface.[24] It proves to be irrelevant and misleading when used to explain the decision-making processes, the structures of power and opposition, and the stories of the rise and fall of the so-called parties in Egypt.

Political power during this phase remained concentrated in the hands of the president. Every criticism toward the government was viewed as communist subversion and was more or less violently suppressed. In May 1971 Sadat liquidated the government's left wing, where opposition to his political and economic concept was developing (a "corrective revolution"). Until 1975 almost all the seats in the one-chamber Parliament had only to fulfill acclamation tasks.

After Egypt's partial victory in the 1973 October war, which strengthened the government's position somewhat, Sadat initiated the process of formal liberalization, which was intended to transform the system into a sort of pluralistic democracy. Egypt was, thus, to become a stable political system; an important prerequisite for the engagement of foreign capital in developing countries. So-called platforms were set up within the ASU that were designed to constitute the basis for future parties.

In the 1976 National Assembly elections, the platform of the middle, the Egyptian Arab Socialist Organization (a government group integrating the rising bourgeoisie with landholders and upper-level bureaucrats), won 28 seats; the Left platform, the Organization of the National-Progressive-Unionist Assembly (so-called loyal Left), composed of leftist intellectuals, Nasserist forces, and union leaders, won two seats; and the right platform (middle-level bureaucrats, technocrats, and medium-capital owners) won 12. In the 1979 elections, these platforms, which Sadat had declared to be parties after the 1976 election, were changed into the National Democratic party (government party), which was said to have won 302 out of 382 seats; the Socialist Liberal party, which won 3 seats; and the Socialist Workers party, which won 26 seats. Owing to their offensive behavior toward the government, however (the Socialist Workers party rejected the Opening Policy and attacked the Camp David accords), all but two of their representatives were expelled from Parliament (and these last two were removed in 1980 as well).[25]

In spite of the establishment of three parties, the government did not tolerate any opposition. It maintained monopoly control over the media and manipulated the press to tell the people that while the new democracy in Egypt did not presently measure up to the Western model of democracy, it nevertheless held this model as its long-range goal. Therefore, the people would have to be content for the time being with this transitional form.

The process of democratization that was begun in 1976 was stopped in early 1977, however, as a result of the January unrest. Emergency decrees were issued that suspended the right of political action—a prohibition of demonstration and assembly and severe punishment for spreading rumors and inciting the people against the government either verbally or in writing, for example, through political songs—and redefined the status and conditions for action of leftist parties in a restrictive mode. These laws enabled the government to suppress every type of dissidence, not to mention active opposition to its policies (including arbitrary arrest of anyone suspected of opposition and the imposition of life imprisonment for involvement in illegal organizations).[26]

Leftist papers were prohibited, and leftist intellectuals were removed from their positions in universities and the media. As far as it was not liquidated or forced to submit to the regime, the leftist movement was crippled through the almost ritually repeated arrest of its at most 200 active members. Even the traditional bourgeoisie was temporarily denied its action base through the June 1977 law that prohibited the reestablishment of pre-1952 parties, because the regime feared the possible establishment of a new center of oppositional power against it by these forces. However, in 1978 the traditional bourgeoisie was permitted to rebuild the Wafd party (which was the organ of the bourgeoisie from the 1920s till its dissolution in the wake of the 1952 coup) under a new name. This was a reward for its positive reaction to Sadat's late 1977 Jerusalem trip and to the peace course initiated by it.

The peace policy was not very well received in Egypt (except by those groups within the ruling class that expected to secure advantages through it). Of course, this necessarily led to the government's attempts to compensate by a suppression of the right- and left-wing opposition. Even the New Wafd party, which in spite of its support for the peace policies continued to compete with the regime for power, was not, under these conditions, destined to last very long. Three months after it had been constituted, it dissolved itself again under the pressure of a regime that did not tolerate real competition.[27]

In view of these inhibiting conditions and the facts that Parliament was at no time able to exercise any legislative functions; that all elections, referenda, opinion polls, and the like that had anything to do either with Sadat's person or his program were never passed with less than 99 percent of the votes given in his favor; and also that every form of opposition continued to be severely suppressed, it cannot possibly be maintained that political life in Egypt had been democratized under Sadat. Given the increasing state control over the political processes after 1978, even favorably disposed observers could not continue to maintain their assertion that the trend to democratization was constant. However, for them, this "does not completely negate the effects of the initial moves" for "once political freedoms are extended they cannot readily be retracted."[28] But these initial moves had exhausted themselves in formal measures with no effect on the distribution of power. Political freedoms had at no time been meaningfully extended in Egypt.

The campaign to renew political life in Egypt, by which the Sadat regime proclaimed the establishment of the rule of law and the

state of institutions as well as its intention to orient its activities in
the future according to the provisions of the constitution, turned out
to be a campaign of polemic, lacking any implication with regard to
the change of the existing political order. This served the purpose
of making the Sadat regime stand out against the Nasser regime on
the propaganda level with respect to its attitude toward the people's
participation in political power. Because the Sadat regime was not
able to improve the material living conditions of the majority of the
population, it had at least to portray itself as the defeater of Nasserite
totalitarianism. This, however, did not mean that Sadat, by any measure he took during the 11 years of his rule, did anything to alter the
power structures that Egypt had inherited from the Nasser regime
and, in the final analysis, even from the regime that the latter had
superseded. [29]

Neither did the existing power structures change in any substantial way as a result of the growth of the religious movement in Egypt
during the Sadat regime. It cannot be denied that this religious-po-
litical movement enjoyed ever-increasing support after 1975 in the
underprivileged classes, who possessed no legal means for articulating
their interests. In spite of this fact, however, the movement did not
constitute a homogeneous force that could have seriously questioned
the regime's authority.

The Muslim Brothers formed a conservative tendency within the
religious movement, but had been outflanked in the 1970s by more
fanatic religious currents. Although this organization had indeed
reached a relatively high development standard, it became at no time
strong and organized enough to be able to challenge successfully the
concerted power of all the regime-sustaining forces. The various
fanatic splinter groups were also not capable of coordinating their activities against the regime; these were an outgrowth of the Muslim
Youth associations, which in the early 1970s, had been instigated at
universities and factories by Sadat and supported by Qaddafi to combat leftist influence there. After the disruption of relations between
Egypt and Libya, their members went underground, from where they
committed sporadic terroristic assaults like the one on the Technical
Military Academy in 1974, the kidnapping and killing of a minister for
religious affairs in 1977, and finally, Sadat's assassination in October
1981. They could provoke temporary unrest, but were never able to
initiate any sustained significant upheavals or changes in the social or
domestic political spheres. [30]

Moreover, the religious establishment and the oppositional religious forces near to the regime were always willing to curb their
potential opposition against it, if it granted them freedom of action
with regard to some cultural and social issues and the pursuit of their
purely religious interests. From the mid-1970s on, the Sadat regime

attempted to win these forces over by announcing mostly Islamic in-
spired laws, like prohibition of alcohol selling and consumption in
public, capital punishment for Muslims who dare change their faith,
and in 1979 the declaration of the Islamic legal code as the basis for
legislation in Egypt. The religious opposition movement cannot be
said to have any significance as a force that could compete for power
with the regime and its supporting class. It actually did nothing but
symbolize the growing rejection of Sadat's internal and external policies by an increasingly disenfranchised and powerless people.[31]

The Sadat regime, after the abolition of Arab Socialism, could
not supply a legitimizing ideology—one that was coherent to a certain
extent and that could have mobilized the impoverished population.
The regime aimed at perpetuating and maintaining equilibrium, an
aim definitely not appropriate as the leitmotif of an ideology about
which people who are suffering from existing conditions can be expected to become enthusiastic. In principle, the regime rejected
left-wing ideologies; therefore, the only choice it had, in order to
avoid the formation of a dangerous ideological vacuum in Egypt, was
to fall back on Islam. Although the organized religious forces tended
to oppose the Sadat regime, as they had opposed the Nasser regime
before, they were nevertheless probably the only forces upon which a
status quo regime in Egypt could potentially rely. They did not attack the given regime because it maintained the status quo, but rather
because it did not maintain it properly or because it was not willing
to reestablish the status quo of the past. Moreover, Islam was a
most suitable official ideology, because, at this point in recent history, it had played a significant role as a factor in stabilization of
the ruling order by stressing the values of obedience toward the
rulers and of submission to fate. Unlike the religious establishment
in Iran, the religious establishment in Egypt was closely connected
to their respective regime. Generally, it helped the government to
neutralize the potential of social conflict, rather than mobilize the
population toward overcoming the established social and political order. Although in times of crisis fanatic religious groups had sprung
up who aimed at doing this, such groups constituted, on the whole, a
marginal force that had no influence whatsoever on the domestic political processes. The main current of the religious movement was
thus formed by the conservative forces and the forces of the religious
establishment, which Sadat intended to galvanize for the purpose of
stabilizing his government.

Immediately after his inauguration, Sadat released the impris-
oned activists of the religious movement from custody, declared
himself the title of "Devout President," and presented his policies as
the result of divine inspiration, which he was sure to have received.[32]
After 1972, when unrest among students and workers began to grow,

he activated the alliance with the Muslims (without, however, thereby "retraditionalizing" the system) to create a counterbalance to the apparently growing forces of left-wing opposition. The Sadat regime granted the religious forces great freedom of action. The Muslim Brother's paper al-Da wah, which was not licensed, was published at a large circulation rate without any interference on the part of the government, whereas the licensed leftist, very popular, al-Ahali was repeatedly seized after its publication and finally prohibited. But Sadat kept a close eye on these activities to make sure they could be limited in case they would grow to be a danger to the stability of the regime. The regime actively supported the more mystic factions of the religious movement (the Sufis), while at the same time it attempted to restrict those of the more earthbound fanatic forces (especially after the ties to Libya were disrupted and closer ties to the West were being formed).

This attempt, however, was unsuccessful, the reason for its failure being primarily that the deteriorating economic situation and its implication of loss of individual and collective future perspectives led forcibly to the multiplication of the fanatic elements within the religious movement.

When he proclaimed Islam as the official ideology in Egypt, believing himself able to control these elements, Sadat in fact conjured up forces for which he was no match. Even if these forces would not be able to overthrow the regime, given their own organizational shortcomings and the ruling class's fast grip on power, the fanatic forces did (as far as their means would allow them) spread a degree of terror in Egypt. Sadat finally became a victim of this politically limited but specifically potent force. The Coptic minority has also been particularly exposed and subjected to harassment and worse by Islamic fundamentalists. (In the eyes of the Muslim fanatics, the Coptic minority is the representative of the Judeo-Christian civilization that they consider the source of all the evils they are struggling against.)[33]

A difference between the Nasser and the Sadat regimes other than a return of the privileges to the ruling class became apparent in a changed style of government after Nasser's death. Although both regimes were equally dictatorial, the Nasser regime presented itself as bureaucratic, ascetic, and concerned about social improvement and the acceleration of national capital accumulation, while the Sadat regime tried to appear as liberal but was, in fact, only ostentatious, being mainly preoccupied with its attempts to secure the redistribution of the national resources by which capital owners were to benefit most. At the same time Sadat tried to compensate for the reduction of social welfare, the deteriorating living conditions, and the abandoning of overall development goals by granting formal civil rights to the population. This was a shallow gesture at best in a regime that sought not to relinquish but to extend the various mechanisms of sociopolitical control.

Apart from the change in government style, nothing occurred under the Sadat regime that had any altering effects on the structures existing since Nasser's time. The peasants remained unable to organize; the workers were, with regard both to their organizational capacities and to the degree of development of their consciousness, not in a position to form a significant pressure group. The professional associations were coopted by the regime to serve only in technical and administrative capacities. Pressure groups of any kind were nonexistent or irrelevant, and the oppositional forces possessed no means to exert pressure on the government.

Even the role of the army, as the most important social and political force supporting the regime, did not change substantially. On the whole, the army did maintain its allegiance to the regime, although the foreign policy of the Sadat regime was considered to be the opposite of Nasser's foreign policy of fighting Israel and Western imperialism, as represented by the United States. In anticipation of potential opposition or upheavals, after 1973 Nasserite army commanders were replaced by officers of undoubted loyalty, most of whom were trained in the United States. Some high-ranking opposition officers were transferred to civil posts, some were sent abroad as ambassadors or military advisers in Arab and African countries, and some, the most serious cases, were liquidated.[34]

Although the Sadat regime was keen on portraying itself as liberal, it nevertheless fell back on the security service to stabilize the ruling order. Trying to repudiate the Nasser regime (not because its nature was different from that of the Sadat regime but because it aimed at somewhat different objectives), Sadat declared in 1971 "the end of police rule and of arbitrariness in wielding power," which were supposed to be characteristics of the Nasser regime. At the same time he announced the "beginning of the rule of law and justice."[35] The proclaimed change was simply little more than another difference in style. From this point forward, the regime only made less clumsy use of the coercion apparatus.

Very few things had changed to alleviate the conditions under Sadat that had made such a security apparatus a necessity in Egypt during Nasser's rule. The economic situation continued to deteriorate, and the discontent continued to increase among the population. Power continued to be concentrated in the hands of a ruling minority that exclusively controlled the decision-making apparatus and did not allow the interests of the majority, which remained excluded from power, to be articulated. The social welfare expenditures continued to be insufficient and even decreased relatively under Sadat. In view of these conditions, Sadat's regime could not eliminate or reduce the coercion apparatus, the heart of which was the security service. Rather, it developed and extended it (with the help of the United States) until it became one of the world's most effective security services.[36] Although it could not prevent the assassination of Sadat (usually the security service was not allowed to be present when Sadat met the army, which was the case on the day of the assassination), the security service, nevertheless, has succeeded to this point (and most probably will succeed in the near future) in stabilizing the ruling order in Egypt.

Considering the above, the best estimate is that under Mubarak not too much will change of a basic nature, either in the internal political conditions in Egypt or in its external relations. Sadat's assassination did not initiate any social processes that could possibly lead toward an overthrow of the existing order because it was not the result of a broad popular movement that integrated relevant and well organized social forces. The act was the effect of an impulse on the part of a fanatic splinter group that was not backed by the majority of the population and that was based neither on a coherent political program nor on sufficient power to destabilize the regime. In other words, the assassination of Sadat did not reveal any alternative perspective to the regime as such. The factors stabilizing the power of the ruling class have continued after Sadat's death, and no political force capable of challenging the position of the ruling class seriously has appeared on the scene. Owing to the circumstances, as already described, in 1981 there was no possibility for a comeback of a Nasser-like or an even more radical regime.

The new president could only be a protagonist of the status quo. Exactly like Sadat, Mubarak was inaugurated on the basis of elections that supplied him with the usual 98 percent of the votes. Like his predecessor, he proclaimed a presumptious political program announcing actions to eliminate poverty, parasitic wealth, corruption, illiteracy, and religious fanaticism. As in Sadat's case, one can easily predict his chances to realize this program as being rather slim, considering the objective conditions under which he will be forced to act.[37]

Again, like his predecessor had done before him, Mubarak changed the style of government upon assuming command. He forbade demonstrations of servitude and public reverences, which Sadat was so keen on, and he appeared to be less addicted to publicity. He is said to have made it clear to his staff that he expects hard work rather than vain eloquence and dry reports on progress so far achieved instead of verbose, emotional explanations of progress aimed at.

Essential structural changes directed at restricting the increasing anarchy and collapse in economic life are no more present in the mind of the new regime than in the case of the Sadat regime. There are no policies aimed toward the elimination of the parasitic rich class from the positions of power through which it has been able to secure the redistribution of the national resources to its advantage. No efforts are foreseen to begin rational planning for the distribution of resources and of the production process, to initiate policies reversing the still deteriorating production-consumption ratio, or to correct the relative decline of the production capacity and of the capacity for supplying the population with the needed basic consumption goods. The government's range of action under Mubarak, furthermore, is just as tightly restricted by the scarcity of resources, the dependence on foreign resources, and the interests of the ruling class and its ability to have its way with regard to economic and political issues, as it was under Sadat.

With respect to foreign policy, Mubarak may deviate from Sadat's course rhetorically. In fact, however, he continues to follow a basically similar course, and considering Egypt's objective situation today, it seems he has no other choice but to do so. According to Mubarak, Egypt belongs neither to the West nor to the East, but is rather an independent, Arab, Islamic, African country.[38] Whereas Sadat never ceased stressing Egypt's close ties to the West, Mubarak has attempted to give some new definition to Egypt's foreign relations, although they do not affect in the least the real conditions to which the system is subjected and that the regime cannot change. The Mubarak regime is attempting to overcome Egypt's isolation within the Arab world by drawing nearer to the Arab status-quo regimes. This route may prove to be a short-term way to avoid internal unrest arising from the crisis that the Egyptian people have seemed to be suffering because of the conclusion of the separate peace treaty with Israel and the abandonment of the Palestinian issue by the Sadat regime. This identity crisis has repeatedly found expression in domestic political unrest.

Apart from these rather more symbolic gambits, however, Egypt's relations to Arab and Islamic, not to mention African, countries cannot possibly have any consequence with regard to changing the country's basic structural situation and plight. Owing to its increasing economic difficulties and to the fact that its attempts to achieve emancipation and ascent in the international hierarchy failed in the mid-1960s, Egypt became necessarily a status quo system, totally dependent on industrialized countries. It lost, in the course of this process, the capacity to form presently or in the predictable future a union with any Arab, Islamic, or African country, which alone might constitute the foundation for a policy aimed at reducing dependence on the more developed systems.

Egyptian foreign and domestic policies under Mubarak are still determined by the very same factors that had determined them under

Sadat. Mubarak's possibilities for alternate actions have not widened compared to Sadat's, but tend rather to grow narrower. President Mubarak is left with few options other than to continue essentially on the same course that Sadat in his turn had little choice but to adopt.

NOTES

1. A general discussion of this dramatic shift in direction is found in P. J. Vatikiotis, A History of Egypt from Muhammad Ali to Sadat (Baltimore: Johns Hopkins University Press, 1980), pp. 406-11. It is developed at considerable length in Salua Nour and Carl F. Pinkele, The Costs of Interdependence: Egyptian Politics from Nasser to Mubarak (in preparation).

2. For a more in-depth discussion of this dimension of Sadat's policies, see the Nour and Pinkele chapter, "Camp David and After: Foreign Policy in an Interdependent Environment," in this volume.

3. M. Kerr, "The Political Outlook in the Local Arena," in The Economics and Politics of the Middle East, ed. A. S. Becker et al. (Rand Corporation, 1975), p. 52.

4. M. Hussein, La lutte des classes en Egypte, 1945-1970, 2d ed. (Paris: Francois Maspero, 1971), p. 190.

5. "Bundesstelle für AuBenhandelsinformation, Länderberichte: Ägypten," 1977, p. 43.

6. J. Tucker, "While Sadat Shuffles: Economic Decay and Political Ferment in Egypt," MERIP Reports, no. 65 (March 1978), p. 3; Area Handbook for Egypt, 1976, p. 292.

7. A. Farhat, Misr fi zil al-Sadat, 1970-1977 [Egypt under Sadat, 1970-1977] (Beirut: Farabi, 1978), p. 85; A. Shohayeb, Muhakamet al-Infitah al-Iqtisadi fi Misr [Open door policy on trial] (Beirut, 1979), p. 216; and Neue Züricher Zeitung, July 25-26, 1976.

8. "Class Roots of the Sadat Regime: Reflections of an Egyptian Leftist," MERIP Reports, no. 56 (April 1977), p. 4; Neue Züricher Zeitung, July 25-26, 1976.

9. Tucker, "While Sadat Shuffles," p. 3; Shohayeb, Open Door, p. 96.

10. Area Handbook for Egypt, 1976, p. 256; S. Shahin, "Fordernde und hemmende Faktoren der Industrialisierung in Ägypten seit 1952" [Factors promoting and hindering industrialization in Egypt since 1952] (Ph.D. diss., Bochum University, 1980), p. 104.

11. "Ägypten im Polypengriffe der Burokatie: lastende Erbschaft aus der Nasser Ara," Neue Züricher Zeitung, July 2, 1975; "Ägyptens (fast) unlösbare Probleme," Neue Züricher Zeitung, October 27, 1977.

12. N. Choucri and R. S. Eckaus, "Interactions of Economic and Political Change: The Egyptian Case," World Development 7 (1979): 783-97.

13. W. Stockklausner, "Ägyptens Massen wollen ihrem Führer nicht mehr folgen; nur die neureiche Händler-Kaste hat von Sadats neuer Politik protitieren Können" [Egypt's masses would not follow their leader any more: Only the new-rich trader-caste could profit from Sadat's new economic policy] Frankfurter Rundschau, January 28, 1977; Steuershcraube, "Inflation und Lohnstopp in Ägypten" [Tax increase, inflation, and stop of wage increases in Egypt], Neue Züricher Zeitung, May 7, 1977; and "Ägyptens Entwicklungsengpasse" [Egypt's development bottlenecks], Neue Züricher Zeitung, January 16, 1982.

14. Shohayeb, Open Door, p. 64.

15. Ibid., pp. 142-47.

16. Ibid., pp. 86-130, 172-202; A. el-Gritly, Chamsah Wa Ishrun Aman: Dirasah Tahliliyah fil Siyasat al-Iqtisadiyah fir Misr, 1952-1977 [Twenty-five years: a study of the economic policies in Egypt, 1952-1977] (Cairo, 1977); Tucker, "While Sadat Shuffles," p. 4; and "Ägyptens Entwicklungsengpasse."

17. "Ägyptens (fast) unlösbare Probleme."

18. Tucker, "While Sadat Shuffles," p. 4; el-Gritly, Chamsah Wa Ishrun Aman, pp. 142-58; Shohayeb, Open Door, p. 342; The Economist, June 28, 1975, p. 42; Neue Züricher Zeitung, January 23-24, 1977. Between 1973 and 1978 Egypt received an average of $3.2 billion in aid from Arab states and $1 billion from the United States and the World Bank. However, in 1977 the Sadat government was urgently in need of $12 billion to avoid an economic crisis; for this see Stockklausner, "Ägyptens Massen."

19. Tucker, "While Sadat Shuffles," p. 4.

20. A Hussein, "Al-Duyun al-Charigiyah—Hatta la Tuhaded Istiqlaluna al-Iqtisadi" [The external debts—As not to threaten our economic independence], al-Taliah 2, no. 10 (October 1975): 69-74.

21. Neue Züricher Zeitung, November 3, 1979; T. Wybranier, "Der Frieden fordet seinen Preis: Ägyptens wirtschoftliche Lage nach dem Arabischen Boykott" [Peace has its price: Egypt's economic situation after the Arab boycott], Frankfurter Allgemeine Zeitung, June 16, 1979.

22. Quoted in Tucker, "While Sadat Shuffles," p. 4.

23. These points are developed more completely in Nour and Pinkele, The Costs of Interdependence.

24. Choucri and Eckaus are representative commentators for the viewpoint that Sadat was liberalizing and democratizing the Egyptian system.

25. "Sadat's Kampf mit der Hydra" [Sadat's battle with the Hydra], Woltnoche, September 9, 1981, p. 3; "Bundesstelle für Au-Benhandelsinformation, Lander-Kurzberichte Ägypten," 1979, p. 7.

26. Tucker, "While Sadat Shuffles," p. 7; "Class Roots of the Sadat Regime," pp. 3-11.

27. M. C. Aulas, "Egypt Confronts Peace," MERIP Reports, no. 72 (November 1978), pp. 3-11.

28. Choucri and Eckaus, "Interactions," p. 792.

29. See Nour and Pinkele, The Costs of Interdependence; and Area Handbook for Egypt, 1976, p. 205.

30. "Les échecs de la démocratisation et le lourd heritage de M. Moubarak" [The failure of democratization and the heavy inheritance of Mr. Mubarak], Le monde diplomatic, November 1981, p. 5; and Tucker, "While Sadat Shuffles," p. 6.

31. D. Strothmann, "Der Adler zeigt seine Krallen" [The eagle shows its claws], Die Zeit, November 27, 1981, p. 7; M. S. Ahmed, "L'Egypt réduite du silence" [Egypt reduced to silence], Le monde diplomatic, October 1981, p. 6.

32. "Les échecs de la démocratisation," p. 5; M. S. Ahmed, "Normalization des relations avec Israel et crise d'identité en Egypte" [Normalization of the relations to Israel and identity crisis in Egypt], Le monde diplomatic, November 1981, p. 4.

33. Ahmed, "Le'Egypt réduite," p. 6.

34. "Les échecs de la démocratisation," p. 5; further discussion of this is to be found in Nour and Pinkele, The Costs of Interdependence.

35. Sadat's dialogue with the Egyptian Student Union on February 1, 1977, published by the Ministry of Information, Cairo, 1977, pp. 42-55.

36. "Les échecs de la démocratisation," p. 5.

37. Strothmann, "Der Adler," p. 7.

38. Neue Züricher Zeitung, November 10, 1981.

16

THE DYNAMICS OF
CHANGE AND DEVELOPMENT
IN JORDAN

KAMEL S. ABU JABER

INTRODUCTION

Partly because of the hold the goal of Arab unity has had upon
the Arab imagination, including Jordan's leaders, the country's political climate has remained somewhat unstable. The tension between
Jordanianism and Arabism, between <u>wataniyyah</u> and <u>qawmiyyah</u>, the
latter the ideal of Arab nationalism and unity, is as strong as it is
elsewhere in the Arab countries. Pride in being local is strong, yet
it is a pride that is constrained with the qualification of being Arab.
The local interests, achievements, or styles of life that have taken
root in the various Arab countries in the last few decades are miti-
gated by the hold the ideal of Arab nationalism and unity still has on
the minds and hearts of the Arab masses and, to some extent, on their
leaders. In Syria, Jordan, Tunis, or the Sudan, a person is proud
of being first Syrian, Jordanian, Tunisian, or Sudanese. He is also
proud of being Arab. In the person's self a struggle is at work as to
whether his first loyalty is to Syria or Morocco or to the Arab nation. Abroad, when asked, an Arab would say, almost apologetically,
"I am an Arab from" Morocco or Lebanon or Yemen.
    The diversity of Arab nations, the various styles of life that
have arisen in each of these nations, and their recent historical experiences, including their different recent colonial pasts, aggravate
the tension-guilt feeling between local and pan-Arab nationalisms.

And it is thus that the goal of Arab unity, the central theme of Arab nationalism, has become a destabilizing element rather than a rallying point. This is also true for Saudi Arabia, Algeria, or Iraq. Local nationalism, different colonial pasts, and the accompanying vested interests that have arisen in each Arab country are a constant cause for further inter-Arab and intra-Arab elements of strain.

In varying geographical degrees, virtually all Arab countries border the rim of the Mediterranean, are affected by it, and also interact with it. Geography, however, is only one factor, whose influence, like that of common language, heritage, and religion, is weakened under the impact of other seemingly more forceful and immediate modern issues. All Arab countries are developing in that they share the aspiration for a better standard of living, a more meaningful sense of direction, and thus, ultimately, a better quality of life economically, socially, and politically.

Much of the thought and attention given in this chapter to Jordan applies also to the rest of the Arab countries. This is the case notwithstanding their varied development experiences, methods, styles, orientations, ideologies, and political systems. Whether republics, shaykhdoms, sultanates, kingdoms, or emirates on the left, right, or center, the unifying theme of these countries is change. Directed change, now called development, is their basic and also most important contemporary feature. What applies to Jordan can also be generalized, with some reservations and in varying degrees, to all.

FORCES OF CHANGE

Most of the Arab countries border on the Mediterranean or are affected by its proximity. The proximity is not in geographical terms alone. From ancient times, the Mediterranean has acted as a medium for interaction between the states lying on its shores. Its northern European-shore nations were affected by nations on its southern shore and vice versa. In the modern era, the dynamic force of ideas that led to the development of Europe, the West, was bound to have its impact on the Arab nations as well.

Westernism, meaning to many in both East and West "modernism," was one such leading force to induce change in the Arab countries. Often in previous centuries, this Western force for change was not a cultural or gentle exchange between equals but was of a more violent nature, a technologically superior force imposing or attempting to impose itself on its weaker adversary. W. C. Smith postulates that it is as if there were inherent jealousies and deep-seated animosities between the northern West European nations and the southern nations of the Mediterranean.[1] Often those negative contacts were intensified by religious fervor and zealotry. From Crusader times through to Portuguese, later French, British, Italian, and Spanish imperial expansion attempts, the contact, well into this century, took by and large violent military overtones. Smith also stresses that Islam itself was and remains a potent force and that Islamic society was never a passive recipient of outside influences only. [2]

Thus, in modern times the Arab society's contact with the West was not that of an underdeveloped partner seeking technology, knowledge, and ideas, nor was it a contact between equals. Rather, it was, as it still remains, a less-advanced society dealing with a basically superior military and technological adversary. For the Arabs, their newly discovered weakness in the modern era was shocking on both the individual and the national levels. The shock was the more severe since Muslim society had always thought itself superior. [3] Essentially, the West was and remains to be perceived as a threat to Muslim-Arab society. [4]

From Muhammad Ali of Egypt to the present day officers-leaders of every Arab country, including Jordan, an active national search for military strength has been commenced. The inner questioning has continued, and the process of evaluation and reevaluation of what went wrong, what is at fault, and how it can be righted is still in progress. The national hurt, rather insult, at finding so much weakness has caused the contemporary soul-searching that one finds in Jordanian, indeed Arab, art, literature, poetry, and other facets of life.

Initially, and while some officer-leaders still continue the effort, an attempt was made to create an efficient military subsystem isolated from the other segments of society, the idea being in some quarters that there is nothing wrong with the society except its military weakness. That problem corrected, so the notion goes, everything will continue in rhythm. Hindsight, however, indicates that one cannot create such a subsystem in contrast to what is around it and keep it isolated for a very long time. Eventually, the newly trained military men themselves begin to question their surroundings. It is not very long before they begin to think they have answers. Eventually they begin to act upon their newly found answers and the cycle of coups d'etat commences.

Surely, however, the response to the Western stimulus is not the only force for change. Western (maybe we should begin to call it modern) influence is not only military. Modern ideas coming through educational institutions and eventually the mass media and mass communications are other forces for change. In Jordan, as is the case elsewhere in the Middle East, the desire for change becomes an irresistible force of its own. And, while the government continues to induce and foster change, the idea generates its own momentum.

It is difficult now to isolate which or what changes have been initiated as a response to this or that pressure. Did the government foster or sponsor this project or was it done in response to popular pressure? It is often very difficult indeed to identify the motive for change. By this time, the process has become native and not solely in response to outside challenges. It is native in the sense that native groups begin championing a cause and defending it as a personal and national necessity.

CULTURAL-RELIGIOUS FACTORS

Westernism (modernity) caused a multiplier effect in the process of change and development. Mass communication and physical movement bring the realization that poverty, for instance, is not a condition to be endured but a problem to be tackled. The middle class—white collar workers, intelligentsia, professionals, an entrepreneurial class, bureaucrats, the military—becomes a potent force in the life of the society. Residing in urban surroundings, they induce the changes that finally make the triumph of the town over the rural area a reality.

The fatalism that once was is no longer here. There is a recognition that backwardness, poverty, ignorance, or disease are not God-ordained. Shaykhs and planners extol the virtues of striving to improve personal as well as national conditions. Jordan, part of the Islamic heritage of the Mediterranean, adheres to that heritage. The glorious Koran states that "Verily, God does not change the condition of a people until they change their own condition."[5] Individual change is an absolute necessity for social change, and the "paramount requirement for the modernization of any society is that the people themselves must change."[6] It is recognized that the process of modernization must work itself through the change and adjustment of hundreds of thousands of individual human lives that, in turn, as Lerner puts it, "induces different dilemmas of personal choice."[7]

The impetus for change generated not only national soul-searching of hitherto unproportioned dimensions but also a psychological process of equal proportions on the individual level. Eventually, the realization dawns that change is not an exclusive response to extraneous factors. Change, development (for that is what is meant by change), is seen to be a good in itself. Soon various Hadiths, sayings of the Prophet, are cited to encourage and extole change. It is not long before it is recognized that judgments and laws change with the changing of times: a recognized principle of the Sharia, Islamic law. Commenting on these developments, Sharabi writes, "in recent years even the narrow area of 'personal status'—marriage, divorce, inheritance—left under the jurisdiction of the Sharia courts has been breached and for all practical purposes brought under civil jurisdiction."[8] In Jordan, in fact, it was so with a comprehensive civil code passed in 1976.[9]

Change or perish. This is a principle recognized and accepted by all Middle Eastern governments. That is why government has become the greatest catalyst and motivating force for change.[10] It is thus that change throughout the Middle East generally, and in Jordan in particular, has been by and large controlled and induced change. Unlike the Western experience, where change came as a result of the uncontrolled forces unleashed before, during, and following the breakdown of the feudal system, change in the Middle East came as a result of the deliberate will of the decision maker, the state. Kemal Atatürk put it most eloquently when he said that "nations striving to advance with a medieval mentality and primitive superstitions are condemned to perish, or at least be enslaved and humiliated."[11] He acted on that realization. Atatürk could be speaking for any contemporary Middle Eastern leader. Change in the West was evolutionary, spanning centuries, although at times there too it was painful. In the West change followed the introduction of new modes of production, ideas, and tools; in Jordan it was sudden, encapsulating time and patterned after the positive will of the policy maker. In most instances it was planned, deliberate change. Also, unlike the Western experience, the new social classes that have arisen—the intelligentsia, labor, the mercantile or industrialist classes—came as products, often by-products, rather than as the initiators of change. Again, unlike the Western experience, change came by the will of the ruling elite rather than against it. That is perhaps one explanation for the lack of associated political and social institutions in the Middle East.

Institutions as corporate bodies grow with the growth of experience, traditions, and customs. They cannot come about by the fiat of the leader, who in most cases maintains a tight grip on everything. This, too, helps explain the apparent absence not just in Jordan but elsewhere in the Middle East of grassroots groupings, political parties, or pressure groups.

The state is personalized and identified with the leader, and so are even the political parties. Often even the ideological movement, which by definition should not be so, is also personalized. The anonymity of the state, a necessary ingredient in the building of institutions, is lacking. As with King Abdullah (1921-51), now King Hussein dominates Jordan's political and social life, for example. The state identifies with them rather than the other way around. This is true of most if not all other Middle Eastern countries.

This pattern helps explain why countries in the Middle East have failed to produce competitive, organized political parties of traditional Western dimensions. Because of the absence of institutional avenues for change, whether socioeconomic or political, Arab political parties, in order to survive, have had to become all-encompassing transnational movements whose ideological overtones touch on every facet of a person's life. Pan-Arab, or transnational on the political level, they advocate fundamental changes touching the very heart and soul and fabric of life. As such, in Jordan as well as elsewhere in the Arab world, they clashed with and eventually became the enemy of the state. If these parties caused any change to occur, it was in their negative presence rather than by the platforms they advanced. The state had not only to match them but often to surpass their demands.

ISRAEL

It is difficult to ascertain whether Israel has been an indigenous or exogenous force for change in Jordan; indeed, for the general Arab world. Its proximity, its seeming democratic system, the military threat it poses, the anguish and the frustration it continuously causes remain forces for change of unfathomed proportions. On both the individual and the national levels, its efficient, often merciless, military machine causes limitless and agonizing soul-searching. How do you deal with such an adversary, who talks peace and refuses to reconcile? An adversary who looks democratic and, in fact, is otherwise? An adversary who refuses to define himself, his borders, his intentions, or his ambitions? An adversary who talks a great deal but refuses to communicate? An adversary who, with ancient hatreds, a negative ideology, and many platitudes, captured the imagination of the Western world and who has, for all practical purposes, made the West its tool of policy? These questions and many others of similar hue are constantly at work in Jordanian-Palestinian, indeed all, Arab minds. From the officer to the housewife, the high school pupil, his teacher, the doctor, the lawyer, or the engineer, the unfound answers add further frustrations that, in turn, become self-propelling motives for change. The disparity, whether real or imagined, between what was the glorious past, and what is the dismal and weak present, furthers the frustration. It is thus that Israel, recognized by the Arabs as a daily aggression on their lives and their future, has become a monumental force for change. It is impossible to state how much change it has induced thus far, or whether the quality or direction of that change can be quantified with certitude.

The forces of change identified to this point, whether external or internal, have caused the quality and the quantity of life to change in Jordan. Education, health, welfare, other services, and planning have made Jordan a welfare state. The late nineteenth century Muslim reformer and intellectual Jamal al-Din al-Afghani realized the potency of education for change. He recognized the weakness as owing to ignorance and lack of knowledge. He stated, "The nations of Christendom are able to over run the nations of Islam by their science, the source of all power."[12] Jordan's leap from its past primitive or semiprimitive socioeconomic conditions has been not only sudden, but of lasting effects. From 1948 when Israel came into being until 1982, a little over three decades have elapsed, yet the Jordan of 1981 bears little resemblance to that of 1948. From a semiprimitive socioeconomic society to a modern consumer society, the development has been great.

THE PALESTINIANS

The Palestinians, a by-product of the creation of Israel in 1948, and their later swelling numbers owing to the continued Israeli attacks, have provided a terrific motivating force for change in Jordan on both the personal and national levels. Long subjected to the indignity of immediate and direct British heavy-handed rule, and having come in touch with the newly arrived waves of European Jewish immigrants into Palestine, the Palestinians were sophisticated and aware, although a very embittered and frustrated people. They were forced to open their land to Jewish immigrants and had to watch, helplessly, as British manipulation touched their lives and destinies. Their anger was physically transferred with them to Jordan. Initially, a mass of destitute refugees, their hard work and ambitious nature, once given a chance, began to change their status and the prospects of Jordan. Their arrival in Jordan expanded the horizons of this country, its population, its resources, and its international stature. Jordan is no longer that quiet, serene, small country on the sidelines of international politics.[13]

The Palestinians brought with them to Jordan a healthy respect for modernity, knowledge, and an awareness of the twentieth century. Their frustration and anger were also accompanied by the characteristics of hard work and achievement. Politically mature, they began placing demands on the machinery of the state for services, job opportunities, facilities, and other amenities of life. There is no doubt that their advent in Jordan has helped change its political, social, economic, and even psychological composition. Initially very angry and alienated, Palestinians soon began to take root and participate in the development of the country.[14]

SYMPTOMS OF CHANGE: WHAT HAS CHANGED?

The previous discussion has concentrated on basically two points: an attempt to identify the forces, whether internal or external, leading to change; and an attempt at demonstrating macrolevel change in terms of socioeconomic snapshots of Jordan between 1952 and 1979. In the present section, an attempt will be made to identify the nonphysical, imperceptible, and qualitative changes; changes that have not yet been subjected to systematic examination by either Arab or non-Arab scholars or policy planners and makers.

There is no question that the physical side of life, the obvious and the quantifiable, has changed. This can be seen with the naked eye. Side by side with minarets one can see the newly erected factory chimneys, laboratories, and university towers. In most cases, the shape of the landscape itself is changing. New farming techniques, urban areas, forestation, irrigation canals, highways, and state and private buildings are springing up every day.

Concomitant with these observable, quantifiable changes have been changes in attitudes, relationships, manners, and mores. It is difficult to quantify these changes, many of which are neither admitted nor recognized by the changing or changed individual. Suddenly a Bedouin must choose between his precarious life of freedom and a more secure and settled life in urban or village surroundings; a mother has to ask herself whether to permit her daughter to go to college; and the student or the young man must choose between a secular approach to life or piety. The choices are infinite because the combinations and the sources of knowledge are infinite too. Concerning the influence of the mass media on choices, Halpern eloquently comments that "the radio, movies, newspapers, and books allow a young man for the first time to choose his intellectual and spiritual brothers. Modern scientific thought makes possible, indeed required a reexamination of one's traditions and values."[15]

The changes that will be described here are among groups and within groups as well as within the individual self. What happened to some of the "old" groups, the tribe, the extended family? What happened to the time-honored "old" values? Have they changed? How? Are they being replaced? With what? Because the very process of change itself is a continuous one and because Jordan's society is undergoing the process of change, it is difficult to gauge the depth or, indeed, the direction of change. It is pertinent to remark at this juncture that once the scent of modernity grips a society it seems very difficult, even impossible, to halt the process, let alone reverse it.

Four categories of such subtle changes will be discussed here: (1) style of living, (2) relationships, (3) tastes and manners, and (4) values.

At the outset of this analysis, a few observations seem to be in order. First, while the above categorization may seem somewhat arbitrary, it was a necessary first attempt at isolating and identifying, in a systemic manner, changes hitherto not thoroughly researched by scholars. Second, some of these categories are interdependent and overlap. Third, a reminder: recall an earlier remark concerning the readiness of Jordanian and Muslim society to accept change. This, however, should be qualified. There exist, and no doubt will continue to exist, certain groups that resist change.

Resistance to change has taken a variety of forms in Islamic society in modern times. Political parties and intellectual movements, mostly of a conservative nature, have existed in an attempt to preserve the status quo, and in some instances to turn back the clock. In the past two years, events in Saudi Arabia and Iran attest to the strength and depth of the feelings against change or its style. Fundamentalist groups will no doubt continue their attempts at resisting change. In Jordan, the style of the regime and its political acumen have thus far succeeded in avoiding a clash. In fact, the Jordanian regime has had to ride two horses and has done so perhaps a little more successfully than many of the other regimes in the area. One horse is conservative, adhering to certain religious, cultural, and social principles, and the other is that of modernity with all its implications, from color television stations to colleges and universities, as well as the modern amenities of life.

The above attempt at reconciling the old with the new, or at least at achieving a modicum of peaceful coexistence between them, has been lumped together under the slogan "al-asaleh wa al muasarah," ("heritage and modernity"). 16 The slogan is no doubt true, although it seems at times a bit of a psychological cane upon which to lean until something new and acceptable emerges. Yet, in fact, something new has emerged already. That "something" is neither old nor new; it is a little of each—a wedding of tastes, colors, sounds, and smells. A medley, a hybrid of both al-salaf ("the heritage") and the new. It is a compromise between time and space. Eclectically, one chooses what is deemed desirable from the heritage and attempts to reconcile it with contemporary society. Is the attempt futile? To outsiders, the attempt may seem to be futile and a waste of time. To one undergoing change, it makes the process a bit more merciful and palatable until the next change, and yet the next compromise comes along and must be made.

The changes that have taken place and those that Jordanian society is now undergoing are in all fields of life: economic, social, and political. These changes touch the very fabric of the life of the state and the individual. In all these fields, the trend is toward more rationalization of life. Each change is, in itself, a shock that is soon followed by other moves and shocks. The atmosphere is still uncertain, yet there is no doubt that people, and not just the decision makers, want these developments in spite of some resistance on the part of some groups. The interaction of these developments, all having come at the same time and with deliberate intensity, is causing great pressures and imbalances on group and individual lives and relationships and ultimately a change in values, mores, manners, and style of living and making a living. Again, Issawi relates all these changes to "Western influence which accelerated the transformation . . . and the dissolution of its (Arab) communal and organizational ties and their replacement by individual contractual relationships."[17]

CHANGING STYLE OF LIVING

Perhaps the most significant aspect for the development of Jordan during the past few decades is its rapid movement from a semi-private society to the consumer- and service-oriented society of the present, without having passed through the various agonizing stages of the industrial or even the agricultural revolution. The race with local and regional ideologies necessitated that the Jordanian government encourage a welfare consumer society, and the quicker the better. Professor Charles Issawi, again speaking of the Arab world, states that "Western influence accelerated the transformation from a subsistence to a market economy."[18]

Only a generation ago, four-fifths of Jordan's population lived either a pastoral existence or in small towns and villages.[19] The change to urban-oriented living is both quantitative and qualitative, as well as psychological. Changes in clothes have been accompanied by changes in consumer habits, food consumption, calorie intake, and attitudes toward life. Today, Jordan's population is 70 percent urbanized. The seasonal rhythm of life has changed to the month, the week, even the hour of overtime work. Punctuality is one habit that is indicative of the depth and breadth of change. There now exists a growing healthy respect for both time and space. Punching a clock, which is mandatory in some places of work, instills respect for time. Distance once almost immaterial, since one had plenty of time, is also receiving the respect it deserves. Efficiency becomes a value in itself and a merit deserving attention and glorification. The competition alone, faced in an urban setting, necessitates respect for efficiency, time, and space. And although a new arrival to the city carries with him or her a certain amount of loyalty to kith, kin, and tribe, the very move to the city has by definition removed him many steps from his traditional surroundings. The loosening of the ties is no longer a matter of choice; rather, it is a process dictated by the new circumstances of life. In this new urban environment, a person's family or tribe, for all its prestige, cannot sustain the job, though it may initially help in getting it. The individual alone must ultimately prove personal merit and capability to keep and later to improve job and income status. Surely under these new circumstances one's own interests become paramount over those of the clan. The loyalty to the clan takes a secondary and eventually a much lesser position. The labor union, the professional association, the club, or the new group begins slowly to replace old loyalties and connections.

In the city, the peasants learn new jobs, habits, and eventually attitudes. By definition, their move to the new setting has already marked them as changed people, otherwise, Why would they move? It is they or their children that eventually learn new trades, new attitudes, and new life-styles; theirs or their children's newly acquired knowledge, of necessity, generates new demands. No longer can they tolerate the old deprivations nor will they accept them as inevitable. A bad situation is not the God-ordained state of affairs to be endured forever without question. One can do something to help oneself and improve one's situation.

From fellah, or peasant, to consumer society, many things are lost and others are gained. As Daniel Lerner put it, "Secular enlightenment does not easily replace sacred revolution."[20] New insecurities, fears, desires, and comforts work themselves into people's lives. The old habits and beliefs linger yet a little longer, adding further imbalances. There is an amount of hesitation, tentativeness, perhaps lack of self-confidence coupled with a mood of repulsion-attraction that is profound.

The domesticated animal disappears completely in the new environment and, sooner or later, is replaced by the wheel. A bus, bicycle, or eventually one's own car becomes the mode of movement and transportation. The wheel signifies machine, and machine signifies a certain attitude of efficiency and immediate punctuality.

The replacement of the seasonal by much shorter spans of time signifies not only a new style of living but also new attitudes. Sharecropping and seasonal existence are replaced by cash economy and salary. Salary means some wisdom to be acquired in home economy and eventually planning. Planning for one's or one's family's future becomes a standard feature of life. New, more complicated and sophisticated division of labor becomes the new mark of life, even within the family unit itself. The new economic standard or life-style based on cash income and cash expenditure of shorter durational periods, is a new reality. One is no longer in the comfortable traditional surroundings of rural existence. Rather, one hardly knows one's neighbors, their cares, wants, or ambitions. These new neighbors, themselves new arrivals, are not beholden or obliged to be nice; new connections must be made as well as arrangements to generate new desires and friends within the new environment. Like personal, rather than communal income, prestige, standing, and connections also are personal, dependent on skill and capability in handling new situations.

CHANGING RELATIONSHIPS

Our discussion of the changing relationships focuses on three levels: the national level, with the state acting as an orchestrating coordinator of change; second, the changing relationships between groups, the old group and the groups that have newly arisen, and the variety of tensions that have developed and the compromises and adjustments that have taken place; and third, the structure and relationships within the family itself as a basic unit in the social order and the variety of changes and adjustment it has undergone.

The state was once a mere policeman, regulating sociopolitical traffic with hardly any interference in its citizens' affairs. It used to inspire awe and fear with the popular saying "The sultan is he who is far from the sultan" expressing people's desire to stay away from it. The welfare state is now a reality of life in Jordan. Reasons for this are varied, not the least of which is the desire to maintain social order in a region and a century where the style has been very active. In fact, as an all-encompassing state, its activities leave little room for privacy or private initiative. Partly in response to popular demand and partly to outbid rival political movements alive in Jordan and the region, the state is omnipresent in health, education, welfare, the mass media, and the economy.

A sense of belonging, of loyalty to the state, which is essential to the process of nation building, does not come automatically, especially in a country like Jordan. The government has had to work for it. This is especially the case in Jordan, beset by internal frictions—tribal, familial, and Jordanian-Palestinian rivalries, in addition to the general tension accompanying socioeconomic development. Blessed with longevity, continuity of leadership, and a leadership that is also pragmatic and moderate, Jordan's national development has been rather smooth and evolutionary. In contrast to some of the neighboring countries, Jordan's experience has been rather peaceful, with less social dislocation and political discomfort than other regimes in the region. Its style also can be characterized by a paternalistic approach in the political field. King Abdullah first and King Hussein after him have spoken frequently of the "Jordanian family," with the leader as its head. The concept was honed to fine proportions to oil the joints of development and to smooth over the rough edges and dislocations that exist in an essentially uncertain and fragmented society. [21]

Under its own direction, the state has encouraged the development and the rise of new groups and associations. Initially not intended to replace the old groups, by force of circumstances these new groups and associations are ever so slowly, yet surely, replacing the old groups. Modern social security regulations replace family and clan as a source of personal and economic security. On April 6, 1977, King Hussein addressed a letter to the prime minister and the cabinet asking them to begin laying down the groundwork for the enactment of a comprehensive social security law that since has been enacted. In his letter, the king wanted a law "guaranteeing the individual worker and the good citizen comfort and peace of mind in case of illness, old age, and his family a decent living."[22] The extended family, the tribe, and the village elders as a form of social organization and security are being replaced with bureaucrats, professional associations, and modern business and interest groups. Lawyers, doctors, engineers, pharmacists, business groups, chambers of commerce, industry, the labor and trade unions and associations, teachers, and intelligentsia leagues have arisen to demand the loyalty of the individual. Surely it is recognized that belonging and loyalty to these and similar groups are on a more rational basis than the earlier loyalty to family, clan, or tribe, which used to come about automatically.

Thus far the two types of organization, the old and the new, coexist side by side. Depending on the circumstances and proximity or distance from urban life and organization, the old groups are steadily, however, losing ground. In fact, a Jordanian now almost apologetically admits to belonging to some clan or tribe. It is a recognized fact that urbanization and industrialization tend to weaken family relationships and to change their previous configurations. "As industry begins to hire and classify workers on a skill basis, and as the personal aspect diminishes in agriculture, clan and family organization becomes less important as units of production."[23] The new system of groupings and memberships is slowly replacing the old. The new groups' greatest advantage is that they can be defended as a more rational form of organization. Also, loyalty is not automatic but pragmatic, with service rendered on a mutual basis to both the individual and the group.

Within the family, great changes have also taken place. Mention has already been made of the weakening and, in many cases, the disappearance of tribal and familial ties.[24] In fact, tribal and extended-family ties are invoked on only very rare occasions, with the mass of people still at the stage of wondering whether they should take them seriously or not, nostalgia notwithstanding. In fact, the cellular family is becoming the focus of attention. Under the impact of education and modern economy, the larger unit is no longer tenable or desirable. More and more, individuals are thinking in terms of wife and children, not even of father and mother. Otherwise, Why is there need in Jordan and elsewhere in the region for old peoples homes? The family now is "more egalitarian and more democratic,"[25] although respect for the elders is still invoked as a cherished value.

Even the mechanics of life within this cellular family are changing in response to the demands of modern life. The cellular family is small and moveable or easier to move and manipulate; its activities and future are mutually discussed and planned only within its confines. It has to exist independently of the extended group. Its head and, increasingly, the mother or other male and female components must hold a regular job with regular income. The children are raised with the idea that each is an individual unit having to become educated so as to compete for better opportunities. No longer can an individual depend or even hope to depend on others to live.[26] The cellular family is now in the process of becoming supreme, a trend that is essentially not found unwelcome by most and is thought of as modern or "good" by many.

The prestige of the extended family in this situation takes secondary position. Holding individual urban jobs matters, rather than the previous communal work and living. That is, the very step of life has changed for the family. The rhythm is shorter, perhaps more rational, and certainly more uncertain, with the individual standing, as in other modern societies, almost alone to face the vicissitudes of life. The individual cannot, as in earlier times, depend on others but must work instead for personal goals. Achievements or failures are individual, as are prestige and standing in the community.

The interaction among groups, within groups, and also within the state as a whole is not as tense or as complicated as many outsiders would imagine. Surely however, there is a search for smoother transition to make the potentiality an actuality; a search that the new system of authority is now seeking.

CHANGING TASTES AND MANNERS

It is difficult to ascertain which changed first. Was it tastes and manners, or was it the value systems and values? Surely the two are so intertwined it is difficult to separate. Yet the division here is necessary for the sake of clarity and simplification. Values change according to new situations, styles of living, and circumstances and also in response to newly acquired knowledge. Or is it vice versa? It is a process in which values change as residence, occupation, and style of life change.

Now that the rhythm of life is no longer seasonal, depending on nature or the whim of the landlord, life is geared to accompany the new economic situation. In urban surroundings, with its new distribution system, the problem of food itself is no longer seasonal. One can think of shorter, more manageable and perhaps more rational periods of time for storage and food consumption. Three or four decades ago, the average Jordanian household kept most of its <u>mooneh,</u> food provisions and dry goods, stored for the entire year. Most of the time only fresh meat, fruit, or vegetables were bought. Today very few rare families care to do so, partly for economic reasons, but more important, perhaps because of the availability of such provisions owing to modern techniques and also storage problems. The supermarket or the grocery store is right around the corner, and with the new division of labor, one does not have to spend an unduly long time preparing for storage of large quantities of provisions for the family's needs. It makes it easier now that the pattern of income is of shorter, more manageable spans of one-month and, in some cases, one-week periods.

As taste becomes more developed and the appetite appreciates more, one is no longer satisfied with the previously few stored staple items. Nor is one satisfied to spend his or her time, as was previously customary, in preparing them. Suddenly, there is time for leisure, however meager the income, and this tends to generate new demands and needs. With the wheel becoming part of life, picnics become a respected way of spending leisure time. Vacations too eventually become a necessary way to spend spare time.

The change in food and how leisure time is spent is accompanied by a change in dress. Western clothes slowly but surely have replaced traditional garb. Whereas previously a woman or a man owned few items of clothing, all in the time-honored traditional patterns, now a person's wardrobe, especially in the major cities and towns, will include several items. Changes in food and clothing are soon followed, or rather accompanied, by changes in the field of entertainment. Entertainment becomes a necessity for better-quality living, so much so that in the same household one may find several types of music heard and appreciated, ranging from the very traditional to the classical and semi-classical music of the West today.

There now exists more variety not only in what one wants but also in its supply. The more Jordan's society becomes exposed to modernity, the more complicated and sophisticated its pattern of life becomes. Jordan's exposure to modernity through the mass media adds a certain degree of frenzy to the intensity of doing what the average Jordanian thinks to be the right things. First, consumption, and later, conspicuous consumption among the middle and upper-middle classes becomes the pattern of the day. Capitalist in orientation and taste, people's demands for keeping up with their neighbors, bettering them still, becomes the new pattern of life.[27] The changes inherent in this type of largely unproductive society are indeed great. The conformity in demand for goods and services is one of the distinguishing traits of contemporary Jordanian society. Changing tastes, like the changing values themselves, are becoming more materialistic, often with the goods or service demanded not so much for their intrinsic value as for their conspicuous display for others to see. This is in stark contrast to the traditional pattern of consumption and expenditure and attests to the new-found relaxation and perhaps security of the modern era.

CHANGING VALUES

With changing life-styles, relationships, tastes, and manners, values once cherished are now changing or have changed altogether. The change, as stated elsewhere in this chapter, was initially induced by the state. Now it has generated its own seemingly irreversible momentum. While some attempts have been made by scholars, decision makers, and planners, by and large the changes in the value system have come about without preplanning and forethought. Health, education, welfare, modern facilities, amenities, and even highways have been thought out and planned. Yet little thought was given by planners and decision makers to which new values should replace the old. While the state attempts to instill new values—values like the worth and merit of women and manual labor, respect for space and time—these came piecemeal, and there was no comprehensive plan as to what was really desired. Nor does it seem that such a comprehensive plan is contemplated. The new values that we now see emerging came as concomitant companions to changes in related fields. Modern medicine replaced folk medicine; the scientific approach slowly entered the society through the process of education and is now replacing transcendental, superstitious, even sorcery practices. The tea, coffee, or palm reader may still exist side by side with the modern laboratory, yet the mystic is steadily losing ground to the logical-scientific method.[28] Materialistic middle-class values, acquisitive and to some Jordanians even capitalist values, of necessity are slowly replacing old fatalistic attitudes.

It has been discovered, as education, the mass media, seminars, and symposia explain, that man has something to do with his destiny. The Islamic predilection for a compromise between this world and the next makes the process of change in this direction easier. As the popular Arab saying goes, "Live in this world as if you will live forever, and for the next world as if you shall die tomorrow."

Government encourages frugality, economy, and savings. While its reasoning may be different from that of the individual, the end result is to encourage individual savings for investment. Housing projects encourage an already-strong penchant for the acquisition of private property. Private savings for the assurance of a personally better and more secure future have replaced the communal-tribal concept of security. To further this process, the government encourages the institution of savings plans and retirement schemes among private enterprise firms and itself has instituted a comprehensive national security scheme. One's personal labor and savings for the future, supplemented by government services, have come to replace the earlier dependence on family in sickness or old age. Resolution of disputes, especially in important blood cases that may lead to feuds, once almost totally left for tribal justice is now carried out in courts. Tribal formalities have still to be observed in serious cases, but, in fact, they are more or less mere formalities replaced by state justice.

Now everyone is taught to depend on his own resources and resourcefulness. Nepotism is still strong, yet it is weakening each day with the emergence of a system based on merit. The expansion of education and job opportunities and a finer division of labor are speeding up the emergence of meritocracy. One teaches one's children to depend on themselves, while at the same time teaching them to continue their respect and care for the elderly in the family. Slowly, however, and with the emergence and supremacy of the cellular family, the elders are removed, or remove themselves a step away.

In most cases, now, neither father nor married son expects to live in the same household anymore. Respect for the elderly and a certain amount of traditional values are taught to the young, not out of nostalgia but out of genuine concern for certain honored customs and mores. Among these values is respect for religion; religiosity is paramount, for it is not easy to turn away from religion in a society whose speech is still much interlaced with popular and religious sayings. Yet even here change is taking place. As early as 1958, Harris remarked, "The force of religion remains strong, but the secular influences of the town are bringing into question the traditional meaning."[29]

The concept of shame is still very strong, especially as it relates to the females in the family.[30] Family honor, very much related to the female members of the family, is still exceedingly important even among the educated and the highly educated. It is not uncommon to hear a professor remarking that he had to return from Germany or the United States because he could not accept his daughter

courting with boys. Yet, while the code of honor has seemingly remained rigid in this area, it has relaxed elsewhere. In 1958 Raphael
Patai, writing about Jordan, observed that "the employment of women
away from their homes, increased contacts between unrelated men and
women with a resulting loosening of the rigid sexual mores."[31]

Women becoming educated and now participating in gainful employment outside the home are accepted facts. In fact it has been
encouraged by the government, which, since 1976, has held several
conferences especially to focus on this issue and to further draw more
women into economic and social activities. Women's emancipation
is no longer an issue in Jordan. It has already become an accepted
fact, with women found in factories, offices, business firms, the
medical and legal professions, and in almost every walk of life.
Women were granted the right to vote and to be elected to Parliament
as of 1974. Three women participated in the First Consultative Council, while four are members in the current, Second Consultative
Council, serving as Jordan's Parliament. Since 1979, one woman
has been a member of the Jordanian cabinet and holds the portfolio
of Minister of Social Development.

The previous discussion portrays both the breadth and depth of
social attitudes and change toward women in the last three decades.
In fact, it has become almost a shame for young educated females
not to seek gainful employment. It should by now have been noticed
that no mention was made of the veil; the reason is that it no longer
exists in any significant manner at all. The walls that once existed
between the female and the male society as exemplified by the veil
have been disappearing rapidly.

CONCLUDING REMARKS

The earlier catalogue of the forces for change and the later
description of the symptoms of change in this chapter give a rather
intense, perhaps confusing picture of a society in transition, a society
in the process of becoming something else the nature and the general
contours of which are not yet very clear. In spite of the frustration
and the agony caused by the proximity of the Palestine problem and
the accompanying dislocations that this has caused, and in spite of
the hardships of development and change on the individual and even
the group level, the new estrangement and uncertainty are demanded.
It is curious that the price does not seem to be too high even on the
individual level. Perhaps the pragmatism and the humane approach
of the Jordanian experiment, lacking as it does the use of force in
any form, made the process more palatable.

At first it was the state, the decision maker that realized that
change or perish is a fact of life in the twentieth century. Later, the

desire for change filtered even to the level of the individual, and the process became self-propelling. A few remarks concerning individual response to change seem to be appropriate here.

The realization that one must change must be somewhat shocking to many. After all, one has lived part of his life in the grip of comfortable and familiar traditions, circumstances, and responses. True, the physical aspect of life may not have been too comfortable, yet psychologically and socially the situation was familiar, resting, and perhaps even appealing. There was, after all, only one source of knowledge: the tradition. This ancestral, traditional source provided acceptable answers to most questions of life. If an answer was not found, that fact itself was explained in fatalistic terms that themselves were accepted and, in a sense, comfortable too. One does not have to think too much nor worry unduly under these circumstances. In the changing society of today's Jordan there are many answers; some traditionalists would say too many answers, some of which are undesirable, and too many sources of knowledge (Right, Left, Center) that are too confusing and contradictory still.

With the emergence of the achievement-oriented society, the cellular family, cash economy, and cash jobs, the individual too begins to emerge. The suddenness and the speed of change add another dimension to the situation. Is it easy to break away from the old familiar ties? Should one move to the city? Educate his daughter? Permit her to work with men? These and similar questions piled on top of other worries in new and mostly unfamiliar surroundings cause further imbalances or strains on the already strained self. The individual soul-searching, evaluation, and reevaluation become a constant process as each step toward further change is taken. "Am I doing the right thing?" one asks oneself, one's friends, and God. The answer is never sure, for no one knows the situation better than the person asking the question and he cannot seem to provide himself with answers. He simply plods along.

The uncertainty, both individual and national, causes nervousness until each individual appears to be a bundle of contradictions and complexes. The individual moodiness reflects itself in national crises whose intensity are further deepened with the frustrations caused by the failure to deal adequately with Israel. Wavering between what was, what is, and what is to be, the atmosphere, though at times bewildering, is hopeful, looking toward the future rather than facing the past. A new identity is emerging forged by a multiplicity of factors all in the direction of creating a new sense of belonging. The state, the architect of this process of nation building, is accelerating the process.

The changes are still uneven among groups, between rural and urban areas, even within sectors of the same city, yet they are proceeding to further shatter a way of life that once was. If anything, hindsight in the Arab region and in Jordan in particular shows that modernity is a way of life, that you cannot introduce change into one segment of the society or into one of its aspects without having it spill into other sectors and spheres of life. As Halpern once remarked, it is a "change in what men believe, how men act, and how men relate to each other."32 Finally, while the Jordanians have accepted or seem to accept the tangible ingredients, tools, ideas, methods, and styles of twentieth-century civilization, there still lingers some hesitancy, some questioning. For, indeed, in the process of change, "Modernization must be thought of not as a simple transition from tradition to modernity but as part of an infinite continuum from the earliest times to the indefinite further."33

NOTES

1. W. C. Smith, Islam in Modern History (New York: New American Library, 1959), pp. 76, 109, 111; B. Ward, The Interplay of East and West (New York: W. W. Norton, 1962), p. 54.
2. Smith, Islam, pp. 14, 21.
3. See J. C. Harrison, "Middle East Instability," Middle Eastern Affairs, no. 5 (1954), p. 76, where he states that the intellectual foundations of Arab life in modern times were found to be inadequate. This was shocking to a society that thought itself superior.
4. Smith, Islam, p. 64.
5. The Koran 13:11.
6. M. Milikan and D. L. M. Blackmer, The Emerging Nations (Boston: Houghton-Mifflin, 1961), p. 23.
7. D. Lerner, The Passing of Traditional Society: Modernizing the Middle East (Glencoe, Ill.: Free Press, 1958), p. 44.
8. Hishram Sharabi, "Islam and Modernization in the World," in Modernization of the Arab World, ed. J. H. Thompson and R. D. Reischaner (New York: Van Nostrand, 1966), p. 29.
9. The Jordanian Civil Code was published in the Official Gazette, August 1, 1976.
10. See Charles Issawi, "Social Structure and Ideology in Iraq, Lebanon, Syria and the UAR," in Modernization of the Arab World, ed. J. H. Thompson and R. D. Reischaner (New York: Van Nostrand, 1966), p. 141.
11. M. Halpern, The Politics of Social Change in the Middle East and North Africa (Princeton, N.J.: Princeton University Press, 1965), p. 36.
12. Quoted in H. Sharabi, Mugadamah li dirasat al-mujtama al-arabi [Introduction to the study of Arab society] (Beirut: al-Dar al-Muttuhidah Lial-Nashr, 1975).

13. Glubb describes it thus, "From 1932 to 1948 the whole of Jordan was one of the happiest little countries of the world." J. B. Glubb, A Soldier with the Arabs (New York: Harper & Brothers, 1957), pp. 21-26.

14. R. Patai, The Hashemite Kingdom of Jordan (Princeton, N.J.: Princeton University Press, 1958), p. 50.

15. Halpern, Politics, p. 29.

16. For further elaboration of this point, see Kemal Abu Jaber, "Social Change in the Arab World," Majallat al-ulum al-ijtima'yyah, fa al-watan al-arabi [Journal of the social sciences] no. 1 (April 1979): 119-34.

17. Sharabi, Modernization, p. 16.

18. Ibid.

19. As late as 1947 Amman, which now with its surrounding suburbs contains over 1.1 million inhabitants, had a population of no more than 30,000.

20. Lerner, Passing, p. 43.

21. See Patai, Hashemite Kingdom, pp. 76-78.

22. See Al-shab, April 7, 1977.

23. Y. T. Ismael, Governments and Politics of the Contemporary Middle East (Homewood, Ill.: Dorsey Press, 1970), p. 106.

24. Halpern, Politics, pp. 28-29.

25. Ismael, Governments, p. 106.

26. For a negative view of the traditional Arab family, see Sharabi, Introduction, pp. 31-57.

27. Ibid., pp. 61, 68, 70, 79, and 126.

28. Ibid., pp. 72-73.

29. G. L. Harris, Jordan: Its People, Its Society, Its Culture (New York: Grove Press, 1958), p. 65.

30. Sharabi, Introduction, pp. 43-45.

31. Patai, Hashemite Kingdom, p. 89.

32. Halpern, Politics, p. 3.

33. C. E. Black, The Dynamics of Modernization (New York: Harper & Row, 1967), p. 54.

17

THE DYNAMICS OF
ISRAELI ATTITUDES TOWARD
THE PALESTINIANS

RUSSELL A. STONE

The future of the Palestinians is an issue of major concern in the Middle East and throughout the Mediterranean world. It is a focus of the continuing conflict between Israel and the Arab states. The Palestinian presence in Lebanon contributes to the continuing instability and violence in that country. The peace process between Egypt and Israel is now focused on the issue of Palestinian autonomy, for which there is not even a common definition and in which the Palestinians themselves are not currently participating. Palestinians live in many countries around the Mediterranean, where their role varies from that of a small ethnic community to an important component of the skilled manpower available for development, depending on the country involved. In every instance, however, the influence of the Palestine national movement is felt, whether in political pressure, propaganda, terrorism, or active community organization in support of the cause of a Palestinian homeland.

To some extent, the Palestinians can be considered a nation without a state, one of the many examples of nonstate nations that exist throughout much of the Mediterranean area and, indeed, throughout the world. While the phenomenon itself is not new, there has been a recent increase in analytic attention paid to such groups. [1] The Palestinian Arabs have been included among the nonstate nations analyzed in one collection of studies, and in the same book Israel was discussed as a former nonstate nation. The characteristics of such groups include a common ethnic, racial, or linguistic identity and a common sense of national unity and separateness among the people involved; but host country unwillingness to recognize the legitimacy of claims to a separate national existence for the "submerged nations." In some cases, only one host nation is involved (the Welsh in England), and in other cases the nonstate nation is submerged in more than one state (the Basques in Spain and France; the Kurds in Syria, Iraq, Iran, Turkey, and the Soviet Union). There is a further distinction between nonstate groups for which a state with similar characteristics does exist somewhere, although not in the territory they claim to be theirs (Greeks and Turks in Cyprus, Croats and Magyars in Austria, Albanians in Italy), and groups that do not have a homeland of like people anywhere (Armenians, Kurds). In many cases, the peoples of nonstate nations are suspected of separatism and, indeed, are often involved in acts disloyal to the countries in which they live, ranging from maintaining a separatist culture through religion, education, and/or language to terrorism in support of their national cause or in the name of more general dissident goals. Some nonstate nations seek a form of recognition short of full independence, such as autonomy within a federal system. Bertelsen claims the Palestinians are distinct in that they seek complete sovereignty. [2] The uniqueness of the Palestinian stand is challenged by a reviewer who claims other groups, such as the Croats and Kurds, also want complete independence. [3] The nature of the Palestinian sovereignty goal must also be questioned, for it is stated differently by various Palestinian groups, and its very indistinctness in terms of territorial aspirations and coexistence (or not) with Israel is a continuing source of uneasiness and opposition to the idea by Israel.

The existence of more ethnic groups and nascent national movements than there are sovereign states in the world can hardly be questioned. It is likely that such issues will continue to be sources of conflict within countries for the foreseeable future. Issues such as minority group rights and pluralism on various criteria (language, ethnicity, religion, national background, and so forth) will inevitably be a part of continuing political and social analyses of the Mediterranean area, no less than the rest of the world. However, the Palestinian issue is of particular importance because of the growing amount of attention and legitimacy accorded it, and because of its centrality to war-and-peace issues in the area, such as the conflict in Lebanon, terrorism within Israel, and the Israel-Egypt peace process, which has brought about the return of Sinai territory to Egypt and is currently in second-stage negotiations focusing on Palestinian autonomy. The possibility of future war(s) in the area and superpower involvement in military conflict makes the topic one of worldwide importance.

An irony of the situation recognized by Bertelsen is that Israel was a nonstate nation, much like the Palestinians today, before the state was founded in 1948.[4] That the Palestinians find themselves in conflict with the Jewish state, competing for some (or all) of the same territory, and following a path of building a national movement in ways similar to those pursued by Israelis earlier in the century makes the situation unique and fraught with unusual volatility, even for national struggles. Part of the irony is that Israelis can understand what the Palestinians are going through, although few Israelis sympathize with the Palestinian plight enough to support it openly. From the Palestinians' point of view, it is difficult to imagine a more formidable adversary than a nation that has successfully undergone a similar process within living memory, and so understands the strategy and tactics so well. These include shaping of public opinion within the national group, political organization, recruiting support in the world system, military activity, and even terrorism for national political ends.

Many Israelis claim that the situation of the Palestinians is nothing like that of the Jews on one of the criteria mentioned above. The claim is that Palestinians do have a state—that Jordan is a Palestinian state. Whether this is true or not depends upon how a state is defined in ethnic, national, or population terms. The majority (about 60 percent) of the population of Jordan is Palestinian.[5] However, Jordan has never agreed to view itself as a Palestinian state, and in fact, the Jordanian army fought against Palestine Liberation Organization (PLO) forces in September 1970 at the same time that many prominent civilian Jordanians, including many holding political office, were (and are) Palestinians. Jordan had sought to be recognized as the legal representative of all the Palestinian people, but at the Arab summit meeting in Rabat, Morocco, in 1974, King Hussein acquiesced to the pan-Arab commitment to view the PLO as their sole legal representative.[6]

The plight of the Palestinians and their future prospects have been affected by a series of major events, beginning with the war in 1948 that followed Israel's declaration of independence. Large numbers of Palestinian Arabs became refugees, having fled their homes during the fighting. The war in 1967 created a new wave of refugees as Israel occupied the West Bank of the Jordan River, which already contained large concentrations of 1948 refugees, most of whom moved east across the Jordan. Between 1967 and 1970 the Palestine Liberation Organization worked intensively to create a quasi-state within Jordan among the concentrations of Palestinians along the east bank of the river, which eventually constituted such a strong threat to King Hussein's regime in Jordan that he launched a bloody attack against the PLO-controlled areas of Jordan. The PLO then moved the major locus of its activities to Lebanon, along with a new wave of refugees from Jordan into Lebanon. In the ensuing years,

the Lebanese regime was destabilized and the country has been en-
gulfed in civil war since the mid-1970s. The Arab-Israeli war of
1973 did not directly involve the Palestinians, as neither Lebanon
nor Jordan took part in the fighting to any great extent. However,
that war did drive home to Israel the potential military threat posed
by improving Arab armies. And foremost among the reasons for
Arab hostility toward Israel is the continuing plight of the Palestinians.
It has come to affect internal stability within Arab states as well as
being a unifying political issue among countries that otherwise have
found unity difficult to attain.

Shortly after the 1973 war, a time series of public opinion data
began to be collected from representative samples of the urban Jewish population of Israel regarding Israeli perceptions and attitudes toward the Palestinians. The questions were included from time to
time in questionnaires of the Continuing Survey of Social Problem Indicators. The survey is conducted on a regular basis by the Israel
Institute of Applied Social Research and the Communications Institute
of the Hebrew University in Jerusalem. It has been in operation since
1967 and is an omnibus survey that contains questions on a large number of issues, including a full range of social indicators of well-being
plus questions commissioned from a variety of private, government,
and media sponsors. Questions on the Palestinians, the subject matter of this chapter, appeared in the survey at various times during
the period 1973-80. During that time, the Continuing Survey conducted
polls approximately 30 times per year (biweekly, or more frequently
if the volume of commissioned questions or important events warranted
more frequent surveys). Each poll was taken of a randomly selected
representative sample of the adult Jewish residents of the four largest
urban areas in Israel (Jerusalem, Greater Tel Aviv, Greater Haifa,
and Beersheba). Sample sizes were in the range of 500 to 600, and
all respondents were interviewed personally in their homes by trained
interviewers, assuring quality data from respondents and close conformity to the sample design. Several times a year the sample size
was tripled for greater accuracy, and the sampling frame was expanded to include the small towns and rural areas of Israel as well.
The results showed little difference between urban and rural views on
subjects related to foreign affairs, such as the questions discussed
here.

Not every question is included on each survey, but whenever a
question is asked, care is taken to phrase it in exactly the same way
as on previous questionnaires. Then, as questions are repeated over
time, the distributions of responses can be plotted as a time series,
which shows how attitudes of the population sampled (from independent)
random samples—these are not panel data) vary over time or change in
relation to important events. The Continuing Survey data set is unique

for its frequency and comprehensiveness. It provides an opportunity to study the structure and dynamics of attitudes on important social issues, [7] and how the Israeli public in general reacts to major events in the Middle East. [8] The nine questions regarding the Palestinians discussed here were each asked at least nine times, with some included in the survey as many as thirty times during the seven-year period being considered. The data series are plotted on time-series diagrams labeled with the major events that occurred in Israel and the Middle East during the period, to help with orientation to the relevant history.

Three items are included in each of the three figures in this chapter to permit comparisons among related attitudes as they change over time. This grouping divides the chapter into three sections: the first dealing with Israelis' attitudes toward the Palestinian nation, the second addressing the more specific issue of a Palestinian state, and the third examining Israeli perceptions of Palestinian leadership and representation. Thus, the Continuing Survey data permit a unique opportunity to examine the attitudes of members of an existing state (Israel) toward the nonstate nation (Palestinian Arab) that aspires to gain territorial and political concessions from it.

ATTITUDES TOWARD A PALESTINIAN
NATION AND ITS RIGHTS

Many factions within the Arab world use the phrase "regaining the legitimate rights of the Palestinian people" as the goal of the Palestinian national movement. The phrase is deliberately ambiguous, with different groups defining "legitimate rights" in different ways. What does the phrase mean to Israelis? Do they view it as constituting a threat to their own national existence? To evaluate these questions, the Continuing Survey has asked the following item 12 times beginning in early 1974: "To what extent do you agree with the following idea? When Arabs talk about 'regaining the legitimate rights of the Palestinians,' their true intention is to destroy the state of Israel." Responses to the item could be (1) definitely agree, (2) agree, (3) disagree, or (4) definitely disagree. The proportion of respondents who agreed with the statement (those giving answer 1 or 2) is plotted as the upper line in Figure 17. 1 (RTSDSTR on the legend).

During the four years following the 1973 war, between 70 and 80 percent of Israelis interviewed believed that Palestinian rights meant to Arabs the destruction of Israel. Then, when the question was asked again in 1979, after a break of almost two years and one and one-half years after Egyptian President Sadat went to Jerusalem, the proportion had dropped to from 62 to 64 percent. Thus, the major-

FIGURE 17.1

Israeli Attitudes regarding a Palestinian Nation and
Palestinian Rights

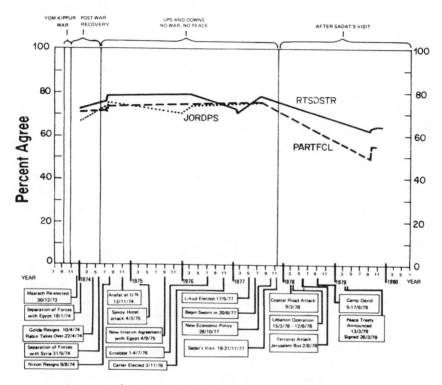

ity of Israelis do believe that Palestinian rights means a threat to
Israel's existence, but the proportion is decreasing slightly over
time. By late 1977 over one-third of Israelis felt Palestinian rights
did not necessarily constitute a mortal danger to national survival.
This reduction in fear of destruction can be attributed in part to the
peace process, in terms of which Israel as well as Egypt have undertaken to deal with the problem. On Israel's part, this implies that
Palestinian rights can be considered and negotiated in political terms.
This is implicit recognition by Israelis that the Palestinians are at
least a social entity, although the nature of the entity remains unde-
fined from the Israeli point of view.

The second full-length time series in Figure 17.1 is the proportion of respondents who agreed with the statement "The 'Palestinian Arab Nation' is an artificial concept that has only emerged in
recent years due to developments (events) in our area" (PARTFCL
on the legend). While close to three-fourths of respondents agreed

with this statement between the 1973 war and Sadat's visit, after the conclusion of the Egyptian-Israeli peace treaty the proportion had dropped to around 50 percent. Thus, a growing proportion of Israelis, close to half, recognized the Palestinians as a distinct national group at the end of 1979. Ironically, this puts them in a similar position to the Jews of Israel before the creation of that state—a people without a homeland.

Finally, a shorter time series in Figure 17.1 shows the proportion of respondents who agreed that "Jordan already fulfills the role of a state for the Palestinian Arab people."[9] This time series (JORDPS) closely followed PARTFCL until the Sadat peace initiative, when it was discontinued. That is, the proportion of Israelis who perceived that Jordan already served as a Palestinian state was similar to the proportion who did not recognize a separate Palestinian entity before Sadat's visit. When the other two questions in Figure 17.1 were asked again, after Camp David and the peace treaty with Egypt, the question on Jordan as a Palestinian state had become less pertinent. Although many Israelis still would agree with the item, as the idea had been supported by many Labor party leaders over the years, Begin's Likud government had agreed to consider a plan for Palestinian autonomy as part of the peace negotiations. There has been much debate, with no agreement to date, on the nature of Palestinian autonomy, but the very concept is recognition by the Israeli government of the existence of a Palestinian national entity.

Overall, Figure 17.1 suggests that there is a slowly growing segment of Israelis, although still a minority, who recognize that the Palestinians are a social entity, the granting of whose rights might not necessarily mean the destruction of Israel. This could be considered evidence, however weak, of a growing acceptance of the Palestinian Arab nation by Israeli public opinion.

Here and throughout this chapter, the distinction between indicators of public attitudes and government policy must be borne in mind. Political leadership must always be sensitive to public attitudes and major shifts in opinion in a democracy with high levels of individual involvement such as Israel. However, the issues considered in this chapter are the subject of much disagreement within the country, and many factors other than public opinion go into official decision making. We can identify temporal relationships between changes in public opinion and changes in government policy, but from the data presented here we cannot assess the extent to which public opinion affects government policy or vice versa. A related study has shown that when the government of Israel does take decisive action, even on controversial issues, public opinion tends to view government performance favorably.[10]

ISRAELI PERCEPTIONS OF A PALESTINIAN STATE

There are numerous proposals regarding the form of a Palestinian political entity, including absorption into Jordan and/or other
Arab countries (Lebanon was a possibility until the mid-1970s); forming a separate province within an Arab country or within Israel (making Israel a binational state—some Palestinian groups have used the
term binational secular state); autonomy in its various forms, as is
currently being discussed in Israel and in negotiations between Israel
and Egypt; and a separate, independent Palestinian Arab state. Of
these concepts, the most extreme, from the Israeli point of view, is
a Palestinian state, while for many Palestinians it is the only acceptable outcome of the conflict. In light of this, the Continuing Survey
has asked three questions regarding a Palestinian state in recent
years.

The first to be posed began shortly after the 1973 war and asks
whether Israelis should face up to the possibility that a Palestinian
state may emerge. The item reads: "Do you agree that Israel should
accept (live with) the possibility that a Palestinian state will be established between Israel and Jordan and in the Gaza Strip?" The proportion of respondents who agreed with the statement is plotted as ACCPS,
the lower line in Figure 17.2.

Throughout the period, fewer than 20 percent of respondents
felt that Israel should face the possibility of a Palestinian state, and
no obvious trends in the data or reactions to events are identifiable.
The proportion rose slightly in mid-1979 as Sadat-Begin talks moved
beyond the peace treaty negotiations to tentative discussions of the
Palestinian question. The series then fell again as the talks achieved
little. There may be a rising trend in mid-1980 as the series ends,
which would be consistent with previous indications that slightly more
Israelis are recognizing a Palestinian Arab national entity.

A correspondingly large majority of Israelis (80 to 90 percent)
answered "yes" or "definitely yes" to the following question: "Do
you think a Palestinian state in Judea and Samaria will endanger the
existence of the state of Israel?" (Judea and Samaria refer to the
area of the West Bank. The terms were adopted by Begin after the
Likud was elected in 1977. They are the biblical Israelite tribal
names.) This attitude is also very stable over time. The lowest
point, in September 1979, corresponds to a high point on the ACCPS
series, indicating consistency of responses to the two items despite
their different phrasing and direction of response ranges.

The third time series in Figure 17.2 was begun after the Camp
David summit in September 1978. The question asks if the respondent
believes a Palestinian state will be created. "In your opinion, as a
result of the agreement signed and the developments which have taken

FIGURE 17.2

Israeli Attitudes toward a Palestinian State

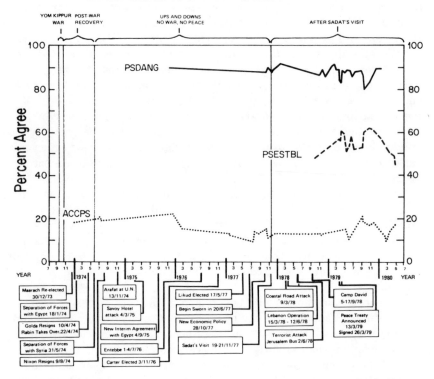

place consequently, in the end will a Palestinian state emerge in the West Bank and the Gaza Strip?" (PESTBL on the legend). The proportion of respondents answering affirmatively varies between 45 and 60 percent, with more variability than the other two series on this figure but no change in overall level. There may be a declining trend toward the end of the series.

This finding reveals some anomaly or contradiction in Israeli attitudes toward a Palestinian state. Although the vast majority of respondents believe such a state will endanger the existence of Israel and that Israel should not accept the possibility of such a state, approximately half believe a Palestinian state will emerge, presumably against the desire of Israel and contrary to its security interests. The anomaly is not a single-time finding. It emerges repeatedly in 17 surveys taken over a two-year period. Also, each of the three time series in Figure 17.2 is very stable over time, showing no change in level.

Finally, comparing Figure 17.1 with Figure 17.2, the most general finding is that although Israeli attitudes became slightly more accepting of some sort of Palestinian national entity, attitudes toward a state per se were consistently negative.

RECOGNITION AND REPRESENTATION

A key issue regarding the nature of a Palestinian entity centers around two questions: Who represents the Palestinian Arab people? and Should the representatives be recognized by Israel? Again, the findings are a mix of consistency and anomaly. More Israelis believe that Yasser Arafat represents the Palestinians rather than King Hussein of Jordan. This can be seen by comparing the YARPALS and KHRPALS time series in Figure 17.3. YARPALS plots the proportion of affirmative responses to this question: "Do you believe that Yasser Arafat (leader of the PLO) represents the majority of Arabs interested in a Palestinian entity?" KHRPALS is the proportion of respondents who believe that "Hussein (King of Jordan) represents most of the Palestinians." The findings are consistent between the end of 1973 and Sadat's visit in 1977. Both time series are stable, as is the relationship between them. More Israelis perceived that Arafat represented the Palestinians than believed that Hussein did. The anomaly is that although around 70 percent of Israelis believed Jordan fulfilled the role of a Palestinian state (at least until 1977—see JORDPS in Figure 17.1), only around 20 percent believed that King Hussein actually represented the Palestinians. Thus, most Israeli respondents recognized, at least implicitly, that the Palestinians did have a problem of representation. Furthermore, the proportion recognizing Arafat as the representative of the Palestinians increased in the last three years of data, rising to over 50 percent after Camp David and the peace treaty.

At this same time, a third question was begun: "In your opinion should Israel recognize the PLO as representing the Palestinians?" (RECGPLO in the legend). Never more than 20 percent of respondents answered yes to this question in 1979–80, as indicated by RECGPLO in Figure 17.3. Thus, at the same time that more Israelis were coming to believe that Arafat represented the Palestinians, very few felt that Israel should be recognizant of his organization.

Finally, close comparison of the RECGPLO plot in Figure 17.3 with ACCPS in Figure 17.2 during 1979 and early 1980 reveals that the (always small) proportion of respondents who felt Israel should accept the possibility of a Palestinian state matched the percentage who felt that the PLO should be recognized. Both time series rose, then fell simultaneously during 1979, remaining always at about the same level relative to each other. This perhaps suggests that once Israelis can bring themselves to accept the idea of a Palestinian state, they

FIGURE 17.3

Recognition and Representation

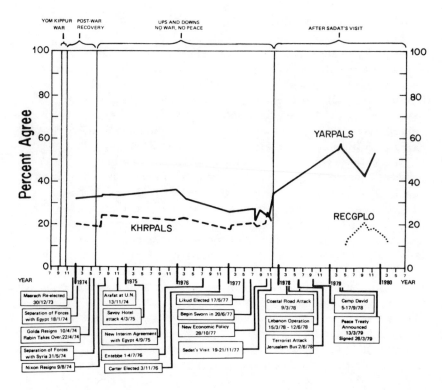

might also accept the PLO as a political organization within that state. However, in the absence of confirmatory correlation analysis, this suggestion is purely speculative.

CONCLUSION

The concept of a nonstate nation is useful for interpreting Israeli public attitudes toward the Palestinians, for there are differences in level and degree of change in attitudes between questions that refer to a Palestinian national entity and those that specifically refer to a Palestinian Arab state. Overall, most of the nine attitudes toward the Palestinians examined over a seven-year period by means of public opinion polls were rather stable, and not very supportive of Palestinian Arab political rights. However, some change does appear toward increased Israeli acceptance of the idea that the Palestinians are a separate national entity and that their rights might not imply the destruction of Israel. The increased acceptance of the idea of a Palestinian nation appeared after Sadat's visit and the peace treaty with Egypt, and the proportion of respondents accepting the idea that a Palestinian Arab nation is not an artificial concept was larger than the proportion who felt that Palestinian rights did not imply the destruction of Israel. A majority of respondents were still wary on the latter issue.

Attitudes toward a Palestinian state were more negative and unchanging than when the terms nation or entity were used. Seldom more than 20 percent of respondents said Israel should accept a Palestinian state, and a corresponding 80 percent felt such a state would endanger Israel's existence. Despite this, by 1978-80, half of the respondents or more perceived that a Palestinian state would emerge eventually in the West Bank and Gaza. The conditions under which this might be acceptable to Israelis have been explored in a previous study.[11] In return for a full and acceptable peace, more than half of the Israelis questioned in the period 1977-79 were willing to give up at least part of the West Bank and the Gaza Strip. These territories, occupied by Israel during the 1967 war, are inhabited primarily by Palestinians, with a growing minority of Israeli settlers. The proportion of Israelis willing to make territorial concessions to the Palestinians in return for peace grew steadily but slowly after the 1967 war from lows below 25 percent to 1979 levels around 60 percent. This despite continued expansion of Israeli settlement in the occupied territories, which reflects the continuing disagreements among Israelis regarding territories and peace. It also highlights the problems inherent in relating public opinion to government policy, warranting caution in interpreting the data in this chapter.

However, it must be noted that the same study showed that change in attitudes toward relinquishing the Sinai in Egypt in return for peace preceded the Egyptian-Israeli peace treaty by many years.[12] Throughout the period since the 1967 war, more Israelis had been willing to return part or all of the Sinai in return for peace than were willing to return the West Bank and Gaza, and the proportions regarding the Sinai rose from below 50 percent after the 1967 war to around 80 percent between 1973 and 1977, and even higher after Sadat's visit. Thus, evidence from both public opinion and government policy shows past Israeli willingness to return occupied territory in return for peace. Then, more recently, public opinion indicators regarding territorial return in the West Bank and Gaza grew more flexible after the 1973 war, so it is not inconceivable that a change in government policy could take place on those territories as well. There would be some public support for it.

This speculation must be treated very cautiously, however. It is clear that no concessions to the Palestinians would be forthcoming if not accompanied by a peace agreement. Also, the experience of actually returning the Sinai to Egypt has created bitter dissension within Israel, which may make government leaders more wary about future concessions. The treaty with Egypt has also spurred supporters of expansion within Israel to increase settlement activity in the West Bank and Gaza, which will complicate matters in the future. It must further be reiterated that the survey data presented here ceased in mid-1980, and events since that time may well have affected public attitudes.

Finally, the findings do reveal some internal inconsistencies among Israeli public attitudes toward the Palestinians. A large majority feel that a Palestinian state will endanger the existence of Israel, but more than half perceive that a Palestinian state may be created in the West Bank and Gaza anyway. More than 70 percent felt that Jordan already fulfilled the role of a Palestinian state, but only around 20 percent perceived that King Hussein, the ruler of Jordan, represented most of the Palestinians. A larger percentage of respondents, which began at 30 percent and increased over time to above 50 percent, felt that Yasser Arafat, leader of the PLO, represented the majority of Palestinians; but only around 15 percent felt that Israel should recognize the PLO as representing the Palestinians. In the final analysis, we must conclude that much of the Israeli public does understand the plight of the Palestinians as a nonstate nation and is aware of the distinction between a Palestinian Arab national entity and a state per se. They are more sympathetic to the former concept, but are both fearful and wary of the Palestinian national movement.

NOTES

1. Judy S. Bertelsen, The Palestinian Arabs: A Nonstate Nation Systems Analysis, Sage Professional Papers in International Studies, vol. 4, no. 02-043 (Beverly Hills and London: Sage, 1973); Charles R. Foster, Nations without a State: Ethnic Minorities in Western Europe (New York: Praeger, 1980); and Vatro Murvar, "Submerged Nations Surface in Research," Contemporary Sociology 1, no. 1 (1982): 14-16.

2. Bertelsen, Palestinian Arabs, p. 247.

3. Murvar, "Submerged Nations," p. 15.

4. Bertelsen, Palestinian Arabs, p. 247.

5. Naseer H. Aruri and Samih Farsoun, "Palestinian Communities and Arab Host Countries," in The Sociology of the Palestinians, ed. Emile A. Nakhleh and Elia Zureik (New York: St. Martin's Press, 1980), p. 117.

6. Ibid., p. 130.

7. Shlomit Levy and Louis Guttman, "On the Multivariate Structure of Well-being," Social Indicators Research, no. 2 (1975), pp. 361-78; and idem, "The Structure and Dynamics of Worries," Sociometry 38, no. 4 (1975): 445-73.

8. Russell A. Stone, Social Change in Israel: Attitudes and Events, 1967-1979 (New York: Praeger, 1982).

9. The complete wording of the JORDPS item is lengthy and complex, which together with changing events, accounts for its being dropped from the Continuing Survey after 1977. In full, it reads: "As is well known, 'Palestine,' according to the British Mandate, includes both banks of the Jordan (river). The Kingdom of Jordan includes half the area of 'Palestine' according to this definition, so there are those who claim that Jordan already fulfills the role of a state for the Palestinian Arab people. Do you agree?" The response range, as for all items considered here, is a four-point scale ranging from definitely agree to definitely disagree.

10. Ibid., chap. 10.

11. Russell A. Stone, "Israeli Public Attitudes to War, Territories and Peace, 1967-1979" (Paper presented at the annual meetings of the Middle East Studies Association, Washington, D.C., 1980); and idem, Social Change, chap. 1.

12. Ibid.

18

TOWARD AN UNDERSTANDING OF MEDITERRANEAN LEGAL CULTURE IN THE CONTEMPORARY WORLD CAPITALIST ECONOMY

JOHN R. SCHMIDHAUSER
LARRY L. BERG

It is the purpose of this chapter to assess the significance of Mediterranean legal culture for world economic systems frequently described or suggested in dependency-theory conceptual frameworks.

Most conceptual frameworks that have been developed to map the purposes and characteristics of judicial systems and legal professions are based upon the fundamental assumption that these institutions evolved or were created in order to institutionalize conflict resolution based upon fair and objective procedure. Indeed, this assumption is an integral part of the received tradition in most law school training in the United States. Conversely, in most revolutionary situations, an existing judicial system or legal profession is frequently characterized by the revolutionaries as an instrument of the regime that must be overthrown or replaced. It is often viewed as unfair, corrupt, or despotic and as a cause of social, economic, and political conflict. Two conceptual frameworks that embody these contradictory assumptions are described and assessed with regard to the impact of Mediterranean legal culture upon them.

The first, a modified Weberian conceptual framework, described below, accepts the assumption basic to the received tradition that a fundamental purpose of judicial institutions and legal professions is conflict resolution based upon fair and impartial procedure. Second, a contradictory conceptual framework was constructed as a corollary to Immanuel Wallerstein's world capitalist economic system perspective and similar perspectives. This corollary rejects this assumption and differs with other Weberian assumptions as well.

One important additional difference in assumptions relates to the determinants of the organization, purposes, and personnel of judicial systems and legal professions. A basic assumption of Weber's model is that such a system is not a product or result of deterministic economic and social forces. Instead, in the Weberian conceptualization, the closer a legal and judicial system comes to attaining the attributes identified in the ideal model, the more optimum the conditions for capitalism. In short, in the modified Weberian hypothesis, the presence of a legal system determines whether capitalism (or, in this adaptation, modernization) may be achieved and operate under optimum conditions. Conversely, in the context of Wallerstein's framework, the organization, purposes, and personnel of court systems and legal professions of nations are determined by seminal changes in political, social, and economic power relationships that establish fundamental parameters of political and military power and economic relations for eras of considerable stability and continuity. [1]

The first of the macroconceptual frameworks utilized in this investigation is one developed from Max Weber's ideal model of a legal system. As noted above, adaptation of Weber's framework to modern conditions, rather than strict application of his original model, is one of the goals of this investigation. Max Weber's intellectual interest in the relationship of legal systems to the development of capitalism provided the basis for his partially completed ideal model of a legal system. [2] In broad perspective, Weber's conception of the legal basis for social control and stability embodies two basic attributes: (1) that legal norms may be distinguished from other social integrative norms because they are enforced and (2) that it is sometimes difficult to distinguish legal mechanisms of social control from nonlegal mechanisms. [3] These attributes were related to his identification of three major bases of political legitimacy—legality, formally correct rules, and accepted procedure. [4] Weber did not posit that capitalism determined the characteristics of Western European legal systems (from which he constructed his ideal model), but he did indicate that an ideal legal system provides the necessary conditions of stability and predictability for the development and maintenance of capitalism. The key criteria employed by Weber for designating such legal systems were "formal" and "rational." The rational criterion was met when a legal system was characterized by objectivity and impersonality in its modes of decision making. [5] Weber classified governmental authority (in his terminology "domination") into three broad groups: traditional, charismatic, and legal. Many decades before linkage politics became academically fashionable, Weber attempted to demonstrate the integration of political structure and legal system. A summary of this is provided in Table 18.1. [6]

TABLE 18.1

Weber's Conception of the Relationship between Political Structure and Legal System

	Types of Domination (Governmental Authority)		
	Traditional	Charismatic	Legal
Obedience is owed to:	Individuals designated under traditional particularistic custom	Individuals considered to be extraordinary and/or endowed with special powers	Enacted rules made in accordance with objective, universal standards
Law is legitimated by its:	Origin in custom and tradition	Origin derivative from each particular charismatic leader	Origin in objective process, determined itself by legal or constitutional standards
Nature of the judicial process and the manner in which judicial decisions are justified is:	Empirical–traditional; case-by-case decision making, but precedent may or may not be invoked; particularistic; justified by custom or tradition	Case oriented but based on revelation; particularistic and often idiosyncratic; justified by revelation	Independent and universalistic; cases decided by formal rules and abstract principles and justified by independence and universality of the decision-making process
Structure of the administration of law is:	Patrimonial; staff recruited through traditional modes, generally ascriptive; duties allocated by discretion of master	Structured administration; often ad hoc selection of staff on charismatic qualifications, with undifferentiated tasks	Bureaucratic; highly organized administration of law by well-trained professionals with a universally delimited jurisdiction
Degree of discretion of ruler(s) is:	High	High	Low
Predictability of rules governing economic life is:	Low	Low	High

Source: This is a modified version of table 2, "Administration, Law and Economic Regulation under the Pure Types of Domination," in David M. Trubek, "Max Weber on Law and the Rise of Capitalism," 1972 University of Wisconsin Law Review (Madison: University of Wisconsin Press, 1972), p. 735.

TABLE 18.2

The Conceptual Model

Components of the Model	Description–Rationale	Strongest Indicator
Nature of the foundation or basis of judicial authority	The judicial system must be protected from short–term political change and external influences in order to ensure its objectivity and independence	Judicial institutions based upon constitutional rather than ordinary statutory authority
Nature of the relationships among major branches of government	The judicial system must be independent of other branches of government or external influences	Constitutionally required functional separation
Nature of the tenure of justices and judges	Judicial independence and objectivity are ensured when judicial tenure is guaranteed	Constitutionally protected tenure for life
Inviolability of judicial salaries	Judicial independence is again safeguarded when salaries are inviolate	Constitutional prohibition on reduction of judicial salaries
Nature of judicial selection standards	High standards are essential to ensure professionally sound and objective decision making	Highest quality professional training, non–ascriptive standards
Nature of the institutional procedures	Predictable institutional procedures are necessary to ensure objectivity and impartiality in decision making	Formal rules, universalistic criteria, and independence
Nature of jurisdiction	Jurisdictional authority should be broad to provide the judiciary with significant governmental authority	Major jurisdiction provided for in the constitution as well as in the statutes
Judicial review	On matters of highest constitutional import, the highest appellate courts must exercise sufficient authority to determine the fundamental interpretation of that constitution	Full establishment and acceptance of judicial review
Compliance with judicial decisions and orders	The judicial system cannot exercise full authority unless other appropriate branches of government enforce its orders and decisions and uphold its authority	Full enforcement of and compliance with judicial decisions and orders
Regime stability	The judicial system is highly unlikely to retain its independence and objectivity in periods of serious political unrest	Fundamental political change by peaceful constitutional change
Distributive justice	The judicial system is highly unlikely to administer justice objectively and independently in an unjust, class–stratified society	High standards and fair practices in economic, social, and political sectors
Probity	The judicial system cannot maintain its independence and objectivity in a society characterized by bribery, nepotism, and conflicts of interest	Prohibition of bribery, nepotism, and conflicts of interest

In order to provide a basis for cross-national and intranational comparative analysis of judicial systems and to identify attributes of institutional stability, independence, and probity, which are conceptually equivalent to elements of Max Weber's ideal legal system, components of the model are identified with a description-rationale and appropriate indicators for each category. They are summarized in Table 18.2.

The emphasis upon the relationship between the characteristics of legal systems and capitalism, which are basic to Weber's ideal model, may appropriately be transposed to a broader linkage framework: the relationship between the characteristics of legal systems and modernization. Among the attributes of modernization commonly considered as significant indicators are rising levels of production and consumption, urbanization, the emergence of a money economy, the development and application of science and technology not only in manufacturing and extractive industries but in health as well, significant advances in education, the development of social and political institutions not based upon kinship, and secularization.[7]

In Max Weber's original model, as in the real world in most of his lifetime, there was no significant formulation or example of the phenomenon of the socialist legal system. Although such a system was, indeed, established near the end of his life, Weber did not modify his model. Logically, the basic attributes of the socialist legal system represent failures to fulfill the requirements of the Weberian model—an unsurprising conclusion, since creation of the optimum conditions for capitalism is a major feature of his model. In this modification and adaptation of the Weberian model, modernization rather than capitalism is deemed the logical end result of the development of an ideal legal system. Presumably, under certain conditions, an advanced socialist legal system as well as an advanced capitalistic legal system might fulfill a number of the conditions required within the conceptual framework. Yet, it is doubtful, in the real world of politics, that such socialist legal systems would meet the requirements related to judicial independence, while it is equally doubtful that their capitalistic counterparts would meet those related to nonascriptive standards or distributive justice. Whatever the empirical evidence may indicate, this adaption and modification of the Weberian ideal model incorporates all legal systems, including the variety of socialist forms that have emerged since 1917.

Although this modification of Max Weber's ideal model of a legal system differs in some respects from conventional conceptions of judicial and legal systems, it shares a fundamental assumption basic to a number of theoretical and descriptive approaches to the comparative study of judicial and legal institutions—that the purpose of these institutions is conflict resolution and the substitution of some

form of adjudication for violence.[8] Indeed, it has been suggested
that this conflict-resolution function is a universal attribute of
courts.[9]

The key components of the world-system conceptual framework
for judicial and legal systems reflect emphasis upon imposed change,
center-periphery relations, and economic as well as political power
conflicts, rather than the conventional basic assumption about conflict resolution. The conceptual model, summarized below in Table
18.3, treats significant elements of the judicial and legal systems of
selected nations and their relationship to the fundamentals of Wallerstein's world-system perspective. In Wallerstein's summation, "A
World System is a social system, one that has boundaries, structures,
member groups, rules of legitimation, and coherence. Its life is
made up of the conflicting forces which hold it together by tension and
tear it apart as each group seeks eternally to remold it to its advantage."[10]

Wallerstein distinguishes three varieties of world systems:
(1) world empires, in which a single political system controls all or
most of the components; (2) the world economy, which has existed for
five hundred years because "the world economy [capitalism] has had
within its bounds not one but a multiplicity of political systems"; and
(3) socialist world government, which has not succeeded in replacing
the second, but would change the system of economic distribution and
political decision making. Within the current world system, the
capitalist world economy, the conflicts and tensions that may change
or even eliminate a national judicial and legal system are distinguish-
able in two broad categories. The first involves the conflicts and tensions that Wallerstein deems characteristic of this world system—
those growing out of the relationship among the dominant nations
(called core states by Wallerstein or center nations by Galtung and
others), the peripheral areas, and third, the semiperipheral areas
between these two. The second category involves the conflicts and
tensions between competing world systems—the capitalist economy
and the socialist world government.[11] At another level of analysis,
systemic division-of-labor considerations are also important components of this conceptual framework. Since the division of a world
economy involves a hierarchy of occupational tasks, "[the] tasks requiring higher levels of skill and capitalization are reserved for
higher-ranking areas."[12]

In terms of the logic of the world-economy conceptual framework, judicial and legal institutions and personnel are determined or
considerably affected by the conflicts and tensions that are characteristic of the capitalist world economy and its political manifestations. Independence from external influences (such as foreign political and/or economic control) would be the monopoly only of the core

TABLE 18.3

The World–System Conceptual Model: General Framework

Components of the Model	Description–Rationale	Strongest Indicators
Origin of organization and purposes of the judicial and legal systems of nations	The genesis of the structure, purposes, and characteristics of judicial and legal systems is found in the requirements of the world capitalist economy (or its antithesis, the world socialist economy).	Core nations develop their judicial and legal systems independent of external influences, but in periphery and semiperiphery nations and territories, the judicial and legal systems are replaced or modified in accordance with core-nation demands, world economy, and/or needs.
Purposes of the judicial and legal system	A major purpose of each national judicial and legal system is the creation of conditions highly conducive to the operation of a capitalist world economy (or a socialist world economy).	The judiciaries and legal professions of core nations and of those periphery and semiperiphery nations and territories that are responsive to the legal needs of the capitalist world economy are heavily oriented toward providing legitimation to the capitalist or socialist world economy, and the heaviest emphasis in law is upon commercial-and property-related issues (or the goals of a socialist society). They are secular rather than religious oriented.
Characteristics of judicial and legal personnel	The capitalist world economy requires that judges and lawyers be cosmopolitan and be part of an international community of legal professionals who maintain and protect the operation and the capitalist world economy or its socialist counterpart and, especially in underdeveloped nations, have more in common with the world economy of legal professionals than with their own countrymen.	In core nations and capitalist or socialist world-economy–oriented periphery and semiperiphery nations and territories, the highest ranking judges and most influential lawyers are products of a few influential schools and constitute a world community of elite legal professionals whose work greatly facilitates the goals of the capitalist (or socialist) world economy.

or center nations. Modification of the judicial and legal structures, purposes, and characteristics of the personnel of periphery and semi-periphery nations or colonial territories would be in accordance with the political and economic needs or demands of the core or center nations. Consequently, the imperatives of the general framework of the world-system conceptual model described in Table 18.3 provide only the most prominent features of that model. A more detailed examination of the major areas of difference between the two competing conceptual frameworks provides the more comprehensive elements of the world-system model as well as an item-by-item comparison. These attributes are summarized in Table 18.4. Because Wallerstein stresses the multiple layers that characterize the world system, references to the requirements of this system pertain to the more complex, competitive needs found at various levels of those multiple layers, such as ruling-class needs or power-bloc demands, among others.

In a number of instances, the attributes of these two conceptual frameworks are directly contradictory, notably items related to (1) conflict resolution, (2) the determinants of the characteristics of judicial systems and legal professions, (3) judicial independence, or the lack thereof, (4) and (5) legal neutrality and objectivity, and (6) the essentiality or expendability of nonproperty human rights. But in others, the differences were more subtle. For example, it has been argued that the world capitalist economy stimulates convergence "by subjecting all societies to the same forces" and stimulates divergence "by creating different roles for different societies in the world stratification system."[13] The homogenizing thrust is generally interpreted as resulting from the needs of the world capitalist market economy, the thrust toward diversity from world division of labor imperatives. Judicial systems and legal professions are uniquely related to the world needs of the capitalist economic system. Therefore, while it might ordinarily be assumed that a branch of a national government (and the professional group most intimately related to the judicial branch) would fulfill a divergent, nationalistic, institutional role, it is hypothesized here that this is the one sector of national governments most likely to fulfill the particular legal needs of the world capitalist economy. The framework and pattern of such legal and judicial fulfillment are set by legal professionals, law schools (or their equivalents), and judicial institutions of the core nations and are generally adopted in the periphery and semiperiphery nations. Consequently, the highest-ranking judges and most successful attorneys in these nations are likely to be cosmopolitans, well trained in the most prestigious of the leading schools of core nations (or in domestic schools modeled after the elite schools of the core nations), and intellectually part of an international community of similar elite

TABLE 18.4

Key Attributes That Distinguish the Competing Conceptual Frameworks

Modified Weberian	Corollary Adapted from Wallerstein
The assumption that judicial systems and legal professions evolved or were created to institutionalize conflict resolution based upon fair and objective principles and procedures is accepted.	Conflict resolution may be a secondary objective, but the primary purpose of judicial systems and legal professions is to fulfill needs of world capitalist economy (or its socialist antithesis).
The characteristics of judicial systems and legal professions are not the products of deterministic economic and social forces.	The characteristics of judicial systems and legal professions are determined by economic and social forces, notably the needs of the world capitalist economy (or its socialist counterpart).
Judicial systems and legal professions and their modes of procedure and decision making are developed and maintained independently of external institutional influences, such as governments, political parties, and religious organizations.	Judicial systems and legal professions and their modes of procedure and decision making are developed and maintained under external institutional influences that seek to maintain the world capitalist economy (or its socialist counterpart).
Decision making is characterized by objectivity and impersonality in its modes	Objectivity and impersonality in decision-making modes may be transcended by the compelling need to fulfill goals of world capitalist economy (or its socialist antithesis).
Law serves as a neutral arbiter of individual relations between equals.	Law protects interests and institutions that are essential to the optimum functioning of the world capitalist economy (or its socialist equivalent).
Legal safeguards for contract and property rights and for noneconomic personal liberties (human rights) are essential.	Legal safeguards for property and contract are essential, but noneconomic safeguards are expendable when the bourgeoisie has achieved effective control of a nation in order to participate in the world capitalist economy. (Tigar and Levy)
Predictability and rationality are deemed essential to a legal system as components of an ideal system by Weber.	The predictability and rationality deemed essential to a legal system by Weber are deemed essential by monarchs (or other nonrepresentative rulers) to encourage commerce and the allegiance of the bourgeoisie to the monarch. (Tigar and Levy)
Judges and elite legal professions are trained and socialized to develop goals of predictability and rationality.	Judges and elite legal professionals are trained and socialized to maintain and develop the world capitalist economy (or the world socialist economy).

jurists and legal professionals. Although many periphery and semi-periphery nations are strongly nationalistic in international policy making and governmental elite recruitment, judicial recruitment often reflects greater emphasis upon the legal needs of the world economic system.

Similarly, the attributes of predictability and rationality extolled by Weber are not unacceptable in the conceptual framework derived as a corollary of Wallerstein's world system. However, such attributes were deemed desirable in the latter for purposes directly related to the development of or maintenance of the world capitalist economy.

The advent of world-economic-system analysis, most significantly identified with Immanuel Wallerstein, has shifted emphasis in a number of important respects from nations to a world-system perspective and from political dominance to economic and cultural influence and penetration. Similarly, such analysis has explicitly rejected modernization as a useless concept. [14] And in contrast to a great deal of the current emphasis upon statistical cross-national comparison, such world-economic-system analysis emphasizes historical, longitudinal, comparative analysis. Such analysis is deemed particularly appropriate in the investigation of both the neo-Weberian model and the corollary derived and modified from the Wallerstein framework. Specifically, a series of time-series investigations of selected core and periphery legal systems should be designed to determine whether they change from traditional to charismatic to rational-legal (or back) in the framework of a Weberian-type model or from precapitalist to capitalist to socialist modes of production and division of labor with role changes from periphery to core (or back) in the context of a framework adapted from Wallerstein.

The modified legal framework based upon world-economic-system perspective, in contrast to that of a modified Weberian model, incorporates the fundamental assumption that judicial systems and legal professions play a very direct role in either a world capitalist economic system or a world socialist economic system. The major elements of such a role are summed up in a preliminary fashion in Figure 18.1.

In accordance with this preliminary theory, the judicial systems and legal professions of core nations enjoy and maintain relative equality of power and influence, while those of semiperiphery and periphery nations do not share such power and influence. The primary task of the judicial systems and legal systems of core, semiperiphery, and periphery nations is the maintenance of either the world capitalist economic system (which currently is most pervasive) or the countervailing world socialist economic system. Although the two systems oppose each other in terms of power, influence, and

FIGURE 18.1

A Preliminary Theory of the Judicial System and Legal Professional Properties of World-System Dynamics

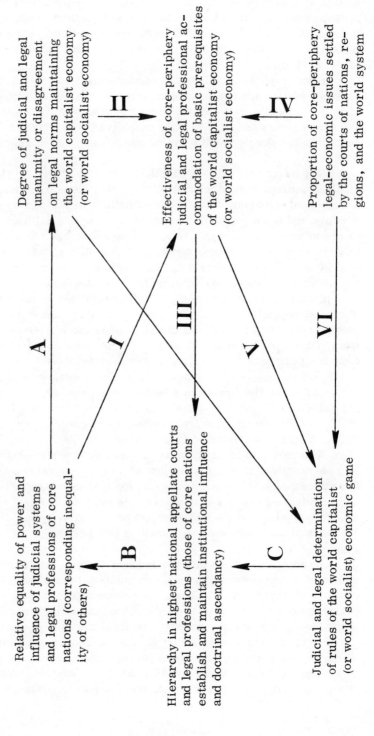

fundamental assumptions, the capitalist and socialist nations each
are characterized by their own principles and procedures that accommodate the fundamental prerequisites of each world economic
system and serve to restrict intrusion of principles or practices that
may be considered detrimental or contradictory to a capitalist or a
socialist legal system. Inga Markovitz's comparative analysis of
several doctrines relating to owning and sharing public offices, housing, and worker compensation developed by courts in East and West
Germany provided several pertinent examples. [15] The hierarchy of
judges in the highest appellate courts and the leaders of the bar associations in the core nations establish and maintain institutional influence and appropriate doctrinal ascendancy. Similarly, legal education within either the world capitalist core nations or the world socialist core nations is oriented to engender the professional behavior,
attitudes, and doctrinal commitments that will maintain the integrity
of each respective system.

What is the relationship of Mediterranean legal culture to these
contradictory models? What are its origins and the major elements
that demonstrate its continuity? In the context of the modern Italian
and Spanish legal systems, do the modern attributes of Mediterranean
legal cultures approximate most closely the neo-Weberian model or
the model based upon a corollary of Wallerstein's world economic
system?

THE MEDITERRANEAN CONNECTION:
HISTORIC AND MODERN

Most modern commentators upon law as an instrument of control rather than of conflict resolution treat such relationships in the
context of the rise of capitalism. However, there is an earlier body
of comparative law investigation and analysis, ignored by most contemporary investigations, that establishes the Mediterranean origins
of law and legal professionalism, its paramount influence in Western
law development, and its ancient and medieval manifestations of law
as an instrument of the powerful.

The conception of Mediterranean civilization as the social and
intellectual setting for the creation of significant social, political,
and economic institutions is central to Fernand Braudel's magnificent
analysis of the Mediterranean world in the age of Phillip II of Spain.
The contributions of several great Italian city-states to the development of the legal profession, banking, and international finance had,
of course, preceded Spain's ascendancy as a transatlantic as well as
a major Mediterranean power. [16] However, the enduring quality of
these social and institutional innovations was, quite appropriately,
emphasized by Braudel. He stated that

the more one thinks about it, the more convinced one becomes of the striking similarities, transcending words,
terminology and political appearances between East and
West, worlds very different it is true, but not always divergent. Experts in Roman law and learned interpreters
of the Koran formed a single vast army, working in the
East as in the West to enhance the prerogative of princes.
It would be both rash and inaccurate to attribute the progress made by monarchy entirely to the zeal, calculations
and denotions of these men. All monarchies remained
charismatic. And there was always the economy. Nevertheless, this army of lawyers, whether eminent or modest, was fighting on the side of the large state. It detested
and strove to destroy all that stood in the way of state
expansion. [17]

Braudel thus emphasized the university of the role of lawyers
in a political world system that both preceded and overlapped the
early stages in the capitalist world system conceptualized by Wallerstein. That capitalist world system derived its major characteristics
from its Mediterranean inventors. Thus, according to Braudel, the
emergence of international financial situations retained the essential
attributes of Genoese origins after transposition to Portugal, the
Netherlands, and ultimately, England. As he put it,

The coming of the age of paper, its extension if not its
first appearance, in fact marked the beginning of a new
economic structure, an extra dimension that now was to
be reckoned with. The Genoese did not, as it is often
nastily assumed, represent the failure of pure finance
and paper money and the triumph of the merchant who had
remained faithful to traditional commerce; it signified
rather the rise of a new capitalism with a different graph-
ical centre of gravity, which had been in the making since
the discovery of America, but which took over a century
to reach completion. Ultimately it marked the victory of
new financiers, the Portuguese money lenders who were
to intervene at Madrid in 1627, and behind them, the heavy
hand of the capitalists of the North. It was in fact one of
the stages in the development of Dutch capitalism, the
superstructures of which, including the most modern form
of credit machinery, which was in place by at least 1609;
the force that was to replace Mediterranean capitalism.
But the old model, patiently assembled over time, was in
every respect a pattern for the new. [18]

The intellectual thrust of Braudel's analysis of the relationship of law and lawyers as the instruments of the powerful, the Mediterranean origins of that relationship, and the concomitant invention and intellectual colonization of capitalism has been largely overlooked by the contemporary leaders of the law-and-development school. They have usefully documented the transportation of Mediterranean legal institutions and procedures from Europe and Latin America, [19] but they have overlooked the rich academic heritage of analysis from the nineteenth and early twentieth centuries, which provided the basis for Braudel's massive modern contribution.

Alexis de Tocqueville's notes on the influence of Roman law in Germany was one such analysis. His interpretation emphasized the enlisting of Roman law in the service of expanding monarchical control and the economic exploitation of the peasantry. De Tocqueville argued that

> at the close of the Middle Ages the Roman law became the chief and almost the only study of the German lawyers, most of whom, at this time, were educated abroad at the Italian universities. These lawyers exercised no political power, but it devolved on them to expound and apply the laws. They were unable to abolish the Germanic law, but they did their best to distort it so as to fit the Roman mold. To every German institution that seemed to bear the most distant analogy to Justinian's legislation they applied Roman law. Hence a new spirit and new customs gradually invaded the national legislation, until its original shape was lost, and by the seventeenth century it was almost forgotten. Its place had been usurped by a medley that was Germanic in name, but Roman in fact.
>
> The Roman law carried civil society to perfection, but it invariably degraded political society, because it was the work of a highly civilized and thoroughly enslaved people. Kings naturally embraced it with enthusiasm, and established it wherever they could throughout Europe; its interpreters became their Ministers or their chief agents. Lawyers furnished them at need with legal warrant for violating the law. They have often done so since. Monarchs who have trampled the laws have almost always found a lawyer ready to prove the lawfulness of their acts. [20]

De Tocqueville's interpretation provided an interesting and challenging assessment of this significant transformation in law, but did not provide a complete conceptual framework. A nineteenth century Italian scholar, Achille Loria, provided a tightly reasoned analy-

sis that integrated historical interpretation of the role of Roman law
with a broadly enunciated conceptual framework for the economic
basis for law.

Loria directly addressed the fundamental question of whether
law is a product of each separate national culture as argued by Mon-
tesqieu and Savigny,[21] or whether it is a product of a rational organization that would create the optimum conditions for capitalism (Weber).
Loria stated that

> if the law then constitutes the sanction that society, or
> more strictly, its ruling classes, accords to existing
> economic conditions, it must then of necessity reflect
> these same conditions, and docilely follow in the train of
> their successive transformations. The law, in other
> words, proceeds from the economic constitution and
> changes as it changes. The theory of Savigny and the
> historical school, which regards the law as the product
> of the national conscience, or the result of the peculiar
> inheritance and habits of a people, is thus entirely er-
> roneous. On the contrary, the legal systems of the most
> widely separated races and nations must be the same
> whenever the prevailing economic conditions are identi-
> cal. On the other hand, every nation must undergo a
> change in its legal system when the onward march of its
> civilization has brought about radical changes in its eco-
> nomic constitutions.[22]

Loria, like Maine and Weber, underscored the overwhelming
influence and significance of Mediterranean legal culture throughout
much of the modern world. The rise of capitalism brought fundamental changes in law. New economic forms related to the emergence of
capitalism, though differing widely from the form

> that prevailing during the feudal period, offered a profound
> analogy to that of the slave economy. Thus though the law
> regulating the labor contract had to be an original creation
> of the new economic system (or at best an elaboration of
> the contract of feudal service), the law regulating the re-
> lations among proprietors could practically be reproduced
> in its classic Roman form. Now it is exactly these rela-
> tions between proprietors that constitute the essential ob-
> ject, and form, as it were, the organic tissue of the law,
> while the relations between property and labor only enter
> in a subsidiary way. Thus the organic and vital side of
> the law could be regulated by the principles of the jus

romanum. The Roman law accordingly emerged from the
tomb where it had so long reposed into the expansion of a
new life. The movement toward this awakening commenced in Italy where the wage economy first began to develop, following the expropriation of the cultivators. The
new and more active economic relations that were springing up in the industrial cities of the Italian peninsula soon
became incompatible with the narrow rigidity of feudal law
and communal customs, and accordingly necessitated the
institution of a legal system more rapid in its workings
and more subtle in its movements, and such a system was
found already elaborated in the Roman law.

Thus legal history shows us that instead of being the
product of abstract reason, or the result of national consciousness, or a racial characteristic, the law is simply
the necessary outcome of economic conditions. For this
reason a definite legal system may pass on from one nation to another and leap from an earlier to a later century,
whenever its corresponding economic system is transmitted from this people to that and from one historical epoch
to another. [23]

Loria thus rejected Savigny and, to some extent, Weber. He
then assessed a variety of judicial institutions and legal customs to,
as he put it, "find additional confirmation of our main thesis." These
included the evolution of the law of the family,[24] of property,[25] of
inheritance,[26] and of contract. [27]

Loria argued that the most direct manifestation of capitalistic
influence in law is in the criminal law. This was true, suggested
Loria, because

crime being a morbid emanation of capitalistic conditions,
tends to interfere with their normal functions, and the punishment of crime is thus the legal means employed to consolidate and protect these same relations. Penal sanctions
have, accordingly, followed the alternate prevalence of the
different forms of ownership and favored the entire evolution of property. Thus an agricultural state metes out its
heaviest penalties to crimes against landed property, while
a commercial state punishes most severely the crime of
issuing false money. Severity against theft, again, is an
indication of the prevalence of movable over fixed property. For this reason primitive Roman law proceeded with
great severity against thieves, while under the code of
Justinian the rigor of the early law was considerably modi-

fied. And in general each state proceeds most severely against the crimes that injure its predominant interests.

But though the law varies thus in its predilections toward different forms of property at different epochs, it is nevertheless always constant in its partiality toward proprietors. It is, indeed, scarcely|necessary to insist upon this point, as the best criminalists have already vigorously denounced the essentially capitalistic character of the law of punishment, with its constant solicitude for the privileges of property and its total abandonment of the poorer classes. To be sure, jurists now recall with indignation that under the Salic law the punishment for the theft of animals was visited more severely upon the poor than upon the rich; some sociologists also regard it as an enormity that savages should punish theft more severely than homicide; and an Italian traveler has recently recounted with horror how theft and brigandage go unpunished among the Somali if committed on a large enough scale. But when we notice what is going on round about us, honesty compels us to admit that, in the matter of legal morality, we Europeans are not much above the Somali. Pelegrino Rossi has, indeed, deplored the fact that in a civilized country like England the indulgence of the law toward assassins should offer so striking a contrast with its severity toward thieves. But the same contrast is to be met with among all modern nations, and the system of punishments generally in force in the most civilized countries of the world certainly deserves no less decisive condemnation. Bismarck also deplored the fact that in matters of money the law shows an absolute rigor, contrasting strangely with its relative indifference to questions of health, life, and honor. The Italian code, likewise, inflicts very severe penalties upon theft and proceeds with vigor against strikers; while it treats with manifest indulgence a large number of crimes especially characteristic of the richer classes.

Citing Vaccaro, Loria concluded that

the office of criminal law up to the present has not been to protect society as a whole with all the various classes that compose it, but more particularly to defend the interests of those under the favor of the constituted political authority, or in other words, the proprietors. [28]

Achille Loria's analysis and his historical examples represent a very striking and well-organized conceptual framework in support of his thesis about the economic foundation of law. Whether the empirical evidence will support Loria's framework is a question that will ultimately be resolved on the basis of conceptually equivalent comparative analysis of the legal and judicial systems of a significant number of nations. But there is little question that these conclusions about the restoration, modification, and adaptation of ancient Roman law are valid. Like de Tocqueville and a number of others, Loria emphasized the political and economic objectives that stimulated the renaissance of Roman law.

The seminal influence of Mediterranean legal culture was thus crucial in the mobilization of Roman law in support of absolute monarchy and in support of the emergence and maintenance of capitalism. With the subsequent decline of absolute monarchy and the emergence of a world capitalist economic system, the mobilization of law in the service of modern, more sophisticated masters has represented the contemporary application of Mediterranean legal culture. For example, Wolfgang G. Friedmann has identified several attributes of the emergent international role of corporations. He argues that "the elimination of differences in status between states and public international organizations, representing public interests and constitutionally responsible to the public, and private corporations, representing private interests and pursuing private objectives, would be subversive of the basic objectives of public international order."[29] Similarly, in the controversies over the interaction of national imperialism and corporate aggrandizement, the use of law as an instrument of those manifestations of power is often central to such debates as illustrated in the testimony of the chief author of a recent Canadian task force report, Economics Professor Melville H. Watkins. He identified the tendency for U.S. corporations to bring U.S. law with them, as a form of "legal imperialism" characterized as "the tip of the iceberg of political control of that manipulation and exploitation which is imperialism proper."[30]

Although the issues have often changed, the significant role of lawyers and legal institutions continues with lawyers, judges, and other legal experts contributing important services, interpretations, and doctrine relating to the redistribution of political, economic, and social power in the world capitalist economic system.[31] Those adaptations of Mediterranean legal culture in the world capitalist economic system are manifestations of the remarkable, albeit highly controversial, persistence of those social institutions.

MEDITERRANEAN LEGAL CULTURE IN ITS
MODERN CONTEXT: THE SYSTEMS OF ITALY,
SPAIN, AND THEIR LATIN AMERICAN PROGENY

Contemporary assessments of major components of Mediterranean legal culture, such as the judicial systems of Italy and Spain,
substantiate the nineteenth-century interpretation of the major influence of Roman law but generally eschew recognition of the relationship
between power and law of the sort emphasized by de Tocqueville or
Loria. Nor do those assessments seriously consider broad conceptual frameworks, whether Weber's, Durkheim's, corollaries of Wallerstein's, or others. For example, Cappelletti, Merryman, and
Perillo's Italian Legal System posited a direct relationship between
law as developed in ancient Rome and the modern legal system of
Italy, stating that "Justinian's codification, as it was interpreted and
developed by Italian jurists from the twelfth century onward, is the
direct source of the Italian legal system." They underscored the uni-
versality of Roman law, indicating that "there was a progressive
growth in the various European countries of a class of jurists with a
common background of study of Roman law at Bologna or at another
school modeled on Bologna." Similarly and concomitantly, "Italy
was the source of the jus commune of Europe," the acknowledged
legal foundation for the civil law system. [32]

The acknowledgment of the Roman-law heritage of Western
legal systems, in general, and contemporary Mediterranean legal
systems, in particular, does not adequately account for the very significant institutional and fundamental doctrinal differences between
the ancient and the modern. For example, Merryman, Clark, and
Friedman document and classify a wide range of Italian and Spanish
judicial institutional and legal culture attributes that actually have
little or no direct relationship to ancient Roman law. Similarly,
these investigators documented the significant adaptation of such
Mediterranean legal institutions and culture in four former Spanish
colonial regions: Chile, Colombia, Costa Rica, and Peru. [33] One
of the most significant of the modern judicial roles, the development
and exercise of the power of judicial review, had no conceptually
equivalent Roman counterpart. Cappelletti and Cohen recognized its
American origins, but also asserted that judicial review is "the result of an evolutionary pattern common to much of the West, in both
civil and common law countries."[34]

The modern judicial institutions of Italy and Spain and those of
five Latin American nations have indeed developed generally similar
court structures. All incorporate characteristics that, at least superficially, resemble the fundamentals of a Weberian-type model.
The authors of these modern works on the judicial systems of Italy,

Spain, and selected Latin American nations made no overt commitment to Weber or any other originator of a conceptual framework for
comparative legal systems. But the main thrust of these descriptive
studies was acceptance of the basic assumptions of the Weberian
model, without any serious consideration of contradictory conceptual
frameworks. The essential elements of the Weberian model shared
by these investigators of the judicial systems of Italy, Spain, and
some of their Latin American progeny are summarized in Table 18.5.

What are the criteria for the classification made in Table 18.5?
The investigations and surveys from which they were derived employed several empirical indicators that approximated key elements
in the neo-Weberian conceptual framework. These indicators included
(1) designations of the precise nature of national constitutional grants
of authority to the judiciary, (2) presence or absence of constitutional
or statutory provision for separation of powers, (3) jurisdictional
grants or limitations, (4) presence or absence of a bill of rights, (5)
limitations on legislative or executive power in relation to the judiciary, (6) presence or absence of machinery for enforcing judgments,
(7) characteristics of judicial and legal elites, (8) presence or absence
of the power of judicial review, and (9) nature of provisions for judicial independence.

The unacknowledged emphasis upon characteristics of a Weberian model may also reflect the frequently held assumption among
many legal scholars that the ideal characteristics of a judicial system
may be achieved by treating a particular system as if it fulfilled the
ideal, even though it fell short of such an achievement. Clearly, a
careful examination of contradictory evidence appears to substantiate
this interpretation. For example, Piero Calamandrei, a distinguished
Italian scholar, recognized that striking contradictions existed between the principles stated in the Italian Constitution and the reality.
He observed that the third paragraph of Article 24 stated that "the
poor are guaranteed through the help of apposite institutions the means
of seeking a remedy and of defending themselves before any tribunal."
But, he noted that "in the case of damages for judicial errors provided
for in the same article, this protection is more a promise than a reality."[35] Although the evidence in the access-to-justice project varies
nation by nation, the comment made by Carlos Alonzo suggests that
this issue may be a universal problem. "Problems of access to justice in today's world are almost identical in all countries. The most
persistent of these constitute the major defects in contemporary civil
procedure: duration and cost."[36]

Mediterranean legal systems function in the context of the major
capitalist world system conceptualized by Immanuel Wallerstein.
Relative failure to provide equal justice to the poor suggests merely
one dimension of judicial systems approximating a conceptual model

TABLE 18.5

Classification of Scholarly Treatments of Mediterranean Legal Systems

Modified Weberian Framework	1	2	3	4	5	6	7
The assumption is accepted that judicial systems and legal professions evolved or were created to institutionalize conflict resolution based upon fair and objective principles and procedures	+	+	+	+	+	+	+
The characteristics of judicial systems and legal professions are not the products of deterministic economic and social forces	+	+	+	+	+	+	+
Judicial systems and legal professions and their modes of procedure and decision making are developed and maintained independently of external institutional influences such as governments, political parties, and religious organizations	+	+	+	+	+	+	+
Decision making is characterized by objectivity and impersonality in its modes of decision making	+	+	+	+	+	+	+

	1	2	3	4	5	6	7
Law serves as a neutral arbiter of individual relations between equals	+	?	+	+	+	?	?
Legal safeguards for contract and property rights and for noneconomic personal liberties (human rights) are essential	+	+	+	+	+	+	+
Predictability and rationality were deemed essential to a legal system by Weber and are components of an ideal system	+	+	+	+	+	+	+
Judges and elite legal professions are trained and socialized to develop goals of predictability and rationality	+	+	+	+	+	+	+

Sources: Column 1—Mauro Cappelletti, John Henry Merryman, and Joseph M. Perillo, The Italian Legal System: An Introduction (Stanford, Calif.: Stanford University Press, 1967), pp. 1, 77, 52; column 2—Vincenzo Vigoriti, "Access to Justice in Italy," in Access to Justice: A World Survey, ed. Mauro Cappelletti and Bryant Garth (Alphenaandenrijn and Milan: Sijthoff & Noordhoff & Givffre, 1978), pp. 649–86; column 3—Kenneth L. Karst and Keith G. Roseann, Law and Development in Latin America (Berkeley and Los Angeles: University of California Press, 1975), pp. 77–240; column 4—Mauro Cappelletti and William Cohen, Comparative Constitutional Law: Cases and Materials (Indianapolis: Bobbs–Merrill, 1979), p. 11; column 5—Rene David and John E. C. Brierley, Major Legal Systems in the World Today (London: Stevens & Sons, 1978), pp. 1–147; column 6—Carlos de Miguel y Alonso, "Access to Justice in Spanish Law," in Access to Justice, ed. Cappelletti and Garth, pp. 844–87; and column 7—John Henry Merryman, David S. Clark, Lawrence M. Friedman, Law and Social Change in Mediterranean Europe and Latin America: A Handbook of Legal and Social Indicators for Comparative Study (Stanford, Calif: Stanford University Law School, 1979).

related to such a world economic system. Private-sector emphasis upon property rights and stability is a significant indicator of the intimate relationship between the world capitalist system and these national judiciaries.

NOTES

1. This portion of the analysis, the development of the two major conceptual frameworks, is derived from John R. Schmidhauser, "Prolegomena for the Comparative Crossnational Analysis of National Judicial Systems and Legal Professions: The Development of Competing Frameworks" (Paper presented at the Conference on Comparative Judicial Studies held at Mansfield College, Oxford University, April 4-6, 1981).

2. This introductory description of the components of Weber's ideal model for a legal system is, of necessity, purposely parsimonious. For a fuller development of Weber's model and its adaptation as a basis for evaluating the U.S. federal appellate judiciary, see John R. Schmidhauser, Judges and Justices: The Federal Appellate Judiciary (Boston: Little, Brown, 1979).

3. See Max Rheinstein, ed., Max Weber on Law in Economy and Society (Cambridge, Mass.: Harvard University Press, 1954), p. 13; and Max Weber, The Theory of Social and Economic Organization, trans. Talcott Parsons (New York: Free Press, 1964), p. 128.

4. Weber, The Theory of Social and Economic Organization, p. 77.

5. Guenter Roth and Claus Wittich, eds., Max Weber: Economy and Society: An Outline of Interpretative Sociology (New York: Bedminister Press, 1968), p. 883.

6. The initial adaptation of this framework was made in John R. Schmidhauser, "Corruption on the Federal Bench: Mapping the Conceptual Framework in the Context of Weber's Ideal Legal System" (Paper delivered at the 1978 meeting of the Midwest Political Science Association, Chicago, April 1978); the broader framework of Weber's ideal legal system was outlined and applied in Schmidhauser, Judges and Justices.

7. Steven E. Beaver, Demographic Transition Theory Reinterpreted (Lexington, Mass.: Lexington Books, 1975), p. 4.

8. See, for example, Richard Abel, "A Comparative Theory of Dispute Institutions in Society," Law and Society Review 8 (1974): 217-347; and Austin Sarat and Joel B. Grossman, "Courts and Conflict Resolution: Problems in the Mobilization of Adjudication," American Political Science Review 69 (1975): 1200-17.

9. Martin Shapiro, "Courts," in Handbook of Political Science, ed. Fred Greenstein and Nelson Polsby (Reading, Mass.: Addison-Wesley, 1975 , p. 322.

10. Immanuel Wallerstein, The Modern World System (New York: Academic Press, 1974), p. 347.

11. Ibid. , pp. 348-49.

12. Ibid. , p. 350.

13. John W. Meyer, John Boli-Bennett, and Christopher Chase-Dunn, "Convergence and Divergence in Development," in Annual Review of Sociology, ed. Alex Inkeles, James Coleman, and Neil Smelser, 3 vols. (Palo Alto, Calif.: Annual Reviews, 1975), 1:223.

14. In Patrick McGowan, "Problems of Theory and Data in the Study of World-System Dynamics" (Paper given on a panel on world system analysis, Twenty-first Annual Convention of the International Studies Association, Los Angeles, March 1980), pp. 15-17, 20; for a fuller discussion of the points of difference between these conceptual frameworks, see for example, Patrick McGowan and Stephen G. Walker, "Marxist and Conventional 'Models' of U.S. Foreign Economic Policymaking" (Paper delivered at the Fifth Annual Hendricks Symposium on U.S. International Economic Policies and Global Scarcities, Lincoln, Neb., April 1980), pp. 1-61.

15. Inga Markovitz, "Owning and Sharing—The Rightholder's Relation to Society in Bourgeois and Socialist Law" (Paper presented at the 1980 Annual Meeting of the American Political Science Association, Washington, D.C., August 1980).

16. Lauro Martines, Lawyers and Statecraft in Renaissance Florence (Princeton, N.J.: Princeton University Press, 1968).

17. Fernand Braudel, The Mediterranean and the Mediterranean World in the Age of Philip II (New York: Harper & Row, 1972), 2:683-85.

18. Ibid. , 1:510.

19. See, for example, John Henry Merryman, David S. Clark, and Lawrence M. Friedman, Law and Social Change in Mediterranean Europe and Latin America: A Handbook of Legal and Social Indicators for Comparative Study (Stanford, Calif.: Stanford University Law School, 1979).

20. Alexis de Tocqueville, The Old Regime and the French Revolution (Garden City, N.Y.: Doubleday, 1955), pp. 222-23.

21. Henry W. Ehrmann, Comparative Legal Cultures (Englewood Cliffs, N.J.: Prentice-Hall, 1976), p. 5. For a contrary view, see Achille Loria, "The Economic Foundations of Law," trans. Lindley M. Keasby, in Formative Influences of Legal Development, ed. Albert Kocourek and John H. Wigmore (Boston: Little, Brown, 1918), pp. 234-35.

22. Loria, "Economic Foundations," p. 240.

23. Ibid., pp. 244-45.

24. Ibid. , pp. 245-48.

25. Ibid. , pp. 248-51.

26. Ibid. , pp. 251-57.

27. Ibid. , pp. 257-60.

28. Ibid. , pp. 264-66.

29. Wolfgang G. Friedmann, The Changing Structure of International Law (New York: Columbia University Press, 1964), pp. 223-24.

30. Cited in Richard Eells, Global Corporations: The Emerging System of World Economic Power (New York: Free Press, 1976), p. 32.

31. See, for example, W. D. Onuf and Robert O. Slater, "Law Experts and the Making of Formal Ocean Policies" (Paper prepared for delivery at the 1974 Annual Meeting of the American Political Science Association, Chicago, August 29-September 2, 1974).

32. Mauro Cappelletti, John Henry Merryman, and Joseph M. Perillo, The Italian Legal System: An Introduction (Stanford, Calif. : Stanford University Press, 1967), pp. 1, 77, 52.

33. Merryman, Clark, and Friedman, Law and Social Change.

34. Mauro Cappelletti and William Cohen, Comparative Constitutional Law: Cases and Materials (Indianapolis: Bobbs-Merrill, 1979), p. 11.

35. Piero Calamandrei, Procedure and Democracy (New York: New York University Press, 1956), pp. 94-95.

36. Carlos de Miguel y Alonso, "Access to Justice in Spanish Law," in Access to Justice: A World Survey, ed. Mauro Cappelletti and Bryant Garth (Alphenaandenrijn and Milan: Sijthoff & Noordhoff & Givffre, 1978), pp. 844-87.

INDEX

ABOUT THE EDITORS AND CONTRIBUTORS

CARL F. PINKELE teaches political science at Ohio Wesleyan University.

ADAMANTIA POLLIS teaches political science at the New School for Social Research.

KAMEL ABU-JABER is Director of the Queen Alia Jordan Social Welfare Fund, Amman, Jordan.

ROBERT BAADE teaches economics at Lake Forest College.

LARRY BERG teaches political science at the University of Southern California.

BARUCH BOXER teaches in the Department of Human Ecology, Cook College, Rutgers University-New Brunswick.

DIETER DETTKE is a member of the SPD parliamentary group in Bonn, Germany.

JONATHAN GALLOWAY teaches political science at Lake Forest College.

GEORGE GINSBURGS teaches law at Rutgers University-Camden.

ALEXANDER KISLOV is Head of Section, Institute of U.S. and Canadian Studies, Academy of Sciences of the U.S.S.R.

GEORGE LENCZOWSKI teaches political science at the University of California-Berkeley.

STEFAN MUSTO teaches political science at the Technical University of Berlin and the German Development Institute.

JOSEPH L. NOGEE teaches political science at the University of Houston.

SALUA NOUR is a political scientist living in Berlin, West Germany.

DEMETRIOS G. PAPADEMETRIOU, Director, Center for Migration Studies, Staten Island, New York.

JOHN SCHMIDHAUSER teaches political science at the University of Southern California.

BETTY SMOLANSKY teaches sociology at Moravian College.

OLES SMOLANSKY teaches political science at Lehigh University.

RUSSELL STONE teaches sociology at the State University of New York at Buffalo.

MARILYN WALDMAN teaches history at Ohio State University.

GABRIEL WARBURG teaches history in the Institute of Middle Eastern Studies, University of Haifa, Israel.

SUSAN L. WOODWARD teaches political science at Yale University.

36,867